Advances in Olericulture

Series Editor
Silvana Nicola, Department of Agricultural, Forest and Food Sciences, DISAFA, Vegetable Crops & Medicinal and Aromatic Plants, VEGMAP University of Turin, Grugliasco (Turin), Italy

The book series Advances in Olericulture provides a state-of-the-art account of research in olericulture, the applied life science of production and utilization of vegetable crops. The series focuses on various aspects of vegetable science and technology covering primarily but not exclusively species where the vegetative organ is the economically important component. The series of books spans current topics from sustainable fertilization to organic production; from open field cultivation to advanced soilless growing techniques; from vegetable seed and seedling physiology to vegetable quality and safety; from environmental stresses to phyllosphere communities interaction with vegetables; from postharvest biology and technology to minimally processing of vegetables. The series is designed to present the most advanced scientific information available linking basic and applied research for serving olericulturists, research workers, teachers and advanced students.

Shashank Shekhar Solankey • Meenakshi Kumari
Editors

Advances in Research on Vegetable Production Under a Changing Climate Vol. 2

Springer

Editors
Shashank Shekhar Solankey
Department of Vegetables & Floriculture
Agricultural Research Institute
Patna, Bihar, India

Meenakshi Kumari
Department of Vegetable Science
Shree Guru Gobind Singh
Tricentenary University
Gurugram, Haryana, India

ISSN 2367-4083 ISSN 2367-4091 (electronic)
Advances in Olericulture
ISBN 978-3-031-20839-3 ISBN 978-3-031-20840-9 (eBook)
https://doi.org/10.1007/978-3-031-20840-9

This Springer imprint is published by the registered company Springer Nature Switzerland AG
The registered company address is: Gewerbestrasse 11, 6330 Cham, Switzerland

Preface

Climate change refers to long-term shifts in temperature and weather patterns which are attributed directly or indirectly to anthropological activities. The changing climatic conditions have resulted in increased atmospheric CO_2 concentration, temperature, and precipitation patterns.

The issue of rising temperature due to increased emission of greenhouse gases such as CO_2, CO, CH_4, NO, N_2O, and SO_2 is a major concern worldwide. Another significant aspect of climate change is the increased frequency of occurrence of extreme events such as droughts, floods, cyclones, rise in sea level, and intrusion of saltwater. The top emitter of CO_2 is China (29%), followed by the USA (14%), the European Union (10%), and India (7%). It was reported that Indian GHG emissions in 2020 (3000 million tonnes) were around three times more than that observed in 1990 (988 million tonnes). Over the past 111 years, India's average annual temperature has risen by 0.46° C.

Vegetables are the cheapest natural source of nutrition, such as vitamins, minerals, fibers, and carbohydrates, and are basic need for our health system as well. Vegetable crops have great potential for higher yield and provide the farmer with substantial compensation as well, since they sell for more return than other agricultural crops. As we know that most vegetables are herbaceous in nature and very sensitive to climate-borne disorders, crop failures, low yields, declining quality, and increased pest and disease occurrences are other major hindrances to the cultivation of vegetables under changing climatic conditions.

Several mitigation and adaptation strategies are essentially required to lessen the severity and sensitivity of these effects owing to climate change and help the farming community to support itself. Additionally, genetically modifying vegetables is a suitable adaptation approach to deal with the negative effects of climate change. A thorough knowledge of the role played by the environment in the conversion of a genotype into phenotype may be obtained via a combined study of genomics and phenomics. Another method of using plant biodiversity to combat climate change is by grafting a sensitive scion cultivar onto a resistant rootstock. Agronomic techniques including mulching, organic farming, agroforestry, and cropping systems that sequester carbon offer a variety of potential options for mitigating the effects of climate change on vegetable production. Protected farming and post-harvest technologies can also be important strategies for overcoming climate change issues.

The book entitled *Advances in Research on Vegetable Production Under a Changing Climate (Vol. II)* is intended to be an authoritative work at an advanced level, accessible to vegetable scientists, plant breeders, biotechnologists, plant pathologists, and entomologist working on climate change, especially with vegetable crops, and in the libraries of all research facilities and enterprises where this recently interesting topic is investigated, studied, or taught. This book comprises 16 chapters will hopefully be beneficial as a part of the scientific basis for understanding the recent advances in research trends in vegetables and on different aspects such as emerging obstacles of vegetable production, nutraceutical properties of vegetables, nutritional stress management, impact on leafy, salad, perennial, underexploited, tuber vegetables, seed production, kitchen gardening, protected cultivation, grafting, molecular breeding, emerging insect-pests and diseases and postharvest quality of vegetables.

The main aim of this book is to provide a thorough, up-to-date account of recent advances in vegetable production under changing climatic scenarios, and we expect that it would be also highly appreciated by the readers like Volume I.

First of all, we want to express our gratitude to *Almighty God* for giving us the courage, ideas, and inspiration to finalize this book. We are thankful to all the subject matter experts for their contribution towards creation of this novel book by providing the innovative information and in-depth scripts which make it possible to complete this novel book. We are really grateful to our family members, whose unwavering support and inspiration helped us to compile this book. We also acknowledge our institutions (Bihar Agricultural University, Sabour, Bhagalpur, Bihar, and Shree Guru Gobind Singh Tricentenary University, Gurugram, Haryana) and express sincere thanks for their kind support and cooperation.

The entire academic staff at Springer Nature deserves praise, particularly Series Editor Dr. Silvana Nicola (Advances in Olericulture); Ms. Deepthi Vasudevan, Project Coordinator (Books); Ms. Melanie van Overbeek, Assistant Editor, Life Sciences (Agronomy); and Mr. Nitesh Shrivastava, Production Editor (Books), who generously supported the compilation and completion of this assignment. We also appreciate the direct and indirect assistance and insightful recommendations provided by our professors, elders, colleagues, and friends throughout the preparation of this book. We would be delighted if readers could offer any worthwhile ideas to improve the value of this book. The editors gratefully recognize the assistance and support received from the different experts, books, journals, writers, publishers, and online publishing platforms whose works were consulted in the preparation of this book. We heartily appreciate Springer Nature for publishing this book in such a meticulous manner.

Patna, Bihar, India — Shashank Shekhar Solankey

Gurugram, Haryana, India — Meenakshi Kumari

December 2022

Contents

About the Editors

Shashank Shekhar Solankey is presently working as an Assistant Professor–*cum*–Junior Scientist (Vegetable Science) at Agricultural Research Institute, Patna (Bihar Agricultural University, Sabour, Bhagalpur, India). He obtained his master's degree in vegetable science from Acharya Narendra Deva University of Agriculture and Technology, Kumarganj, Ayodhya (India), in 2006 and doctorate in horticulture from Banaras Hindu University, Varanasi (India), in 2010. Dr. Solankey has served as SRF as well as Research Associate at ICAR-Indian Institute of Vegetable Research (IIVR), Varanasi (U.P.) from 2010 to 2012. He joined as Assistant Professor–*cum*–Junior Scientist (Vegetable Science) Bihar Agricultural University, Sabour, on 17 September 2012. He has been involved in teaching, research, extension, and training activities at the university. Thereafter, he was deputed at the newly established Dr. Kalam Agricultural College, Kishanganj, under the umbrella of BAU, Sabour, on 7 September 2015 and acted as Nodal Officer of the newly established Horticulture Research Centre, Kishanganj, Bihar, from May 2021 to June 2022. He handled four research projects on vegetable crops as PI/co-PI with the objective of biotic and abiotic stress management as well as quality improvement in solanaceous vegetables and okra at BAU, Sabour. He has been associated with development of two eggplant varieties (Sabour Sadabahar and Sabour Krishnakali) and one technology on "Management of sucking pests in okra." He is now handling two state non-plan research projects titled "Improvement of okra genotypes for YVMV tolerance" and "Collection, evaluation and assessment of feasibility of promising vegetables for

Bihar." Dr. Solankey has supervised four M.Sc. students and also acted as member of advisory committee for seven M.Sc. and three Ph.D. students. He has published 56 research papers, 7 review papers, 1 souvenir paper, 8 edited books, 1 authored book, 50 book chapters, and 30 popular articles. Dr. Solankey is also life member of the Horticulture Society of India, New Delhi; Indian Society of Vegetable Science, IIVR, Varanasi; International Society for Noni Science, Perungudi, Chennai; Society for Scientific Development in Agriculture & Technology, Meerut; and Bihar Horticulture Society, BAU, Sabour, Bihar. He is also a reviewer of the *International Journal of Plant & Soil Science, Vegetos* as well as *Scientia Horticulturae*. Dr. Solankey has been awarded with Best Teacher Award (2016) as well as Best Researcher Award (2016) by Bihar Agricultural University, Sabour. Beside these, he has also been a recipient of 13 other awards and recognitions.

Meenakshi Kumari is presently working as an Assistant Professor in the Department of Vegetable Science at Shree Guru Gobind Singh Tricentenary University, Gurugram, Haryana, India. She has specialization in vegetable science. She acquired her master's degree in horticulture (Vegetable and Floriculture) from Bihar Agricultural University, Sabour (Bhagalpur), Bihar, in 2016, and Doctorate in Vegetable Science from Chandra Shekhar Azad University of Agriculture & Technology, Kanpur, Uttar Pradesh (India), in 2019. She attained first rank in both master's and doctoral degree programmes. She qualified ICAR-ASRB NET examination in the discipline of Vegetable Science in 2017 and qualified CAR-SRF exam in 2016. She was also selected for the DST-Inspire fellowship during her Doctoral studies by the Department of Science & Technology, Govt. of India. Presently, Dr. Kumari is guiding three master's students. She has published 27 research papers (national & international), 5 review papers, 31 book chapters, 3 books, 1 manual, 15 popular articles, and more than 35 abstracts/ extended summaries. Dr. Kumari has been bestowed with the Best Oral Presentation Award (2018), Best Article Award (2018), and Best Thesis Award (2018) as well as has seven other awards and recognitions. She is also life member of the Horticulture Society of India, New Delhi, and the Indian Science Congress, Kolkata, India.

Contributors

Shirin Akhtar Department of Horticulture (Vegetable and Floriculture), Bihar Agricultural College, Sabour (Bihar Agricultural University, Sabour), Bhagalpur, Bihar, India

Ayon Alipatra Department of Agronomy, Dr. Kalam Agricultural College (Bihar Agricultural University, Sabour, Bhagalpur), Kishanganj, Bihar, India

Bahadur Singh Bamaniya Department of Horticulture, School of Life Sciences, Sikkim University, Gangtok, Sikkim, India

Satyaprakash Barik Department of Agriculture and Allied Sciences, C.V. Raman Global University, Bhubaneswar, Odisha, India

D. R. Bhardwaj Division of Crop Improvement, ICAR-Indian Institute of Vegetable Research, Varanasi, Uttar Pradesh, India

Sangeeta Bhattacharyya Division of Agricultural Extension, ICAR-Central Citrus Research Institute, Nagpur, Maharashtra, India

Panchaal Bhattocharjee Department of Horticulture, Anand Agricultural University, Anand, Gujarat, India

Tirthartha Chattopadhyay Department of Plant Breeding and Genetics, Bihar Agricultural College, Sabour (Bihar Agricultural University, Sabour), Bhagalpur, Bihar, India

Sunil Kumar Dash Department of Vegetable Science, Odisha University of Agriculture and Technology, Bhubaneswar, Odisha, India

Erayya Department of Plant Pathology, Dr. Kalam Agricultural College (Bihar Agricultural University, Sabour, Bhagalpur), Kishanganj, Bihar, India

Ayesha Fatima Department of Agronomy, Dr. Kalam Agricultural College (Bihar Agricultural University, Sabour, Bhagalpur), Kishanganj, Bihar, India

Sheikh Mohammad Feroze Department of Agricultural Economics, College of Agriculture, Central Agricultural University, Imphal, Manipur, India

Keshav Kant Gautam Division of Crop Improvement, ICAR-Indian Institute of Vegetable Research, Varanasi, Uttar Pradesh, India

Sarvesh Pratap Kashyap Bio Zenith Scientific Research Pvt. Ltd., Mirzapur, Uttar Pradesh, India

S. Bhargav Kiran Division of Vegetable Science, Indian Agricultural Research Institute, Pusa, New Delhi, India

Anjani Kumar ICAR-Agricultural Technology Application Research Institute (ATARI), Patna, Bihar, India

Meenakshi Kumari Department of Vegetable Science, Shree Guru Gobind Singh Tricentenary University, Gurugram, Haryana, India

Rajesh Kumar Department of Horticulture, School of Life Sciences, Sikkim University, Gangtok, Sikkim, India

Randhir Kumar Agricultural Research Institute (Bihar Agricultural University, Sabour, Bhagalpur), Patna, Bihar, India

Santosh Kumar Department of Plant Pathology, Mandan Bharti Agricultural College (Bihar Agricultural University, Sabour, Bhagalpur), Agwanpur, Bihar, India

Udit Kumar Department of Horticulture, Post Graduate College of Agriculture (PGCA), Dr. Rajendra Prasad Central Agricultural University, Pusa, Bihar, India

Jitendra Kumar Kushwah Department of Agriculture, Government of Uttar Pradesh, Auraiya, Uttar Pradesh, India

Kalmesh Managanvi Department of Entomology, Dr. Kalam Agricultural College (Bihar Agricultural University, Sabour, Bhagalpur), Kishanganj, Bihar, India

Araghya Mani Department of Horticulture, School of Agriculture, Lovely Professional University, Phagwara, Punjab, India

Shubham Maurya Department of Horticulture, Post Graduate College of Agriculture (PGCA), Dr. Rajendra Prasad Central Agricultural University, Pusa, Bihar, India

Pragnya Paramita Mishra Department of Floriculture and Landscaping, College of Agriculture, Odisha University of Agriculture and Technology, Bhubaneswar, Odisha, India

Durga Prasad Moharana Department of Horticulture, Institute of Agricultural Sciences, Banaras Hindu University, Varanasi, Uttar Pradesh, India

Anirban Mukherjee Division of Agricultural Extension, ICAR-Research Complex for Eastern Region, Patna, Bihar, India

Arindam Nag Department of Extension Education, Dr. Kalam Agricultural College (Bihar Agricultural University, Sabour, Bhagalpur), Kishanganj, Bihar, India

Abhishek Naik Marketing Department, Ajeet Seeds Pvt. Ltd., Kolkata, West Bengal, India

Menka Pathak Department of Vegetable Science, College of Agriculture, Odisha University of Agriculture and Technology, Bhubaneswar, Odisha, India

Madhuri Pradhan Department of Agriculture and Allied Sciences, C.V. Raman Global University, Bhubaneswar, Odisha, India

K. Prasad Department of Horticulture, Tirhut College of Agriculture (TCA), Dholi, Dr. Rajendra Prasad Central Agricultural University, Pusa, Bihar, India

M. Prashant Department of Entomology, University of Horticultural Sciences, Bagalkot, Karnataka, India

Vijay Laxmi Rai College of Agriculture Campus, (Archarya Narendra Dev University of Agriculture & Technology, Ayodhya), Kotwa, Azamgarh, Uttar Pradesh, India

Rajiv Department of Vegetable Science, Chandra Shekhar Azad University of Agriculture and Technology, Kanpur, Uttar Pradesh, India

Pankaj Kumar Ray Krishi Vigyan Kendra (Bihar Agricultural University, Sabour, Bhagalpur), Saharsa, Bihar, India

Ramnath K. Ray ICAR-Research Complex for Eastern Region, Patna, Bihar, India

K. Madhusudhan Reddy Department of Horticulture (Vegetable and Floriculture), Bihar Agricultural College, Sabour (Bihar Agricultural University, Sabour), Bhagalpur, Bihar, India

Anirudhha Roy Department of Agricultural Economics, Dr. Rajendra Prasad Central Agricultural University, Samastipur, Bihar, India

Pinaki Roy National Agricultural Science Fund, Krishi Anusandhan Bhawan-I, ICAR-Indian Council of Agricultural Research, New Delhi, India

Joy Rudrapaul Department of Horticulture, Post Graduate College of Agriculture (PGCA), Dr. Rajendra Prasad Central Agricultural University, Pusa, Bihar, India

Bholanath Saha Department of Soil Science, Dr. Kalam Agricultural College (Bihar Agricultural University, Sabour, Bhagalpur), Kishanganj, Bihar, India

Sushanta Saha Directorate of Research, Bidhan Chandra Krishi Viswavidyalaya, Kalyani, West Bengal, India

Jyoti Prakash Sahoo Department of Agricultural Biotechnology, College of Agriculture, Odisha University of Agriculture and Technology, Bhubaneswar, Odisha, India

Surabhi Sangam Department of Horticulture (Vegetable and Floriculture), Bihar Agricultural College, Sabour (Bihar Agricultural University, Sabour), Bhagalpur, Bihar, India

Subhashish Sarkhel Department of Plant Pathology, Dr. Kalam Agricultural College (Bihar Agricultural University, Sabour, Bhagalpur), Kishanganj, Bihar, India

Md. Shamim Department of Molecular Biology and Genetic Engineering, Dr. Kalam Agricultural College (Bihar Agricultural University, Sabour, Bhagalpur), Kishanganj, Bihar, India

Lomash Sharma Department of Horticulture, School of Life Sciences, Sikkim University, Gangtok, Sikkim, India

Kumari Shubha Division of Vegetable Science, ICAR-Research Complex for Eastern Region, Patna, Bihar, India

Anand Kumar Singh Department of Horticulture, Institute of Agricultural Sciences, Banaras Hindu University, Varanasi, Uttar Pradesh, India

D. P. Singh Department of Vegetable Science, Chandra Shekhar Azad University of Agriculture and Technology, Kanpur, Uttar Pradesh, India

Hemant Kumar Singh Krishi Vigyan Kendra (Bihar Agricultural University, Sabour, Bhagalpur), Kishanganj, Bihar, India

Prakash Singh Department of Plant Breeding & Genetics, Veer Kunwar Singh College of Agriculture (VKSCOA) (Bihar Agricultural University, Sabour, Bhagalpur), Dumraon, Buxar, Bihar, India

Raj Narain Singh Directorate of Extension Education, Bihar Agricultural University, Bhagalpur, Bihar, India

S. K. Singh Department of Plant Pathology, Post Graduate College of Agriculture, Dr. Rajendra Prasad Central Agricultural University, Pusa, Bihar, India

Shashank Shekhar Solankey Department of Vegetables & Floriculture, Agricultural Research Institute, Patna, Bihar, India

Barsa Tripathy Institute of Agricultural Sciences, S.O.A. Deemed to be University, Bhubaneswar, Odisha, India

M. A. Waseem Department of Entomology, University of Horticultural Sciences, Bagalkot, Karnataka, India

Sudheer Kumar Yadav Department of Horticulture, Post Graduate College of Agriculture (PGCA), Dr. Rajendra Prasad Central Agricultural University, Pusa, Bihar, India

Abbreviations

%	Per cent
@	At the rate of
$^1O_2$	Singlet oxygen
ABA	Abscisic acid
ADP	Adenosine diphosphate
AICRIP	All India Coordinated Rice Improvement Programme, Hyderabad
ATP	Adenosine triphosphate
AVRDC	Asian Vegetable Research and Development Centre, Taiwan
B	Boron
BSA	Bulked segregant analysis
BSR	Bulked segregant RNA
C	Carbon
CAX1	Calcium exchanger-1
Cd	Cadmium
CDPKs	Calcium-dependent protein kinases
Ce	Cerium
CFC	Chlorofluorocarbon
CH_4	Methane
Cl	Chlorine
CNGCs	Cyclic nucleotide gated calcium channels
CO	Carbon monoxide
Co	Cobalt
CO_2	Carbon dioxide
CPRI	Central Potato Research Institute, Kufri, Shimla
CTCRI	Central Tuber Crops Research Institute, Thiruvananthapuram
Cu	Copper
DAG	Diacyl glycerol
DMSP	Dimethylsulfonium propionate
DNA	Deoxyribonucleic acid
DREB	Dehydration-responsive element-binding protein
EC	Electrical conductivity

ECI	Efficiency conversion of ingested food
eCO_2	Elevated atmospheric CO_2
EMS	Ethyl methane sulfonate
FAO	Food and Agriculture Organization, Rome, Italy
Fe	Iron
FYM	Farmyard manure
GDP	Gross domestic product
GHGs	Greenhouse gases
GM	Genetically modified
GMOs	Genetically modified organisms
GPX	Goaiacol peroxidase
GWAS	Genome-wide association analysis
H	Hydrogen
H_2O_2	Hydrogen peroxide
HO	Hydroxyl radical
HSP	Heat shock proteins
IARI	Indian Agricultural Research Institute, New Delhi
ICAR	Indian Council of Agricultural Research, New Delhi
ICMR	Indian Council of Medical Research, New Delhi
IIHR	Indian Institute of Horticultural Research, Bengaluru
IIVR	Indian Institute of Vegetable Research, Varanasi,
INM	Integrated nutrient management
IPCC	Intergovernmental Panel on Climate Change, Geneva, Switzerland
La	Lanthanum
LEDs	Light-emitting diodes
MARD	Market access for rural development
MAS	Marker-assisted selection
Mn	Manganese
Mo	Molybdenum
Mo	Molybdenum
N	Nitrogen
N_2O	Nitrous oxide
Na	Sodium
NASA	National Aeronautics and Space Administration, Washington, DC, USA
NOAA	National Centres for Environmental Information
NSU	Neglected and underutilized species
O	Oxygen
O_2^-	Superoxide
O_3	Ozone
OA	Osmotic adjustment
^oC	Degree Celsius
OEE	Oxygen evolving enhancer
^oF	Degree Fahrenheit
OH	Hydroxyl radical

PAL	Phenylalanine ammonia lyase
PIP2	Phosphatidylinositol bisphosphate
PLC	Phospholipase C
PNT	Poverty-nutrition Trap
ppm	Parts per million
QTL	Quantitative trait locus
rDNA	Recombinant DNA
RGR	Relative growth rate
Ri plasmids	Root hair inducing plasmid
RNA	Ribonucleic acid
ROS	Reactive oxygen species
RuBisCO	Ribulose bisphosphate carboxylase
RWC	Relative water content
Se	Selenium
Seq	Sequencing
Si	Silicon
SLAF	Specific-locus amplified fragment
SNP	Single nucleotide polymorphisms
SO_2	Sulphur dioxide
SPM	Sulfitation pressmud
SPS	Sucrose phosphate synthase
TFs	Transcriptions factors
Ti plasmids	Tumor-inducing plasmid
UNFCCC	United Nations Framework Convention on Climate Change
UV	Ultraviolet
Va	Vanadium
WHO	World Health Organization, Geneva, Switzerland
WUE	Water usage efficiency
Zn	Zinc

Chapter 1
Advances in Research Trends in Vegetables Under a Changing Climate: A Way Forward

Shashank Shekhar Solankey, Meenakshi Kumari, Hemant Kumar Singh, Pankaj Kumar Ray, Shirin Akhtar, and Bholanath Saha

Abstract Change of climate is a never ending process but from last few decades the swiftness of alteration causes vulnerability in atmospheric temperature (heat, cold), rainfall (flood, drought), greenhouse gasses (CO_2, CH_4, N_2O, O_3) and intensity of extreme events, that is basically due to direct and indirect human interferences. Due to change in climate, agriculture food products are significantly impacted particularly succulent crops like vegetables, which are extremely vulnerable to change of climate. Vegetables has vast production capacity to fulfill the daily requirements of growing global population, but the present climate change severely affect the vegetable productivity. Hence, the scientists are thinking to reduce the speed of climate change and to work on the adaptation and mitigation strategies. Studying how climate change affects the growth, development, production and

S. S. Solankey (✉)
Department of Vegetables & Floriculture, Agricultural Research Institute, Patna, Bihar, India

M. Kumari
Department of Vegetable Science, Shree Guru Gobind Singh Tricentenary University, Gurugram, Haryana, India

H. K. Singh
Krishi Vigyan Kendra (Bihar Agricultural University, Sabour, Bhagalpur), Kishanganj, Bihar, India

P. K. Ray
Krishi Vigyan Kendra (Bihar Agricultural University, Sabour, Bhagalpur), Saharsa, Bihar, India

S. Akhtar
Department of Horticulture (Vegetable and Floriculture), Bihar Agricultural College, Sabour (Bihar Agricultural University, Sabour), Bhagalpur, Bihar, India

B. Saha
Department of Soil Science, Dr. Kalam Agricultural College (Bihar Agricultural University, Sabour, Bhagalpur), Kishanganj, Bihar, India

S. S. Solankey, M. Kumari (eds.), *Advances in Research on Vegetable Production Under a Changing Climate Vol. 2*, Advances in Olericulture,
https://doi.org/10.1007/978-3-031-20840-9_1

quality of agricultural products is therefore urgently needed. At present a number of climate tolerant varieties have been evolved in various vegetable crops and with the help of recent molecular techniques the task of gene identification, genetic modification and varietal evolution is little bit easier for the olericulturists. Agronomic practices that preserve water and vegetable crops from harsh environmental conditions must be constantly developed and made broadly accessible to farmers in impoverished countries. A successful expansion plan that considers technological, social, and political aspects must be in place.

Keywords Climate change · Heat · Drought · Salinity · Greenhouse gasses · Adaptation · Mitigation

1.1 Introduction

Farmer struggle to cultivate crops as a result of the changing environment, which eventually puts tremendous strain on the ability to supply food for the world's expanding population. Unpredictable weather patterns and unstable atmospheric temperatures are results of climate change. Numerous international researchers have issued warnings in this direction, stating that if we do not implement eco-friendly agricultural practices, climate change could cause the world's food supply to decrease, making it impossible for us to produce high-quality food. However, affluent countries contribute significantly more to greenhouse gas emissions than under developed nations do. The Food and Agriculture Organization of the United Nations (FAO) estimates that global food production will need to increase by 70% by 2050 in order to feed an extra 2.3 billion people.

This issue is brought on by the atmosphere's buildup of greenhouse gases like carbon dioxide (CO_2), methane (CH_4), nitrous oxide (N_2O), and ozone (O_3), which are produced by humans as a result of long-term, intensive industrial expansion and high-consumption lifestyles in industrialized nations. India requires a national strategy to first adapt to climate change and then improve the ecological sustainability of India's growth path while working with the international community to collectively and jointly address this challenge.

Vegetables are thought of as a wholesome food and an essential component of the human diet. In addition to being rich in vitamins and minerals, vegetables also bring in more money than traditional food crops (Solankey et al. 2021a). Despite the fact that vegetable production has increased much more than the 15 million tonnes of 1950 to 191.77 million tonnes in 2021, moreover in fiscal year 2022, the total production of vegetables was estimated to be at approximately 200 million metric tons (Anonymous 2022).

More vegetables must still be grown to provide at least 300 g of vegetables per person everyday. Indian is the biggest producer of okra and ginger, while it comes in second place for potato, onion, cauliflower, brinjal, cabbage production. Cultivation of varieties which are both high yielding and have increased nutritional content with resistance to a variety of biotic and abiotic stresses can help to achieve our goal. Vegetable breeding combined with molecular techniques has a bright future in developing vegetable varieties with high amounts of nutraceuticals and bioactive chemicals appropriate for different markets. However, it is crucial that nutrient-rich foods are consumed to promote health security and bio fortified vegetable varieties be bred to meet the food needs of an ever-growing population.

Additionally, the distribution and quality of natural resources are impacted by climate change, which has a negative impact on people's ability to support themselves. The globe may be in serious danger due to the anticipated changes in climate because of an economy that is heavily reliant on its natural resource base and sectors that are vulnerable to them, including forestry, agriculture, and water. Food production has increased dramatically as a result of the green revolution, which also improve global food security. The natural resource basis for agriculture has been destroyed worldwide as a result of intensive agricultural production practices, endangering future output. Revolutionary discoveries made in the 1970s and 1980s fuelled expectations of significant improvements in agriculture and sparked interest and investment from entrepreneurs. These predictions are already coming true in the market because ongoing improvements in knowledge, technology, and business experience. Due to the effects of climate change and the rising competition for land, water, energy as well as farmers in developing countries will need to triple food production in order to fulfil demand over the next 40 years. Plant biotechnology and the application of molecular markers in conventional plant breeding techniques have established the foundation for molecular plant breeding, an interdisciplinary approach to crop improvement in the twenty-first century, much like recent advances in genetic engineering. Although molecular plant breeding techniques are constantly changing and are of great interest to agricultural specialists worldwide (Nelson et al. 2007; Lörz and Wenzel 2005; Varshney et al. 2006; Eathing ton et al. 2007; Mumm 2007). In 1994, the first wave of products in biotechnological application in vegetables was made available in pilot test markets. By using biotechnology tools some traits like postharvest life, quality of processing (deep red colour) was increased in vine ripe tomatoes, novel virus resistance in squash and insect killing protein in potato were increased. The Flavr Savr® tomato, which is reported as a superior vine-ripened flavour, was one of the first products to receive both popular attention and regulatory scrutiny. Alarming impact of change in climate are predicted. By 2030, due to change in climate 22 million people, the bulk of whom are farmers could be surviving in the extreme poverty. Variability in production is projected to rise when droughts and floods occur more frequently. Consider the possibility that, by 2100, grain productivity may drop by 10–40% while prices could rise by 29%. Every degree Celsius increase in temperature results in a 4–5 million tons decline in wheat production (Mohanty 2021).

Productivity of crop could be dropped by 10–40%. The rising temperatures would raise the amount of fertilizer which is needed by crop to achieve the same production goals thus increasing the emissions of greenhouse gas. The spawning, migration, and harvest of fish are anticipated to be affected by rising sea and river water temperatures. By 2030, coral reefs will begin to deteriorate. Increased animal needs for food, shelter, and energy would have an impact on milk output. There is a glimmer of optimism among these alarming figures. The soil holds the key role as a carbon sink, the soil of the earth absorbs and holds onto greenhouse gases. Carbon is released when soil is disturbed. Agriculture and the food value chain are responsible for 25% of all greenhouse gas emissions. Climate-smart farming techniques assist minimize soil disturbance, reduce greenhouse gas emissions, and maintain the soil's fertility (Mohanty 2021).

Lower intake of unprocessed meals but higher consumption of ultra-processed foods, particularly in tropical areas, were linked to decreases in rainfall and rise in temperature. Change in climate affects a wide range of health outcomes.

1.2 Effect of Climate Change on Vegetable Production and Its Management Techniques

Vegetables are usually succulent and sensitive plants therefore, severely affected by minor changes in the climate pose several challenges and negative impacts upon both quality as well as yield (Solankey et al. 2021b). The impact of various climatic threats and their mitigations are as follows:

1.2.1 Temperature

The main consequence of climate change that harms vegetable output is fluctuations in daily mean maximum & minimum temperatures since many physiological, biochemical, and metabolic processes in plants are temperature dependent (Abewoy 2018). In tropical and desert regions, the prevalence of high temperatures affects the production of vegetables (Abdelmageed et al. 2004). The plant's morphological, physiological, biochemical, and molecular responses are significantly altered by high temperatures, which has an impact on the plant's growth, development, and yield. Hazra et al. (2007) outlined the signs that high temperatures in tomato plants cause failure in fruit set, which include development of abnormal flower, poor dehiscence, viability and production of pollen grains, dropping of buds, abortion of ovules, minimum availability of carbohydrate & abnormalities in reproduction. In a similar manner, temperatures exceeding 25 °C have an impact on pollination and fruit setting in tomato (Abdelmageed et al. 2004; Solankey et al. 2017). In addition, high temperatures can significantly diminish tomato yield since they result in smaller, lower-quality fruits and reduced fruit set percentage (Bhardwaj 2012). Increased temperature after pollination in pepper decreased setting of fruits, demonstrating that fertilisation is vulnerable to extreme heat stress. However, pre-anthesis temperature exposure had no effect on pistil or stamen viability (Erickson and Markhart 2012). High temperatures results in flower and fruit dropping, abortion of ovule in green chilli (*Capsicum annuum*) while affect the red colour development in mature fruits of chilli (Arora et al. 2010). Additionally, seeds of *Citrullus lanatus, Cucurbita pepo, Cucurbita maxima* & *Cucurbita moschata* will not germinate above 42 degree Celsius. Seeds germination of *Cucumis sativus* and and *Cucumis melo* var. *cantalupensis* is significantly inhibited at 42 and 45 °C, respectively (Kurtar 2010). In melons, fluctuations of temperature results in delay ripening and also decrease the sweetness of fruit. Low yield of warm season crops such as *Benincasa hispida, Lagenaria siceraria* & *Cucurbita moschata* is a result of poor production of female flowers and also due to vegetative growth increased under warm humid climate (Ayyogari et al. 2014). The author also reported that high temperatures have resulted in increased abscission of floral buds, blooms, and immature pods, as well as decreased pod size and number of seeds per pod. When cole crops are produced for vegetables, high temperatures results in bolting, which is undesirable (Abdelmageed et al. 2004).

1.2.1.1 Plant Response to Heat Stress

Even slight temperature fluctuations can be detected by plants (Steven 2008). This process involves a large number of molecules, including transcription factors, chaperones, and osmoprotectants (Jacob et al. 2017). ER-UPR, Cyt-UPR, and membrane cyclic nucleotide gated calcium channels (CNGCs) appear to be heat stress sensors (Hasanuzzaman et al. 2013; Singh et al. 2020). Decreased rate of photosynthesis is caused by heat stress because it lowers internal CO_2 concentration, stomatal conductance, and leaf water potential (Hasanuzzaman et al. 2013). Heat stress, however, also has an impact on photosynthesis by altering stromatic reactions, grana and thylakoidal membrane architecture, as well as the grana and stromatic responses. Photosynthetic pigments are diminished by the lipid peroxidation of membranes, and PSII activity (Fv/Fm) is similarly decreased. Additionally, a decrease in Rubisco subunit proteins was observed under heat stress (Sage et al. 2008) and of the creation of starch and sucrose (Hasanuzzaman et al., 2013). Additionally, the ROS singlet oxygen ($1O_2$), superoxide radical (O_2), hydrogen peroxide (H_2O_2), and hydroxyl radical (OH^-) are produced during heat stress (Hasanuzzaman et al. 2013). Oxidative stress harms cell membranes by resulting in lipid peroxidation, protein breakdown, and the inhibition of root development after heat stress.

This happens as a result of the root cells' increased O_2 levels (Hasanuzzaman et al. 2013). Heat stress can have an impact on a plant at different phenological phases. However, the high temperature during blooming can entirely stop all grain production, while it can cause blossoms to drop off, give birth, or generate sterile flowers during the reproduction stage.

At different phenological stages, heat stress can have an impact on a plant. While it can cause flowers to droop, abort, or produce sterile blooms during the reproduction stage, the high temperature during blooming can completely cease all grain production (Maheswari et al. 2012). A shift in meiosis in both the male and female organs, which decreases pollen fertility and results in aberrant ovule and stigma development, is what causes heat-induced sterility in flowers (Cao et al. 2008). The rise in ethylene under high temperature provides an explanation for how heat affects pollen sterility. The metabolism of starch appears to be inhibited by ethylene, which results in less grain filling and sterile grain. The effect of high temperatures that is most concerning overall is yield decline. It has been demonstrated that tolerance species are less impacted by cereal pests than sensitive ones, with a 1000 grain weight reduction of up to 8% in sensitive rice species and up to 4% in tolerant species (Bashandy and El-Shaieny 2021).

1.2.1.2 Plant Heat Stress Defence Mechanisms

Living things have been classified as either psychrophiles, which prefer growth temperatures between 0 and 10 °C, mesophyles, which prefer growth temperatures between 10 and 30 °C, or thermophyles, which prefer growth temperatures between 30 and 65 °C. (Zróbek-Sokolnik 2012; Giordano et al. 2021). However, based on how they respond to high temperatures, the different species have been grouped into sensitive, tolerant, and resistant species, and their adaptation to high temperatures has been split into avoidance and tolerance (Hasanuzzaman et al. 2013). Stress brought by the heat that has similar effects as drought. The morphological and physiological avoidance methods that plants use when they are under heat stress include changing the direction of their leaves, closing their stomata, increasing their density, and changing the lipids in their membranes (Hasanuzzaman et al. 2013). To shield their leaves from heat, plants can also produce tomentose hairs, which grow thickly on their surfaces and cuticles. To prevent too much light from being reflected, leaves are either oriented parallel to the sun's rays or they roll up to limit water loss (Zróbek-Sokolnik 2012). Smaller leaves can handle heat better than larger ones because they experience less resistance on their surface as heat is expelled from the inside to the exterior. Additionally, in order to survive periods of high temperatures, plants attempt to complete their reproductive cycle when the weather is favourable. Many plants adapt their photosynthesis to high temperatures via the C_4 and CAM processes, which helps them survive in arid situations (Zróbek-Sokolnik 2012).

Other pathways that might lead to tolerance include those involving ion transporters, proteins from the late embryogenesis abundant (LEA) family, osmoprotectants, antioxidants, and transcriptional regulators (Hasanuzzaman et al. 2013; Singh et al. 2020). Tolerance though, can differ between tissues, organs, and species within a single plant. Other method of avoiding the damaging effects of heat stress is by the creation of heat shock proteins (HSP) (Xu et al. 2013). Five of these proteins work with other chaperones to repair damaged proteins while also being involved in systems for avoiding stress. Their elimination from the plant determines developmental changes (Raza et al. 2020). Under conditions of heat stress, salicylic acid supports the regulation of other stress hormones (Raza et al. 2019). The quantity of ascorbate, glutathione, tocopherols, and carotenes is increased by the plant as well as a number of antioxidant enzymes in response to the oxidative stress brought on by heat stress (Raza et al. 2019). Beyond a certain temperature, antioxidant enzymes lose their ability to function. For instance, it was shown that tolerant genotypes, as opposed to sensitive ones, showed an increase in the activities of catalase, ascorbate peroxidase, and superoxide dismutase under circumstances of heat stress. But at temperatures above 500 °C, these enzymes are inactive (Zeng et al. 2021).

1.2.1.3 Heat Responsive Genes

With a high degree of conservation and fairly predictable patterns, the heat-shock reflex is a biological response that almost all species display. At the transcriptional level, the heat shock response is characterised by a complicated alteration in mRNA and protein production as well as a significant accumulation of a collection of typically conserved proteins.

Genes are activated as a result of heat stress, especially if it losts for a long period, although the Heat Shock Proteins (HSPs) and related transcriptional activators are the main sources of an increased accumulation of transcripts (Guo et al. 2016). The term "HSPs" refers to a diverse class of proteins that was first studied in model organisms. HSPs are divided into five classes based on their molecular weight, though there are other classifications as well. The tiny HSPs are particularly prevalent in terrestrial plants (Waters and Vierling 2020). However, more recent studies in the fields of transcriptomics and functional biology have shown that the relevance of HSPs is far broader than previously thought. It essentially entails having a major impact on the stability and folding of proteins. It includes an essential part of cellular homeostasis under normal circumstances as well as defence against biotic and environmental stress. The fact that HSPs have been used successfully in the vegetable industry to increase post-harvest freezing tolerance is only superficially conflicting (Aghdam et al. 2015). A very extensive multigene superfamily, whose members are categorised based on their intracellular localization and function, encodes HSPs in plants. In plants, HSPs are encoded by a very large multigene super family, whose members are classified according to their intracellular location and function (Wang et al. 2004). The molecular response to heat stress in plants involves genes involved in a variety of processes, such as perception of stress, inhibition of "normal" protein and mRNA synthesis, preservation of cellular functions, development of thermo tolerance, long-distance signalling, and the beginning of morpho-physiological adaptation to long-term stress (Qu et al. 2013). The association of heat inducible genes with primary and secondary metabolism in plants, as well as with basic biological functions including transcription and translation, phytohormome signalling, and post-translational modifications, is not surprising. For heat responsive genes, post-transcriptional regulation, changing inducibility, and transcriptional control are all conceivable (Janni et al. 2020). Additionally, the modulation of the transcriptional regulation in plants under heat stress is governed by the small RNAs (Sunkar and Zhu 2004). Different genes are overexpressed in different crops when they are exposed to heat stress (Plantenga et al. 2019; Aleem et al. 2020; Singh et al. 2020).

1.2.2 Drought

Climate change is anticipated to have a considerable influence on water availability, and high levels of water stress will lower agricultural output in general and vegetable productivity in particular. Drought is the main reason for crop loss in dry and semi-arid areas & significantly lowers down the average yields by more than 50% for the majority of agricultural crops (Sivakumar et al. 2016). Lack of rainfall or inadequate soil moisture can produce drought stress can result in a number of biochemical, plant physiological, and genetic processes that drastically restrict crop growth (Vadez et al. 2012). The seeds germination in vegetable viz. onions and okra as well as the sprouting of potato tubers are negatively impacted by the presence of drought conditions (Arora et al. 2010). The drought environment causes embryo abortion in tomato flowers (Bhatt et al. 2009). Water stress during the reproductive stage was reported to have reduced tomato output by more than 50% (Srinivasa Rao and Bhatt 2012). Water stress during the flowering stage is thought to decrease the amount of nutrients given to the floral organs during assimilation and photosynthesis, which may enhance the pace of abscission. Increased soil solute concentration brought on by drought stress causes water to osmotically drain out of plant cells. Moisture stresses lowers the production of most vegetables by increasing the loss of water in the cells of plants and a variety of biochemical and physiological processes, including respiration and photosynthesis (Abewoy 2018). In addition to slowing down photosynthetic rate by lowering stomatal conductance, drought stress causes metabolic disfunction (Yordanov et al. 2013). When there is not enough water available, photosynthesis & the ability of photosynthesis are diminished. Reduced invertase activity may alter the body's capacity to metabolise sucrose, which could impact ovary growth and hexose concentration (Abewoy 2018).

1.2.2.1 Plant Response to Drought

In response to drought, plants have developed defences that differ depending on the species, the severity, and the length of the drought (Zhu et al. 2020). Numerous physiological traits are linked to water availability and can be employed as indicators of drought stress, such as leaf water potential, osmotic adjustment (OA), maximal quantum yield of PSII (Fv/Fm), water consumption efficiency (WUE), cell membrane integrity, and relative water content (RWC) (Hasanuzzaman et al. 2014; Goto et al. 2021). The decrease in cell turgor pressure and associated processes (such stomatal closure) during drought stress suggest decreased water losses and decreased plant nutrient absorption from the soil (Escalante-Magaña et al. 2019; Taiz and Zeiger 2017). For instance, the tomato stomata close at a water potential between −0.7 and −0.9 MPa, whereas the pepper (*Capsicum annuum* L.) stomata close at a somewhat greater range between −0.58 and −0.88 MPa (Chatterjee and Solankey 2015). Stomata closure decreases transpiration on the one hand while decreasing gas exchange and photosynthetic rate on the other. Other metabolic activities, such as the effectiveness of carboxylation, the quantity and regeneration of Rubisco, and the suppression of PSII activity are also impacted by prolonged drought stress. Since, drought-tolerant species have high WUE and can open their stomata quickly when the water deficit is alleviated, which enable them to fix carbon under stress (Chatterjee and Solankey 2015). Plants can modify OA by accumulating organic solutes in their tissues in order to combat the physiological harm brought on by drought (Taiz and Zeiger 2017). The cellular osmotic potential is decreased by an increase in solute concentration, allowing water to enter the cells and stabilising their turgor. It has been demonstrated that in Brassica species, a high value of OA enables the extraction of water from 90 to 180 cm deep soil layers (Chatterjee and Solankey 2015). Additionally, genotypes with high OA allow for the maintenance of a high turgor pressure even when the leaf potential is below 2.4 MPa (Chatterjee and Solankey 2015). Proline, glycine betaine, sugars (trehalose and sucrose), polyols (sorbitol, mannitol, arabitol, and glycerol), and other low-molecular-weight molecules, such as dimethylsulfonium propionate (DMSP), are among the most often found solutes in water shortage conditions (Escalante-Magaña et al. 2019; Sun et al. 2015). All of these substances serve as osmolytes, allowing the body to absorb water, as well as stabilisers and stress-resistance agents for proteins, cell membranes, chloroplasts, and liposomes (Razi and Muneer 2021). Proline has received a great deal of attention in particular because of its multifunctional role in plants'reactions to stress as a source of nitrogen and energy for cells as well as a radical scavenger (Sun et al. 2015). On the other side, it was discovered that proline contributes to the creation of cell wall proteins, such as extensin, that provide mechanical support under stress. In response to diverse abiotic pressures, glycine betaine builds up in a variety of species (plants, mammals, bacteria, cyanobacteria, and algae) (Escalante-Magaña et al. 2019). Because the light that is caught cannot completely be converted into chemically bound energy under drought stress, the pace of photosynthetic growth is delayed, and the energy surplus causes photo inhibition, which results in a decrease in the maximum quantum yield of PSII reaction centres (F_v/F_m). Several methods, include non-photochemical quenching, photorespiration through the Mehler reaction, and dissipation of non-radiant energy, and chlorophyll content management, can counteract the harmful effects of photo inhibition. Fv/Fm can be used to both identify water stress circumstances and identify genotypes that are resistant to and susceptible to water stress. As an illustration, tomato genotypes tolerant against moisture stress and maintained good PSII activity, leading to better photosynthetic activity than susceptible genotypes (Chatterjee and Solankey 2015). The ratio of photosynthesis (ACO_2) to transpiration (E) over time can be used to define water usage efficiency. Additionally, it shows the proportion of dry matter accumulation to water use throughout the growing season. WUE stands for a genotype's ability to efficiently absorb water from the soil in water-stressed situations. The genotypic differences in WUE, which are based on the ability to use soil water for absorption rather than transpiration, can help distinguish between tolerable and sensitive genotypes (Chatterjee and Solankey 2015).

Stress from the environment can disturb the cell membrane because it changes how permeable it is and causes ions to leak out. The loss of electrolytes from the cell can be used to measure this loss. The electrical conductivity of cytoplasm can be used as a gauge of drought tolerance since tolerant genotypes keep their membranes intact and display less electrolyte leakage than sensitive genotypes do (electrolyte leakage) (Chatterjee and Solankey 2015). Another drought indicator is the RWC, which shows the hydration status of plant tissues under water stress. Although this decline is genotype-specific, it reduces as the water shortage grows (Zhu et al. 2020). Abiotic stressors like drought often result in an increase in the amount of ROS in plant tissues (Sun et al. 2015). When oxygen is reduced by reducing molecules, ROS are created. When a plant is under stress from its environment, its stomata close to prevent water loss and the amount of CO_2 in its leaf's decreases. This causes oxygen to be reduced in its radical forms, including superoxide (O_2), hydrogen peroxide (H_2O_2), and hydroxyl radical (HO), through the action of NADPH, or reduced ferredoxin. It's conceivable for macromolecules like proteins, lipids, and nucleic acids to react with the recently created ROS (Sgherri et al 2018).

1.2.2.2 Defence Mechanisms Against Drought Stress

According to the research, resistance to drought stress is produced by combining three separate defence mechanisms, namely escape, avoidance, and tolerance (Kumar et al. 2012; Giordano et al. 2021). Plants'ability to complete their life cycle before the start of drought conditions allows for the escapement process. Plants mature more quickly as a result of this response, which is caused by the shorter maturation durations for the distinct phenological phases (Kumar et al. 2012). On the other hand, the avoidance approach results in high water potential in plant tissues by increasing water absorption and reducing water losses from cells during the dry periods. This is achieved via a variety of methods, such as decreasing the canopy and leaf area, which decreases transpiration by diminishing the perception of solar radiation. Stomata closure, the production of cuticular wax, and adjustments to root density and length are also a part of this system (Kumar et al. 2012; Giordano et al. 2021). Finally, plants may endure drought stress if they maintain cellular turgor and water loss in circumstances of moisture deprivation and low water potential. This can be accomplished by increasing the cytoplasmic concentration of solutes (such as OA), increasing the flexibility of cell membranes, and shrinking the size of the cells (Kumar et al. 2012). Despite the fact that plants'ability to endure in water-stressed conditions has overall advantages, the adaptation processes may result in undesirable features that have a detrimental effect on crop yield and output. For example, stomatal closure or a reduction in leaf area significantly affect CO_2 absorption, which affects biomass output and total yield (Kumar et al. 2012). While osmolytes accumulation may have a negative impact on the quality of the finished product, osmotic regulation by osmotic adjustment typically entails a significant energy expenditure, reducing the amount of photons available for biosynthetic processes. Therefore, it is imperative to establish the ideal balance between crop performance and drought survival, especially in commercial farming settings.

1.2.2.3 Drought Responsive Genes

Numerous species have been shown to include genes that alter physiological and morphological features in response to drought stress. For instance, the expression of numerous genes' recessive alleles dictates root thickness whereas the activity of several genes and the expression of their dominant alleles defines the length and number of roots (Kumar et al. 2012). The genes involved in the accumulation of solutes (such as the mannitol-accumulating mtlD gene or the proline-accumulating P_5CS gene) useful for reversing the decline in water potential in plants, encode various enzymes required for the production of these molecules.

Several species have previously discovered several of these genes, and their overexpression results in particular responses to drought tolerance (Kumar et al. 2012). DREBs/CBFs and ABF_3 genes that produce transcription factors that provide resistance to cold, salt, and drought stress; The production of transcription factors by the $SNAC_1$ gene increases stomata's sensitivity to ABA, which reduces water loss. When there is dryness, the ERA1 gene decreases stomatal conductance; Mn-superoxide dismutase is produced by the Mn-SOD gene, which confers resistance to several forms of stress; Root development is influenced by the AVP_1 gene; Osmo-tolerance is mediated by the build-up of proline and mannitol via the P_5CS and mtlD genes. Increased photosynthetic rate and tolerance to water deprivation during drought stress are mediated by the $GF_{14}l$ gene; WUE and stomatal conductance are affected by the NADP-Me gene. Under drought stress, the wilty gene contributes to the withering of tomato leaves. (Kumar et al. 2012; Giordano et al. 2021).

1.2.3 *Salinity*

Salinity is a serious problem that inhibits the growth and production of vegetable crops in many salt-affected areas. Throughout the course of a plant's development, many agricultural crops, including the majority of vegetables, are highly sensitive, suffer from excessive soil salt, which lowers output (Abewoy 2018). Because of the comparatively large solute concentrations in the soil, salinity physiologically imposes an initial water deficit, results in ion-specific stressors from changing K^+/Na^+ ratios, and creates a build-up of Na^+ and Cl^- concentrations that are harmful to plants. Salt stress causes the plant to lose its turgor, grow more slowly, wilt, shed its leaves, produce less respiration and photosynthesis, lose its cellular integrity, develop tissue necrosis, and eventually die (Cheeseman 2008). Onions are extremely vulnerable to saline soils, but cucumbers, eggplant, peppers, and tomatoes are just marginally sensitive to them.

In cabbage, salinity significantly lowers the germination rate, germination percentage, length of the fresh shoot and root weight (Jamil and Rha 2014). Because the mechanism in early, growing potato leaves that prevents salt build up from occurring has been disrupted, the combined stress of saline and heat causes vegetative development recovery to fail. The leaf area index and canopy function are adversely affected by this (Bustan et al. 2004). As stated by Lopez et al. (2011), In chillies, salinity lowers down the net absorption rate, relative growth rate, leaf area, and dry matter production.

The author also mentioned that salinity has a greater impact on fruit production per plant than it does on fruit weight. All cucurbits lose fresh and dry weight due to high salt content. Relative water content and total chlorophyll content changes are related to these changes. Salt stress in bean plants inhibits growth and photosynthetic activity, as well as changing the conductivity, number, and size of stomata. Salinity reduces transpiration and cell water potential in bean plants (Kaymakanova et al. 2008). It is well recognised that the excessive saline levels of soil and irrigation water influence many metabolic & physiological processes, which inhibits cell development.

1.2.3.1 Plant Response to Salinity Stress

Dissolved salts in the soil solutions come into direct contact with roots due to the osmotic effect and may hinder plant growth by reducing water potential in leaves and tissues. Plant development and production will be hampered by an excessive concentration of salts since these substances might interfere with key processes including among other things, germination, photosynthesis, nutrition balance, and redox balance (Petretto et al. 2019; Parihar et al. 2015). For instance, salinity may hinder germination because it lowers the osmotic potential of the germination media and inhibits seed imbibition (Parihar et al. 2015) & it also modifies the activity of enzymes necessary for the metabolism of proteins and nucleic acids (Parihar et al. 2015). Species and cultivars utilised, as well as salt levels, all affect salinity's impact on germination differently (Lauchli and Grattan 2007). Salinity and germination rate typically have a negative association, as evidenced by studies on rice, wheat, maize, *Brassica* spp., and tomato (Giordano et al. 2021). Two distinct phases of the salinity influence on plant development can be seen (Parihar et al. 2015). The initial stage of salinity has no detrimental effects on plant growth because Na^+ and Cl^- that enter the xylem are collected in the vacuoles and the meristems continue to grow by feeding through the phloem. Only the decrease in the growth of the leaves and roots is seen during this period (Parihar et al. 2015). During the second phase, salts accumulate in plant tissues, but cells are unable to store them in vacuoles. As a result, the cytoplasmic salt content increases and various enzymes' activities are significantly reduced. Salinity also affects photosynthesis because it lowers plant water potential and produces more chlorophyll. Cl was found to particularly inhibit chlorophyll synthesis, and 490 mg kg^{-1} of Cl in the soil may result in a 10% decrease in crop yield (Parihar et al. 2015). However, there are species-specific differences, and the levels of Cl required for plant growth can vary from 4–7 mg g^{-1} for species that are Cl sensitive to 15–50 mg g^{-1} for species that are Cl tolerant (Parihar et al. 2015). Salt can also lessen the number of carotenoids and xanthophylls as well as the intensity of chlorophyll's fluorescence, as shown in mung beans. However, it was shown that chlorophyll b is more susceptible to a rise in salt than chlorophyll a (Saha et al. 2010). Salinity can jeopardise the nutritional balance of plants because of its adverse impact on nutrient uptake and delivery within the plant (Parihar et al. 2015). Salinity significantly affects pH and redox. Potential of soil solution, which decreases the solubility of micronutrients. However, the species and salinity levels affect the micronutrients status (Parihar et al. 2015). Salinity may reduce the absorption of nitrogen due to interactions between sodium and NH_4^+ or chlorine and nitrate (Parihar et al. 2015). Atriplex griffithii's shoots and leaves showed decreased Ca absorption as a result of excessive salinity, although K levels were steady in the roots and decreased in the leaves. Due to the high salinity, the shoots and leaves of *Atriplex griffithii* exhibited decreased Ca absorption, but the K levels in the roots and leaves were stable (Parihar et al. 2015). Additionally, it was demonstrated that soluble ions (Na^+, Cl^-, SO_4^{--}) in the soil reduced phosphate absorption (Parihar et al. 2015). As the stomata close, an excessive amount of energy is transferred to oxygen, resulting in the creation of superoxide (O_2), hydrogen peroxide (H_2O_2), hydroxyl radical (OH-), and singlet oxygen, high salinity may cause oxidative stress comparable to drought and heat stress ($^1O_2$) (Raza et al. 2020; Petretto et al. 2019).

1.2.3.2 Plant Salinity Defence Mechanisms

Halophytes are referred to as plants that can survive in salty environments. The ability of plants to tolerate salt is influenced by a variety of traits at the genomic, proteome, and metabolomic levels. The plant uses a variety of strategies to deal with high salt concentrations, including reducing, absorbing and moving of salt or compartmentalising and extruding salts from plant tissues. The transcription level of proteins is altered during salt stress, but post-transcriptional alterations are also an issue. The answers provided by genomic and transcriptomic analysis can therefore be completed by proteomic analysis. A proteomic investigation of a protein's function, post-transcriptional alterations, subsequent interactions with other proteins, and distribution in the cell and tissues may all be used to learn more about a protein's function in salt stress tolerance. Different protein types undergo functional group changes during salt stress. Among these are ion transporters, energy-metabolizing proteins, and signal proteins. Saline stress activates some of these proteins, such as the calcium-binding calmodulin and annexin. They are involved in the transmission of abscisic acid signals. The transmission of saline stress signalling is mediated by additional guanosine triphosphate-binding proteins from the Rab family (GTPase). Under saline stress, $OsRPK_1$ protein kinase controls plasma membrane H^+-ATPases and restores ion homeostasis. Additionally discovered to accumulate during salt stress, the glutamate ammonia ligase protein aids with plant tolerance. It contributes to the synthesis of amino acids and the uptake of nitrogen.

Additionally, it was shown that salt stress causes the protein glutamate ammonia ligase, which is essential for assimilating nitrogen and producing amino acids, to build up (Kumar Swami et al. 2011) as well as glutamine synthetase. Several of these proteins are down-regulated in salt-sensitive plants, such as potatoes, emphasising their significance for salt tolerance. Protein breakdown is another feature of the changes observed at the protein level during salt stress.

An illustration of this is the rise in the ATP-dependent metalloprotease FtsH-like protein, which breaks down PSII's D1 core component. Under saline conditions, osmotic stress causes stomata closure, which lowers CO_2 absorption. This decrease affects the proteins that make up the Oxygen evolving enhancer (OEE) complex proteins, Rubisco subunits, and Rubisco activase. NADPH reductase and CP_{47} protein in ferredoxin, which protects D1 cells from salt stress, are additional proteins whose concentrations fluctuate (Parihar et al. 2015). Salinity causes the enzymes necessary for the manufacture of numerous hormones to increase, including the amounts of ethylene, abscisic acid, jasmonic acid, and giberellins (Raza et al. 2020). Lipid metabolism also has a role in salinity. For instance, the decrease of monogalactosyl diacylglycerol synthase, an enzyme of the galactosylglycero lipids of the membranes of chloroplasts and thylakoids, jeopardised the integrity of cell membranes (monogalactosyl diacylglycerol, digalactosyl diacylglycerol) (Parihar et al. 2015).

1.2.3.3 Salinity Responsive Genes

Plants that are tolerant seem to have particular genes that sensitive plants lack. The literature has identified three gene families as being crucial for salt tolerance: (i) genes that control how ions are transported and absorbed; (ii) genes that modulate osmotic pressure (iii) the genes that influence plant development (Munns 2005). Analysis of Arabidopsis mutants prone to high external Na^+ concentrations allowed for the discovery of three SOS genes implicated in salt tolerance (Parihar et al. 2015) (i): SOS_1 encodes a protein kinase that activates SOS_1, SOS_2, and SOS_3. SOS_3 also produces a calcium-binding protein that activates SOS_2, as well as a plasma membrane Na^+/H^+ transporter that transports sodium in the apoplast. Additionally, SOS_4 (a fourth gene) appears to control SOS_1, SOS_2, and the $SCaBP_8$ protein, which is controlled by SOS_2 (Parihar et al. 2015). Not only was salinity tolerance seen in Arabidopsis plants with overexpressed SOS genes, but also a decrease in Na^+ accumulation and an increase in K^+ accumulation. (Yang et al. 2009). SOS_1, SOS_2, and SOS_3 genes have been linked to tolerance to salt stress and a high Na^+/K^+ ratio in Brassica and rice (*O. sativa* L.) plants (Giordano et al. 2021). Other genes produce suitable solutes, osmolytes, or osmo protectors. Four classes of these osmolytes are distinguished: N-containing solutes include sugars like sucrose and raffinose, straight-chain polyhydric alcohols (polyols), such as mannitol and sorbitol, and cyclic polyhydric alcohols (cyclic polyols) (Parihar et al. 2015). The genes involved in plant development are all influenced by hormones, transcription factors, signal molecules, and environmental stressors. Examples of stress sensor molecules that can vary their concentration or proteins that can change their structure in response to drought, salt, and cold include the molecules that move from roots to shoots to create tolerance to salinity stress (Parihar et al. 2015).

1.2.4 Flood

Another important abiotic stress is flood by which the development and productivity of vegetables are negatively affected, which are typically thought of as crops prone to flooding (Parent et al. 2008). O_2 deficit is typically brought on by flooding situations because of the sluggish gas diffusion in water and the O_2 consumption of microbial life and plant roots. The majority of crops are extremely vulnerable to flooding, and tomato in particular exhibits minimal genetic diversity in this regard. Vegetables damaged by flooding are typically harmed because there is less oxidation in zone of root, which hampered aerobic functions. *Solanum lycopersicum* plant that have been flooded build up endogenous ethylene, which harms the plants (Drew 2009). In tomatoes distinctive responses to wet conditions is the quick development of epinasty growth of the leaves, and the significance of ethylene build-up has been suggested. Rising temperature make flooding symptoms more severe, and tomato plants typically wilt and die quickly after a brief flooded interval during a hot season (Kuo et al. 2014). Flooding during bulb development in onion also causes yield lose up to 30–40%. The growth stage, duration, and intensity of the shocks, which collectively account for more than 50% of yield losses globally, determine how plants react to environmental pressures (Kumar 2017).

Vegetable plants' physiology is impacted by flooding. Reduced stomatal conductance is one of plants'earliest physiological responses to soil flooding (Folzer et al. 2006). It increases the amount of CO_2 inside the plant, slows down carbon exchange significantly, increases CO_2 concentration, and ultimately raises leaf water potential. It also causes a decrease in stomata conductance (Liao and Lin 2014). Flooding has a detrimental impact on plant development, both vegetative and reproductive as a result of its deleterious effects on physiological function. Flooding harms sensitive crop plants by causing leaf chlorosis and reducing root and shoot development as well as the build-up of dry matter and overall plant production. Storms can increase the windborne distribution of spores, heat waves and drought can make all the plants more prone to disease & flooding can facilitate the transmission of water-borne infections (Abewoy 2018).

1.2.5 Responses of Insects and Diseases to Climate Change

Additionally, climate change affects insect pests'ecology and biology (Jat and Tetarwal 2012). In several insect groups, higher temperatures encourage reproduction and expedite the conclusion of the life cycle of insects with short life cycles, such as aphids and diamondback moths. Consequently, these are able to produce more generations every year than they often do (FAO 2009). Contrarily, certain insects' life cycles might take a long time to complete.

Some insect species that spend all or part of their life cycle under soil likely to suffer more than insects that are present above the soil surface because soil acts as an insulating medium that tends to buffer temperature changes more than the air (Bale et al. 2002). Insect species migrate to higher latitudes as a result of temperature rise, however in the tropics, greater temperatures may have a negative impact on certain nuisance species. High atmospheric temperatures lead to an increase in insect development and oviposition rates, insect outbreaks, and the introduction of invasive species, while the capacity of fungi for biological control of insects increases, the accuracy of economic threshold levels, insect diversity in ecosystems, and parasitism all decrease (Das et al. 2011).

The growth of the Colorado potato beetle, European corn borer, onion maggot, and cabbage maggot will all be accelerated by rising temperatures (Newton et al. 2011). The breeding season will last longer as a result of rising temperatures, which will further boost reproduction. Insects may be able to reach their minimum flying temperature with increased temperatures, earlier, enhancing their ability to disperse, according to studies on aphids and moths (Zhou et al. 1995). Because stored nutrients are used up more quickly at warmer temperatures, the length of an insect's diapause is shortened. Warming in winter may delay the beginning of diapauses in insects, while warming in early summer may hasten their end, allowing the insects to resume active growth and development. This has a significant consequence that, in a global warming scenario, a temperature increases of between 1 and 5 °C will improve the survivability of insects because of population increase, early infestations, decreased winter mortality, and agricultural loss brought on by insect-pests as a result (Harrington et al. 2010; Abewoy 2018).

1.2.6 Adaptive Management Strategies for Climate Change

The implementation of suggested production processes should be prioritised in order to enhance the efficiency of water consumption and adjust to the dry, hot conditions. According to Welbaum (2015), methods like shifting planting or sowing dates should be used to reduce the predicted spikes in temperature and water stress during the agricultural growing season. Increasing nutrient availability can be done by improving soil fertility and nutrient absorption with soil amendments and changing fertiliser application. The two crucial low-cost interventions are conserving soil moisture reserves and providing irrigation throughout key stages of crop growth (Malhotra 2016). The use of plastic mulches and crop leftovers as mulches, among other crop management technique, helps to preserve soil moisture.

Moreover, some colourful plastic mulches also useful for management of insect-pests and diseases in various vegetable crops under climate change scenario (Table 1.1).

Table 1.1 Use of coloured mulches in vegetable cultivation

Mulch colour	Advantages/ Benefits	Vegetable crops
Transparent	Soil solarization, moisture conservation.	Crop raising can be performed in temperate areas/ season
Black	Preventing the soils from sun light, increases soil temperature, suppress weed growth, reduce moisture loss and increase crop yield.	–
Silver	Increase crop output, deters certain aphids and whiteflies, and lowers soil temperature, all of which help prevent or postpone the spread of insect-transmitted viruses.	Okra, chilli and capsicum
Red	Increasing soil temperature, supress weed growth, soil moisture conservation, increasing crop yield, suppression of nematodes and early blight.	Tomato
Blue	The colour is appealing various thrips species which are responsible for transmission of viral diseases in many cucurbitaceous vegetables and potato.	Cucumbers, summer squash, bitter gourd
Green IRT	Supress weed growth, mild increasing soil temperature.	Muskmelon/Cantaloupe
Yellow	Serves as a trap to attracts whitefly, cucumber beetle, aphids and ultimately prevent damage to the main crop.	–

Modified from Chandra (2009)

Growing crops in raised beds could help in situations where too much soil moisture from a lot of rain becomes a serious issue (La Pena and Hughes 2007). The authors also mentioned that growing vegetables might be done using clear plastic rain shelters, which would have less adverse effect on developing fruits and would also result in less flooding of the fields during the rainy season. During rainy season, vegetables which are grown on raised beds produce more because of enhanced drainage, which lessens anoxic stress on the root system (Welbaum 2015).

Increased ability to withstand stress through to grafting: Vegetable grafting was first used in East Asia in the twentieth century to lessen the impacts of soil-borne diseases like fusarium wilt, which lowers the output of vegetables including tomato, eggplant, and cucurbits (Lee et al. 2008). In Asian countries like Japan and Korea, grafting is recognised as a standard practise in the cultivation of vegetables. This practise is an effective, quick substitute for the generally slow breeding methodology meant to increase the ability of horticultural crops, and vegetables in particular, to withstand environmental stress (Martinez Rodriguez et al. 2010). One of the most effective ways to modify a plant‘s root system and increase its resistance to numerous abiotic challenges is by grafting (Bhatt et al. 2013).

Grafted plants are most commonly used in many vegetable crops to increase tolerance to abiotic stresses such as minimum and maximum temperatures, dehydration, salt, and flooding (Martinez Rodriguez et al. 2010). The production of grafted plants in crops like *Solanum lycopersicum*, *Solanum melongena, Capsicum annuum* and cucurbits (melon, cucumber, watermelon, and pumpkin) has expanded recently due to these advantageous effect of grafting (Lee et al. 2010). Grafting began with eggplants in the 1950s, then cucumbers and tomatoes in the 1960s and 1970s (Edelstein 2004). Yetisir et al. (2006) reported that grafted melons were shown to be highly resistance to salt as compare to melons that are not grafted when grown on hybrid squash rootstocks. However, rootstocks' susceptibility to salt varies widely between species, with rootstocks from *Cucurbita* spp. being maximum tolerant to salt than those from *Lagenaria siceraria* (Matsubara 2012). Because certain eggplant cultivars can withstand drought, eggplant rootstocks may offer protection against soil moisture stress in addition to floods. The plant's ability to withstand heat stress is improved when temperature-sensitive tomato varieties are grafted onto varieties of more heat-resistant rootstock. According to reports, tomato grafted plants grow more successfully compared to ungrafted plants under heat stress. Additionally, the fruit output of the eggplants (S. *melongena* cv. Yuanqie) increased by 10% after they were grafted onto a rootstock that is heat-resistant (cv. Nianmaoquie). Creating climate-resilient vegetables: The most affordable way for farmers to handle the problems of a changing climate is to develop improved, adaptable vegetable germplasm. (Altieri et al. 2015). However, the majority of contemporary cultivars only reflect a small portion of the genetic diversity that is now accessible, including stress tolerance. It's possible that features that might have helped new varieties adapt or tolerate low input and unfavourable circumstances were counter-selected when they were bred, especially for high-input, intense manufacturing systems in industrialised nations. The production of better varieties with a greater capacity for environmental adaptation may result from the identification of specific genetic variation for tolerance to various biotic and abiotic stimuli. The ability to recognise and create genotypes may be achievable with enhanced properties caused by superior allele combinations at several loci. To find out these enhanced genotypes and related features, particularly in related, wild species that thrive in conditions that are inhospitable to the establishment of their domesticated relatives, cultivated variety, improved selection procedures are required. Plants that are native to areas with substantial seasonality offer opportunities to discover the genes or gene combinations that confer such resilience because of their innate capacity to adjust to changing environmental conditions. Due to the genetic and physiologic complexity of this trait, attempts to increase salt tolerance in crops through traditional breeding programmes had relatively little success (Flowers 2008).

Furthermore, tolerance at one stage of plant growth does not always coincide with tolerance at other stages; tolerance is a developmentally regulated, stage-specific phenomenon. Effective screening techniques, genetic diversity, and the capacity to transfer genes to the target species are necessary for successful breeding for salt tolerance. Only a little amount of variety occurs in cultivated species, and the majority of commercial tomato cultivars are only mildly susceptible to increasing salt (Flowers 2008). In both domesticated and wild tomato species, there exist genetic variations for salt tolerance during seed germination. The capacity of tomato seed to germinate quickly under salt stress is genetically regulated, as demonstrated by a salt-sensitive tomato line (UCT5) and a salt-tolerant S. esculentum accession

(PI-174263) hybrid with a narrow-sense heritability (h^2b) of 0.75 (Fooland and Jones 2011). According to various research, the ability of tomato seeds to tolerate salt during germination is regulated by genes with cumulative effects and might be enhanced through directed phenotypic selection (Foolad 2014). Vegetable kinds that can resist high or variable levels of salinity and are suitable for various production settings may arise more quickly if the mechanism of salt tolerance at different growth phases is understood and salinity tolerance genes are introduced into vegetables. Vegetable varieties that can withstand heat, drought, and salt have been produced in India by the ICAR-IARI in New Delhi, the ICAR-IIHR in Bengaluru, the ICAR-IIVR in Varanasi, the CPRI in Shimla, as well as other recognised organisations and universities (Table 1.2).

1.2.7 Plant Biotechnology

Vegetable crop development has until recently, mostly been limited to traditional breeding methods. These programmes rely on both naturally occurring and purposefully induced random mutations, as well as interspecific hybridization of plants with desirable heritable traits. The insertion of additional genetic information may lead to greater taste, heightened sensitivity to environmental factors, or increased resistance to illnesses. The insertion of genetic material (single or multiple genes) and its integration into a recipient cell, which affects the plant genome, is known as plant genetic engineering, also known as genetic engineering of plants. Transgenic plants or genetically modified (GM) plants are those with altered genomes. The direct biotechnological change of an organism's genome is known as genetic engineering, often referred to as genetic modification. The term "genetically modified organisms" (GMOs) refers to creatures whose DNA has undergone a change that does not occur normally. The field is frequently referred to as "modern biotechnology," "gene technology," "recombinant DNA technology," or "genetic engineering." It makes possibility for some genes to be transferred between unrelated species as well as between different creatures. By using such techniques, GM plants are created, which are subsequently utilised to cultivate GM food crops. In recent years, cointegrate and binary vectors, which are modified versions of the *Ti* and *Ri* plasmids of the hairy root disease bacterium *Agrobacterium rhizogenes* and the crown gall bacterium *Agrobacterium tumifaciens*, respectively, have been employed to transmit genes coding for tolerance to biotic and abiotic stressors (Srivastava 2003).

In order to boost agricultural output in unfavourable conditions, advanced technology will be required to augment conventional tactics, which are usually insufficient to prevent yield losses due to environmental constraints. Genes have been identified, and their roles are now known. As a result, it is now feasible to genetically modify genes that are related to environmental stress tolerance. These instruments are expensive, but they also guarantee quicker and even stunning results. Several efforts use these genetic and molecular technologies, with varying degrees of success. Vegetable stress tolerance has not been extensively studied using

Table 1.2 Vegetable varieties with various stress tolerance released in India for cultivation

Cold tolerance	
Tomato	Pusa Sheetal, Pusa Sadabahar, Punjab Tropic, Pusa Hybrid-8
Radish	Pusa Himani
Potato	Kufri Sheetman and Kufri Dewa (frost tolerant)
Heat tolerance	
Tomato	Pusa Hybrid-1, Pusa Sadabahar
Cauliflower	Sabour Agrim
Bottle gourd	Pusa Santusthi
Cucumber	Pusa Barkha, Pusa Uday
Carrot	Pusa Kesar
Potato	Kufri Surya
French bean	Arka Garima
Garden pea	Arka Tapas, Arka Uttam, Arka Chaitra
Radish	Pusa Chetki
Cow pea	Arka Garima
Ivy gourd	Thar Sundar
Sponge gourd	Thar Tapish
Drought tolerance	
Tomato	Arka Vikas, Arka Meghali
Brinjal	Bundelkhan Deshi
Chilli	Arka Lohit
Potato	Kufri Sheetman, Kufri Sindhuri
Sweet potato	Sree Nandini, Sree Bhadra
Cassava	H-97, Sree Sahya
Onion	Arka Kalyan
Dolichos bean	Arka Jay, Arka Vijay
Pumpkin	Thar Kavi
Drumstick	Thar Harsha
Salt/high soil pH tolerance	
Tomato	Sabour Suphala
Eggplant	Pragati and Pusa Bindu
Okra	Pusa Sawani
Musk melon	Jobner 96-2
Spinach beet	Jobner Green
Onion	Hisar-2

Compiled from Singh et al. (2020), Koundinya et al. (2018), Solankey et al. (2015), Kumar et al. (2012)

molecular markers, although attempts are being made to pinpoint the QTLs that underlie this tolerance. In tomato, QTLs for drought resistance have been found. Based on 13C composition, in tomato species *S. pennellii* three QTLs were shown to be associated with water usage effectiveness. When *S. pennellii* was planted in Israel's wet and dry fields, three distinct yield-promoting zones were found, while

Foolad et al. (2010) discovered four QTLs, two of which were provided by *S. pimpinellifolium*, that were connected to seed germination drought tolerance. Additionally, *S. pimpinellifolium* is being researched in order to develop a salt tolerance. Salt tolerance appears to be quantitatively inherited according to QTL mapping. (Foolad 2014). Lin et al. (2006) discovered RAPD (random amplified polymorphic DNA) markers relating to tomato heat tolerance line CL5915 from AVRDC-The World Vegetable Centre, Taiwan. At the Centre, more research is being done to completely comprehend the genetic basis of CL5915's heat tolerance. Studies show that stress tolerance is quantitatively inherited, and that in some circumstances, it is influenced by the stage of growth of the plant. (Abewoy 2018). The researchers noted several limitations in conventional breeding, which can only be solved by developments in contemporary biology. Vegetable crops have been successfully engineered during the past 10 years to possess a variety of qualities, including resistance to biotic stress, quality, and storage life. Some of these features have already been marketed. Today, it is possible to modify vegetable crops for abiotic stress resistance, quality enhancement, therapeutic uses, and industrial applications.

Although transgenic vegetable crop commercialization has advanced at a rather modest pace, transgenic vegetables modified for nutraceutical and medicinal application will soon make a substantial contribution to value-added agriculture. Vegetable crops have been effectively improved through genetic engineering techniques based on the introduction of transgenes. These techniques primarily focused on increasing crop productivity, improving resistance to biotic and abiotic challenges, and producing vegetables with superior nutritional value. There are two fundamental methods for transferring genes into plant cells. These methods include vector-mediated genetic transformation techniques and direct transformation methods. Although several transgenic vegetables have been field-tested, development of transgenic vegetable crops for commerce is still moving at a glacial pace. Only tomato plants with extended shelf life and insect- and virus-resistant potatoes have been commercialized among vegetable crops. One significant drawback is that, even in industrialized nations, public acceptance of genetically modified food is not very widespread. The dissemination of knowledge regarding the benefits and safety of transgenic crops is crucial. In addition to producing more food to fulfil demand, there is a pressing need to use fewer chemicals in agriculture to maintain environmental quality and safeguard biodiversity. Additionally, this would increase the small farmers and others with limited access to resources' economic security.

In the third world, transgenic vegetables will be very helpful in reducing misery. Using modified veggies will make it simpler to defeat deadly illnesses like cholera, hepatitis, and diarrhoea. For transgenic technology to be used and accepted more widely, some application-related obstacles must be overcome. Environmental risks must be thoroughly evaluated on a case-by-case basis, including the effects of transgenic goods on human health and cross-pollination with closely related wild relatives of agricultural plants. Furthermore, there are several concerns about the use of antibiotic and herbicide resistance genes as selectable markers from the perspectives of ecological and human safety. The public may be more accepting of transgenic crops if alternative techniques are used to generate marker-free transgenic plants (De Vetten et al. 2003) building binary vectors or tiny chromosomes to allow for repetitive gene transfers (Goderis et al. 2002). Our capacity to add features with long-lasting benefit may be further increased through improvements in transformation methods for vegetable crops. Molecular markers have been developed to distinguish between various vegetable types in tomato, potato, onion, garlic, and their relatives. Numerous molecular markers have been used for DNA fingerprinting of cultivars and breeding lines in a range of vegetable crops, including *Solanum lycopersicum*, *Phaseolus vulgaris*, *Capsicum annuum*, and *Solanum tuberosum* (Amin et al. 2010). Major disease resistance in tomatoes has been related to molecular markers, including the powdery mildew resistance gene ol-1 on tomato chromosome 6, the tomato mosaic virus, and *Meloidogyne incognita* (Ansari 2015). The MAS has also been used to support backcross breeding operations in the following ways: either by using markers to select for (or against) a certain background genotype, or by using markers to identify the gene that is to be introgress.

Recombination between the marker and the target gene, a small degree of parental polymorphism with different traits, and decreased precision of QTLs as a result of environmental interaction all restrict the use of the MAS in agricultural plants. Unfortunately, there are currently no molecular markers available for a number of significant features that are regulated by several genes or polygenes. Finding answers to these issues is not difficult given recent advancements in both structural and functional genomics. The availability of high-density genomic and physical mapping enables the identification of markers that are physically closer to the target gene and would reduce MAS failure due to genetic recombination. Furthermore, the construction of allele-specific markers would be made feasible by the cloning and characterisation of the target genes, which is conceivable given their location on the linkage map. By using such markers, any chance of the marker-trait relationship breaking would be absolutely eliminated. Additionally, allele mining in the germplasm resources would be made easier by markers based on gene sequences, allowing for the discovery and use of novel alleles in crop development.

1.2.8 Climate Change in the Future and Its Effects on Vegetable Production

The atmosphere contains a lot of greenhouse gases that are persistent. As a result, even if emissions stopped growing, atmospheric concentrations of greenhouse gases would keep rising and staying high for hundreds of years. In addition, assuming concentrations were maintained and the makeup of the atmosphere today remained constant (which would necessitate a sharp decrease in existing greenhouse gas emissions), the temperature of the surface air would rise further. This is due to the fact that it takes decades for the seas, which store heat, to completely react to growing levels of greenhouse gas emissions. Over, coming decades to centuries, the climate will continue to be impacted by the ocean's response to rising temperatures and concentrations of greenhouse gas (IPCC 2013). By 2100, global temperatures are predicted to increase by 0.5 °F on average, reaching 8.6 °F, with increases of at least 2.7 °F predicted in all save the scenario with the most extreme cuts in greenhouse gas emissions.

The average global temperature is predicted to rise by at least twice as much in the coming 100 years as it did in the previous 100, barring the most drastic mitigation measure thought of. Over land than overseas, temperatures at ground level are anticipated to rise more quickly. It is anticipated that some regions of the world will experience greater temperature rises than the global average (IPCC 2013). In most places, the amount of rain that falls during extreme weather is predicted to rise, while storm trajectories are anticipated to move poleward (IPCC 2013). Through the end of the century, global average annual precipitation is anticipated to rise, although regional variations in precipitation volume and severity will be substantial. Precipitation occurrences' average heaviness is probably going to get heavier. High-latitude and tropical regions, where it is also projected that precipitation would rise overall, will be where this will be most obvious. Tropical storm-related wind speeds are probably going to get stronger. Additionally, it's expected that tropical storms will provide more precipitation overall. It is anticipated that annual average precipitation would rise in certain places and fall in others. The projection of regional variations in precipitation under two emission scenarios is shown in the picture to the right (IPCC 2013).

1.2.9 Conclusions

In order to fulfil the demands of a population that is constantly expanding, the world's agriculture, particularly the production of vegetables, is currently facing a challenging position. From a finite amount of land, we must produce more and more food. Increasing biotic and abiotic forces, deterioration in the state of the environment, and the threat of escalating global warming brought on by greenhouse gases all contribute to the problem's aggravation. The scrumptious veggies are highly susceptible to hot, dry, and flooding weather. The impact of climate change on agricultural growth, development, yield, and quality must thus be thoroughly studied. The development of adaption technology and calculating the crops' ability to reduce emissions should also be priorities. Increased temperatures have a negative impact on vegetable crop production and output by reducing crop length, blooming, fruiting, fruit size, and ripening/maturity.

It will be necessary to modify the present vegetable systems to accommodate the expected effects of climate change in order to boost the production and consumption of nutritious vegetables and reduce hunger and poverty in emerging nations. To lessen the effects of climate change on the quantity and quality of vegetable crop production, develop efficient adaptation techniques. Priority should be given to the development of industrial solutions for more effective water use that can be tailored to the hot, dry circumstances. The use of plastic mulches and crop leftovers as mulches, among other crop management techniques, helps to preserve soil moisture. Raised beds allow for crop cultivation, which is a key solution to the issue of excessive soil moisture brought on by heavy rain. Vegetable germplasm with resistance to drought, high temperatures, and other environmental problems as well as the capability to sustain output on marginal soils must be discovered in order to serve as sources of these traits for both public and commercial vegetable breeding programmes. These genetic materials will contain both domesticated and wild accessions with genetic diversity that is not present in currently popular cultivars. To introgress, uncover genes giving resistance to stressors, and simultaneously produce tools for gene isolation, characterisation, and genetic engineering, genetic populations are being produced. Agronomic practises that protect vegetable crops from adverse environmental conditions and conserve water must also be constantly developed and made widely accessible to farmers in impoverished countries. In addition to employing enzymes and antioxidants as defence mechanisms, secondary metabolism chemicals like phenyl propanoids and hormones that cause morphological and physiological changes in response to external stimuli can also help plants adapt to changing environmental conditions. Since all plants possess these defence systems, these reactions might be referred to as "innate tolerances." In addition to displaying these kinds of reactions, certain plants have evolved the capacity to thrive in challenging environments. This reaction reflects the "memory of stress" in plants and can be viewed as a "acquired tolerance." This reaction allows plants to adjust to periods of high stress followed by periods of low stress and to deal with the reappearance of stress more effectively, increasing their resistance when stress returns (Caplan et al. 2019). Extreme weather events have an impact on many parts of the world, not only those with dry or semi-arid climates. Because of how quickly these processes occur, organisms' genetic diversity cannot produce creatures that can tolerate unfavourable environments. In this situation, it is now possible to create more tolerant species using modern breeding techniques. Their practical implementation, however, takes a long time and is always attempting to keep up with the changing environmental factors, including biotic and abiotic stresses. In this case, cutting-edge agricultural practises like the use of bio-stimulants or well-established ones like grafting might give ecologically friendly tools to increase plants'tolerance to abiotic stress in response to escalating agricultural sector needs. As a result, given that vegetable crops encounter numerous and severe abiotic stressors in the actual world, understanding plant defence mechanisms and implementing eco-sustainable farming techniques may be able to help these priceless commodities survive a fast changing environment. Future research is also required to better understand how plants respond to numerous stressors, particularly when heat, drought, salt, and increased CO_2 levels are present together. Additionally, biotic stressors should also be taken into account. Finally, it is important to extensively screen local ecotypes and landraces to find those that can survive environmental stresses in an effort to understand how plants defend themselves.

References

Abdelmageed AHA, Gruda N, Geyer B (2004) Effects of temperature and grafting on the growth and development of tomato plants under controlled conditions. In: Deutsher Tropentag, Rural Poverty Reduction through Research for Development and Transformation; October 5–7, pp. 1–5

Abewoy D (2018) Review on impacts of climate change on vegetable production and its management practices. Adv Crop Sci Technol 6:330. https://doi.org/10.4172/2329-8863.1000330

Aghdam MS, Sevillano L, Flores FB, Bodbodak S (2015) The contribution of biotechnology to improving post-harvest chilling tolerance in fruits and vegetables using heat-shock proteins. J Agric Sci 153:7–24

Aleem S, Sharif I, Etlas A, Mehvish T, Nusrat P, Rasheda A, Muhammad N, Muhammad T, Hussain S (2020) Heat tolerance in vegetables in the current genomic era; an overview. Plant Growth Regul 92:497–516

Altieri MA, Nicholls CI, Henao A, Lana MA (2015) Agroecology and the design of climate change-resilient farming systems. Agron Sustain Dev 35:869–890

Amin A, Razvi SM, Dar ZA, Zafar G, Bhat MA, Mir MR, Wani N, Bhat KA (2010) Molecular markers and their application in improvement of vegetable crops. Int J Curr Res 4:38–48

Anonymous (2022) Production volume of vegetables in India from financial year 2008 to 2021, with an estimate for 2022. Statista Research Department, July 11, 2022, New York. https://www.statista.com/statistics/621210/vegetable-production-in-india/

Ansari AM (2015) Molecular markers in vegetable improvement. Hortic Biotechnol Res 1(1):5–10

Arora SK, Partap PS, Pandita ML, Jalal I (2010) Production problems and their possible remedies in vegetable crops. Indian Hortic 32:2–8

Ayyogari K, Sidhya P, Pandit MK (2014) Impact of climate change on vegetable cultivation-a review. Int J Agric Environ Biotechnol 7:145

Bale JS, Masters GJ, Hodkinson ID, Awmack C, Bezemer TM, Brown VK, Butterfield J, Buse A, Coulson JC, Farrar J, Good JEG, Harrington R, Hartley S, Jones TH, Lindroth RL, Press MC, Symrnioudis I, Watt AD, Whittaker JB (2002) Herbivory in global climate change research: direct effects of rising temperature on insect herbivores. Glob Chang Biol 8(1):1–16. https://doi.org/10.1046/j.1365-2486.2002.00451.x

Bashandy T, El-Shaieny AH (2021) Morphological and molecular marker screening for drought tolerance in Egyptian Jew's Mallow (Corchorus olitorius L.) landraces. Acta Univ Agric Silvic Mendel Brun 69:79–89

Bhardwaj ML (2012) Effect of climate change on vegetable production in india in vegetable production under changing climate scenario

Bhatt RM, Rao NKS, Upreti KK, Lakshmi MJ (2009) Hormonal activity in tomato flowers in relation to their abscission under water stress. Indian J Hortic 66:492–495

Bhatt RM, Rao NKS, Harish DM (2013) Significance of grafting in improving tolerance to abiotic stresses in vegetable crops under climate change scenario. In: Climate-resilient horticulture: adaptation and mitigation strategies, pp 159–175

Bustan A, Sagi M, Malach YD, Pasternak D (2004) Effects of saline irrigation water and heat waves on potato production in an arid environment. Field Crop Res 90:275–285

Cao YY, Duan H, Yang LN, Wang ZQ, Zhou SC, Yang JC (2008) Effect of heat stress during meiosis on grain yield of rice cultivars differing in heat tolerance and its physiological mechanism. Acta Agron Sin 34:2134–2142

Caplan D, Dixon M, Zheng Y (2019) Increasing inflorescence dry weight and cannabinoid content in medical Cannabis using controlled drought stress. HortScience 54:964–969

Chandra P (2009) Use of plastics in production: recent developments. In: Chadha KL (ed) Recent initiatives in horticulture. Westville Publishing House, New Delhi, pp 542–549

Chatterjee A, Solankey SS (2015) Functional physiology in drought tolerance of vegetable crops-an approach to mitigate climate change impact. Clim Dyn Hortic Sci 1:149–171

Cheeseman JM (2008) Mechanisms of salinity tolerance in plants. Plant Physiol 87:547–550

Das DK, Singh J, Vennila S (2011) Emerging crop pest scenario under the impact of climate change-a brief review. J Agric Phys 11:13–20

De Vetten N, Wolters AM, Raemakers K, van der Meer I, Stege R, Heeres E, Heeres P, Visser R (2003) A transformation method for obtaining marker-free plants of a cross-pollinating and vegetatively propagated crop. Nat Biotechnol 21:439–442

Drew MC (2009) Plant responses to anaerobic conditions in soil and solution culture. Curr Adv Plant Sci 36:1–14

Edelstein M (2004) Grafting vegetable-crop plants: pros and cons. In: VII International symposium on protected cultivation in mild winter climates: production, pest management and global competition, pp 235–238

Erickson AN, Markhart AH (2012) Flower developmental stage and organ sensitivity of bell pepper (*Capsicum annuum* L.) to elevated temperature. Plant Cell Environ 25:123–130

Escalante-Magaña C, Aguilar-Caamal LF, Echevarría-Machado I, Medina-Lara F, Sanchez Cach L, Martínez-Estévez M (2019) Contribution of glycine betaine and proline to water deficit tolerance in pepper plants. HortScience 4:1044–1054

FAO (2009) Global agriculture towards 2050 issues brief. High level expert forum. Rome, pp 12–13

Flowers TJ (2008) Improving crop salt tolerance. J Exp Bot 55:307–319

Folzer H, Dat JF, Capelli N, Rieffel D, Badot PM (2006) Response of sessile oak seedlings (Quercus petraea) to flooding: an integrated study. Tree Physiol 26:759–766

Foolad MR (2014) Recent advances in genetics of salt tolerance in tomato. Plant Cell Tissue Organ Cult 76:101–119

Foolad MR, Zhang LP, Subbiah P (2010) Genetics of drought tolerance during seed germination in tomato: inheritance and QTL mapping. Genome 46:536–545

Fooland MR, Jones RA (2011) Genetic analysis of salt tolerance during germination in Lycopersicon. Theor Appl Genet 81:321–326

Giordano M, Petropoulos SA, Rouphael Y (2021) Response and defence mechanisms of vegetable crops against drought, heat and salinity stress. Agriculture 11:463. https://doi.org/10.3390/agriculture11050463

Goderis IJWM, De Bolle MFC, Francois IEJA, Wouters PFJ, Broekaert WF, Cammue BPA (2002) A set of modular plant transformation vectors allowing flexible insertion of up to six expression units. Plant Mol Biol 50:17–27

Goto K, Yabuta S, Ssenyonga P, Tamaru S, Sakagami JI (2021) Response of leaf water potential, stomatal conductance and chlorophyll content under different levels of soil water, air vapor pressure deficit and solar radiation in chili pepper (*Capsicum chinense*). Sci Hortic 281:109943

Guo M, Liu JH, Ma X, Luo DX, Gong ZH, Lu MH (2016) The Plant Heat Stress Transcription Factors (HSFs): Structure, Regulation, and Function in Response to Abiotic Stresses. Front Plant Sci., Feb 9;7:114. https://doi.org/10.3389/fpls.2016.00114. fPMID: 26904076; PMCID: PMC4746267

Harrington R, Fleming RA, Woiwod IP (2010) Climate change impacts on insect management and conservation in temperate regions: can they be predicted? Agric For Entomol 3:233–240

Hasanuzzaman M, Nahar K, Fujita M (2013) Extreme temperatures, oxidative stress and antioxidant defense in plants. In: Vahdati K, Leslie C (eds) Abiotic stress – plant responses and applications in agriculture. InTech, Rijeka, pp 169–205

Hasanuzzaman M, Nahar K, Gill SS, Fujita M (2014) Drought stress responses in plants, oxidative stress, and antioxidant defence. In: Tuteja N, Gill SS (eds) Climate change and plant abiotic stress tolerance. Wiley-Blackwell, Hoboken, pp 209–237

Hazra P, Samsul HA, Sikder D, Peter KV (2007) Breeding tomato (Lycopersicon esculentum Mill) resistant to high temperature stress. Int J Plant Breed 1:31–40

IPCC (2013) Climate change 2013: the physical science basis EXIT. In: Stocker TF, Qin D, Plattner G-K, Tignor M, Allen SK, Boschung J, Nauels A, Xia Y, Bex V, Midgley PM (eds) Contribution of working group I to the fifth assessment report of the intergovernmental panel on climate change. Cambridge University Press, Cambridge/New York

Jacob P, Hirt H, Bendahmane A (2017) The heatshock protein/chaperone network and multiple stress resistance. Plant Biotechnol J 15:405–414

Jamil M, Rha ES (2014) The effect of salinity (NaCl) on the germination and seedling of sugar beet (Beta vulgaris L.) and cabbage (Brassica oleracea capitata L.). Kor J Plant Res 7:226–232

Janni M, Gullì M, Maestri E, Marmiroli M, Valliyodan B, Nguyen HT, Marmiroli N (2020) Molecular and genetic bases of heat stress responses in crop plants and breeding for increased resilience and productivity. J Exp Bot 71:3780–3802

Jat MK, Tetarwal AS (2012) Effect of changing climate on the insect pest population National Seminar on Sustainable Agriculture and Food Security: Challenges in Changing Climate, March 27–28, Hisar, India, pp 200–201

Kaymakanova M, Stoeva N, Mincheva T (2008) Salinity and its effects on the physiological response of bean (*Phaseolus vulgaris*. L.). J Cent Eur Agric 9:749–756

Koundinya AVV, Kumar PP, Ashadevi RK, Hegde V, Kumar PA (2018) Adaptation and mitigation of climate change in vegetable cultivation: a review. J Water Clim Change. https://doi.org/10.2166/wcc.2017.045

Kumar SN (2017) Climate change and its impacts on food and nutritional security in India. In: Belavadi VV, Karaba NN, Gangadharappa NR (eds) Agriculture under climate change: threats, strategies and policies, p 48

Kumar Swami A, Alam SI, Sengupta N, Sarin R (2011) Differential proteomic analysis of salt response in Sorghum bicolor leaves. Environ Exp Bot 71:321–328

Kumar R, Solankey SS, Singh M (2012) Breeding for drought tolerance in vegetables. Veg Sci 39:1–15

Kuo DG, Tsay JS, Chen BW, Lin PY (2014) Screening for flooding tolerance in the genus Lycopersicon. Hortic Sci 17:76–78

Kurtar ES (2010) Modelling the effect of temperature on seed germination in some cucurbits. Afr J Biotechnol 9:9

La Pena RD, Hughes J (2007) Improving vegetable productivity in a variable and changing climate. ICRISAT 4:1–22

Lauchli A, Grattan SR (2007) Plant growth and development under salinity stress. In: Jenks MA, Hasegawa PM, Mohan JS (eds) Advances in molecular breeding towards drought and salt tolerant crops. Springer, Berlin, pp 1–32

Lee SG, Huh YC, Sun ZY, Miguel A, King SR et al (2008) Cucurbit grafting. Crit Rev Plant Sci 27:50–74

Lee JM, Kubota C, Tsao SJ, Bie Z, Echevarria PH et al (2010) Current status of vegetable grafting: diffusion, grafting techniques, automation. Sci Hortic 127:93–105

Liao CT, Lin CH (2014) Effect of flooding stress on photosynthetic activities of *Momordica charantia*. Plant Physiol Biochem 32:479–485

Lin KH, Lo HF, Leo SP, Kuo CG, Chen JT, Yeh WL (2006) RAPD markers for the identification of yield traits in tomatoes under heat stress via bulked segregant analysis. Hereditas 143:142–154

Lopez MAH, Ulery AL, Samani Z, Picchioni G, Flynn RP (2011) Response of Chile pepper (capsicum annuum L.) to salt stress and organic and inorganic nitrogen sources: I. growth and yield. Trop Subtrop Agroecosyst 14:137–147

Lörz H, Wenzel G (2005) Molecular marker systems in plant breeding and crop improvement. Springer, New York

Maheswari M, Yadav SK, Shanker AK, Kumar MA, Venkateswarlu B (2012) Overview of plant stresses: mechanisms, adaptations and research pursuit. In: Venkateswarlu B, Shanker AK, Shanker C, Maheswari M (eds) Crop stress and its management: perspectives and strategies. Springer, Dordrecht, pp 1–18

Malhotra SK (2016) Recent advances in seed spices research-a review. Ann Plant Soil Res 18:300–308

Martinez Rodriguez MM, Estan MT, Moyano E, Garcia Abellan JO, Flores FB et al (2010) The effectiveness of grafting to improve salt tolerance in tomato when an 'excluder' genotype is used as scion. Environ Exp Bot 63:392–401

Matsubara S (2012) Studies on salt tolerance of vegetables, 3: salt tolerance of rootstocks. Scientific reports of the faculty of agriculture Okayama University

Mohanty A (2021) Climate change: Impacts and way forward. January 15, 2021, 5:52 PM IST. TOI/News/Blogs/India

Mumm RH (2007) Backcross versus forward breeding in the development of transgenic maize hybrids: theory and practice. Crop Sci 47(Suppl 3):S164–S171

Munns R (2005) Genes and salt tolerance: bringing them together. New Phytol 167:645–663

Nelson DE, Repetti PP, Adams TR, Creelman RA, Wu J, Warner DC, Anstrom DC, Bensen RJ, Castiglioni PP, Donnarummo MG et al (2007) Plant nuclear factor Y (NF-Y) B subunits confer drought tolerance and lead to improved corn yields on water-limited acres. Proc Natl Acad Sci U S A 104:16450–16455

Newton AC, Johnson SN, Gregory PJ (2011) Implications of climate change for diseases, crop yields and food security. Euphytica 179:3–18

Parent C, Capelli N, Berger A, Crèvecoeur M, Dat JF (2008) An overview of plant responses to soil waterlogging. Plant Stress 2:20–27

Parihar P, Singh S, Singh R, Singh VP, Prasad SM (2015) Effect of salinity stress on plants and its tolerance strategies: a review. Environ Sci Pollut Res 22:4056–4075

Petretto GL, Urgeghe PP, Massa D, Melito A (2019) Effect of salinity (NaCl) on plant growth, nutrient content, and glucosinolate hydrolysis products trends in rocket genotypes. Plant Physiol Biochem 141:30–39

Plantenga FDM, Bergonzi S, Abelenda JA, Bachem CWB, Visser RGF, Heuvelink E, Marcelis LFM (2019) The tuberization signal StSP6A represses flower bud development in potato. J Exp Bot 70:937–948

Qu AL, Ding YF, Jiang Q, Zhu C (2013) Molecular mechanisms of the plant heat stress response. Biochem Biophys Res Commun 432:203–207

Raza A, Razzaq A, Mehmood SS, Zou X, Zhang X, Lv Y, Xu J (2019) Impact of climate change on crops adaptation and strategies to tackle its outcome: a review. Plan Theory 8:34

Raza A, Ashraf F, Zou X, Zhang X, Tosif H (2020) Plant adaptation and tolerance to environmental stresses: mechanisms and perspectives. In: Plant ecophysiology and adaptation under climate change: mechanisms and perspectives I. Springer, Berlin/Heidelberg, pp 117–145

Razi K, Muneer S (2021) Drought stress-induced physiological mechanisms, signaling pathways and molecular response of chloroplasts in common vegetable crops. Crit Rev Biotechnol. Aug; 41(5):669-691. https://doi.org/10.1080/07388551.2021.1874280. Epub 2021 Feb 1. PMID: 33525946

Sage RF, Way DA, Kubien DS (2008) Rubisco, Rubisco activase, and global climate change. J Exp Bot 59:1581–1595

Saha P, Chatterjee P, Biswas AK (2010) NaCl pretreatment alleviates salt stress by enhancement of antioxidant defence system and osmolyte accumulation in mungbean (*Vigna radiata* L.Wilczek). Indian J Exp Biol 48:593–600

Sgherri C, Pinzino C, Quartacci MF (2018) Reactive oxygen species and photosynthetic functioning: past and present. In: Reactive oxygen species in plants: boon or bane – revisiting the role of ROS. Wiley, Chichester, pp 137–155

Singh B, Kukreja S, Goutam U (2020) Impact of heat stress on potato (*Solanum tuberosum* L.): present scenario and future opportunities. J Hortic Sci Biotechnol 95:407–424

Sivakumar R, Nandhitha GK, Boominathan P (2016) Impact of drought on growth characters and yield of contrasting tomato genotypes. Madras Agric J 103:78–82

Solankey SS, Singh RK, Baranwal DK, Singh DK (2015) Genetic expression of tomato for heat and drought stress tolerance: an overview. Int J Veg Sci 21(5):496–515. https://doi.org/10.1080/19315260.2014.902414

Solankey SS, Shirin A, Pallavi N, Meenakshi K (2017) Effect of high temperature stress on morpho-biochemical traits of tomato genotypes under polyhouse condition. Indian J Ecol 44(Special Issue 5):497–502

Solankey SS, Kumari M, Akhtar S, Singh HK, Ray PK (2021a) Challenges and opportunities in vegetable production in changing climate: mitigation and adaptation strategies In: Solankey SS et al (eds) Advances in research on vegetable production under a changing climate, vol. 1. Springer, Cham. pp 13–59. ISBN 978-3-030-63497-1

Solankey SS, Kumari M, Kumar M, Silvana N (2021b) The role of research for vegetable production under a changing climate future trends and goals. In: Solankey SS et al (eds) Advances in research on vegetable production under a changing climate, vol. 1. Springer, Cham, pp 1–12. ISBN 978-3-030-63497-1

Srinivasa Rao NK, Bhatt RM (2012) Responses of tomato to moisture stress: plant water balance and yield. Plant Physiol Biochem 19:36–36

Srivastava DK (2003) Genetic transformation and crop improvement. In: Marwaha SS, Arora JK (eds) Biotechnological strategies in agro- processing. Asiatech Publisher, New Delhi, pp 18–34

Steven P (2008) Temperature perception and signal transduction in plants. New Phytol 179:615–628

Sun C, Li X, Hu Y, Zhao P, Xu T, Sun J, Gao X (2015) Proline, sugars, and antioxidant enzymes respond to drought stress in the leaves of strawberry plants. Kor J Hortic Sci Technol 33:625–632

Sunkar R, Zhu JK (2004) Novel and stress-regulated microRNAs and other small RNAs from Arabidopsis. Plant Cell 16:2001–2019

Taiz L, Zeiger E (2017) Fisiologia vegetal, 6th edn. Artmed, Guelph, p 918

Vadez V, Berger JD, Warkentin T, Asseng S, Pasala R, Rao KPC, Gaur PM, Pooran MJ, Larmure A, Voisin AS, Sharma HC, Pande S, Sharma M, Krishnamurthy L, Zaman MA (2012) Adaptation of grain legumes to climate change: a review. Agron Sustain Dev 32(1):31–44

Varshney RV, Hoisington DA, Tyagi AK (2006) Advances in cereal genomics and applications in crop breeding. Trends Biotechnol 24:490–499

Wang W, Vinocur B, Shoseyov O, Altman A (2004) Role of plant heat-shock proteins and molecular chaperones in the abiotic stress response. Trends Plant Sci 9:244–252

Waters ER, Vierling E (2020) Plant small heat shock proteins – evolutionary and functional diversity. New Phytol 227:24–37

Welbaum GE (2015) Vegetable production and practices. CABI, Wallingford, p 476

Xu J, Xue C, Xue D, Zhao J, Gai J, Guo N, Xing H (2013) Overexpression of GmHsp90s, a heat shock protein 90 (Hsp90) gene family cloning from soybean, decrease damage of abiotic stresses in Arabidopsis thaliana. PLoS One 8:e69810

Yang Q, Chen ZZ, Zhoua XF, Yina HB, Lia X, Xina XF, Honga XH, Zhu JK, Gong Z (2009) Over-expression of SOS (salt overly sensitive) genes increases salt tolerance in transgenic Arabidopsis. Mol Plant 2:22–31

Yetisir H, Caliskan ME, Soylu S, Sakar M (2006) Some physiological and growth responses of watermelon [*Citrullus lanatus* (Thunb.) Matsum. and Nakai] gafted onto *Lagenaria siceraria* to flooding. Environ Exp Bot 58:1–8

Yordanov I, Velikova V, Tsonev T (2013) Plant responses to drought, acclimation, and stress tolerance. Photosynthetica 38:171–186

Zeng A, Xu Y, Song L, Li L, Yan J (2021) Validation of suitable reference genes for qRT-PCR in cabbage (*Brassica oleracea* L.) under different abiotic stress experimental conditions. J Plant Biochem Biotechnol 30:184–195

Zhou X-L, Harrington R, Woiwod IP, Perry JN, Bale JS, Clark SJ (1995) Effects of temperature on aphid phenology. Glob Chang Biol 1(4):303–313. https://doi.org/10.1111/j.1365-2486.1995.tb00029.x

Zhu Y, Luo X, Nawaz G, Yin J, Yang J (2020) Physiological and biochemical responses of four cassava cultivars to drought stress. Sci Rep 10:6968

Zróbek-Sokolnik A (2012) Temperature stress and responses of plants. In: Ahmad P, Prasad MNV (eds) Environmental adaptations and stress tolerance of plants in the era of climate change. Springer, New York, pp 113–134

Chapter 2
Emerging Obstacles of Vegetable Production Due to Climate Change and Mitigation Strategies

Shirin Akhtar, Surabhi Sangam, Tirthartha Chattopadhyay, Abhishek Naik, and Shashank Shekhar Solankey

Abstract Climate change is of pivotal concern of crop growers and researchers in today's scenario. Vegetable crops are somewhat more prone to the vagaries of nature. The increasing global warming, changed precipitation patterns, excess amounts of UV radiation is causing menace to vegetable productivity through altered span and period of crop, reduced pollination and fertilization, thus reducing the yield and adverse effect on other physio-biochemical mechanisms, lowering the quality. The patterns of disease-pest infestation and severity has also been modified providing more threat to the crop. Developing climate resilient varieties by conventional breeding techniques or use of marker assisted breeding and development of transgenics are an important option for coping with the abiotic stresses. Besides, adoption of suitable cultural practices, diversified cropping systems, mulches, precision farming and organic farming may be the adaptation and mitigation strategies that may be adopted to combat climate change.

Keywords Abiotic stress · Adaptation · Mitigation strategies · Resilient varieties

S. Akhtar (✉) · S. Sangam
Department of Horticulture (Vegetable and Floriculture), Bihar Agricultural College, Sabour (Bihar Agricultural University, Sabour), Bhagalpur, Bihar, India

T. Chattopadhyay
Department of Plant Breeding and Genetics, Bihar Agricultural College, Sabour (Bihar Agricultural University, Sabour), Bhagalpur, Bihar, India

A. Naik
Marketing Department, Ajeet Seeds Pvt. Ltd., Kolkata, West Bengal, India

S. S. Solankey
Department of Vegetables & Floriculture, Agricultural Research Institute, Patna, Bihar, India

S. S. Solankey, M. Kumari (eds.), *Advances in Research on Vegetable Production Under a Changing Climate Vol. 2*, Advances in Olericulture, https://doi.org/10.1007/978-3-031-20840-9_2

2.1 Introduction

Vegetables are fresh, edible part of herbaceous plant consumed by humans or other animals as food in either in cooked form or directly as raw, and include all edible plant matter, encompassing the stems, leaves, roots, flowers, fruits and seeds. The total vegetables produced in the world is around 311.82 million tonnes produced from an area of about 21.94 million hectares (FAOSTAT 2021). Vegetables are functional foods with valuable food components involved in body building and healing process. These crops are valuable sources of carbohydrates, vitamins, minerals and other phytonutrients. They help in maintaining the alkaline balance of the body (Hanif et al. 2006). The various vegetable groups, i.e., fruits, seeds, stems, roots, tubers, leafy vegetables, each contributes to diet in its own way (Robinson 1990). Vegetables can be an excellent source for coping up with the malnutrition situation that a huge population of the world is suffering with. However, vegetables are highly influenced by climatic vagaries. The changing climate is bringing up many obstacles in the path of vegetable production.

A long-term change in the existing average weather patterns expressed through the the Earth's local, regional and global climate may be described as climate change. There are a diverse range of effects that are associated with climate change. Since the early twentieth century, human activities have become the major reason behind the changing climate, apart from some natural factors. With urbanisation, there has been more use of fuels including fossil fuels, industrialization, use of appliances that release greenhouse gases, and ultimately has led to trapping of heat due to raised greenhouse gases in the atmosphere, thereby increasing our planet's mean surface temperature. This kind of temperature increment is commonly termed global warming. Primarily due to such human activities, the Earth's global average temperature has seen an increase of about 1 °C or 1.8 °F as compared to the pre-industrial period. At present the decadal increment in temperature is about 0.2 °C (i.e., 0.36°F). It is obvious that the atmosphere, ocean, and land has warmed human due to influence. The last 6 decades has seen an unprecedented rise in temperature. World Meteorological Organization has reported that the mean temperature over 2000–2010 is 0.21 °C higher than that of 1991–2000 and 0.40 °C more than that of 1961–1990 (WMO 2012). The temperature increment in the current decade is also in accelerated mode in the present decade and thousands of species may be lost, i.e., 8% plants if the temperature further increases by 1.5 °C and 16% if there is 2 °C increment, as projected by IPCC (i.e., there will be 1.5 °C increment of temperature within 2030–2052 and more than 3–4 °C by 2100) (Nullis 2018). The constantly rising temperature leads to melting of the polar caps and glaciers, increase of land and ocean temperatures, rise in sea levels, change in precipitation patterns, and frequent extreme events such as droughts, flash floods, hurricanes, wildfires, etc.

2.2 Obstacles of Changing Climate in Vegetable Production

Vegetable crops are very sensitive to environmental vagaries. For optimum yield of these crops, there is necessity of specific climatic requirements, otherwise the productivity and quality both are affected. The several stresses exerted by the changing climate on vegetable crops include extremities of temperature (heat stress and chilling and/or freezing stress), extremities of moisture (drought and waterlogging), soil reactions (acidity, alkalinity), presence of excess salt in soil and irrigation water (salinity), deficiency or excess availability (toxicity) of mineral nutrients, besides change in the severity of disease and pest occurrence.

2.2.1 *Heat Stress*

Wahid et al. (2007) defined heat stress as the rise in temperature beyond a threshold level for a period of time sufficient to cause irreversible damage to plant growth and development. Generally, a rise of 10–15 °C temperature above the ambient condition is considered as heat shock or heat stress. It is a cumulative of factors like intensity of heat (measured by the temperature in degrees), its duration, the rate of increment in temperature and the probability and period of high temperatures occurring during the day and/or the night determining its extent. This kind of stress owing to the high ambient temperatures offers serious setback to the global crop production scenario (Hall 2001).

Like all other crops, high temperature stress is a key setback for vegetable crops. There is tremendous decline in growth above a temperature of 35 °C (Hazra and Som 2006). A range of various morpho-anatomical changes in the vegetable crops is observed which affect the germination of seeds, normal growth of plant, flower drop, viability of pollen, fertilization, setting of fruits, size of fruit, its average weight, quality etc. (Akhtar et al. 2015).

Most vegetables have an optimum soil temperature range of 20–30 °C. The maximum temperature for germination of the warm season vegetables like tomato, brinjal, chilli, cucurbits, beans, etc. ranges between 35 and 40 °C while that of cool season vegetables like peas, onion, cole crops, carrot, beet, etc. is 32–35 °C. The germination is severely hampered when it goes above 30 °C.

Temperature alone or in association with other environmental factors affect the vegetative and reproductive stages of the crops (Abdalla and Verkerk 1968). In vegetable crops, the reproductive phase of vegetables is more high temperature-sensitive as compared to the vegetative phase. An increment of 2–4 °C of optimal temperature affects the process of fertilization and gametic development and thus restricts fruitsetting thereby reducing yields severely (Firon et al. 2006; Peet et al. 1997). The flower bud development is also hampered by high temperatures. High temperature causes shedding of flower and flower bud, deformity in flower, hampers anther dehiscence, decreases production and viability of pollen, causes abortion of ovule, reduces stigma receptivity, decreases carbohydrate availability, etc. and in tomato all of these were reported by Hazra et al. (2007). High temperature often results in bolting in leafy vegetables like palak, and thereby the leaf yield goes down. Similarly, high temperature leads to decreased fruit set and hence the yield is reduced besides the quality in all other vegetables. The normal rate of photosynthesis is hampered at temperatures above optimum, and therefore there occurs considerable loss of potential productivity. During the seed filling stage, even a very short exposure of the plants to high temperatures may accelerate senescence, reduce both seed setting and the seed weight, and thereby diminish the seed yield (Siddique et al. 1999), since under these conditions, it is the natural tendency of the plants divert their resources to cope with the heat stress, and very little photosynthates would be available for reproductive development. High temperature induced sterility may also be observed in many plant species when the heat stress is imposed immediately before or during anthesis.

Temperature has profound effect on the development of the economic part of the vegetables and its quality. The following table shows the impact of temperature on the economic part of different vegetables, its development and quality (Table 2.1).

2.2.2 *Cold/Chilling Stress*

The genetic potential of vegetable crops is severely affected by cold and chilling stress, thereby limiting the productivity. The geographical distribution of plants is hugely influenced by low temperatures which also determines the survival of plants. Plant growth, yield is immensely affected by cold stress which leads to significant crop loss (Xin and Browse 2000). Tolerance level of plants, particularly, vegetables, varies for chilling (i.e., 0–15 °C) and freezing (i.e., lesser than 0 °C) temperatures. Raison and Lyons (1986) termed plants as "very sensitive to chilling" when they showed visual injuries when exposed to temperature of 15 °C and even a little reduction in temperature may lead to even 50% decrease in productivity without any visible symptom of damage. The severe yield reduction or even complete crop failure may be due to delayed maturity or direct damage caused by the chilling temperature.

A wide-ranging variety of cellular components and metabolism are affected by cold stress of different severity varying with its duration, intensity and frequency. Freezing injury is common in plants and the cell membrane systems are the primary site of such injury (Steponkus 1984; Levitt 1980). The chief basis of membrane damage induced by freezing is the severe dehydration associated with it (Steponkus 1984; Steponkus et al. 1993). When temperature goes to lesser than 0 °C, ice is formed in the intercellular spaces since the intracellular fluid has lower freezing point than the extracellular fluid having lower solute concentration. The chemical potential of ice being less than that of liquid water at any given temperature, a drop in the water potential outside the cell occurs due to the formation of extracellular ice. This results in movement of unfrozen water from inside the cell to the

Table 2.1 Impact of high temperature on economic part of vegetable, its yield and quality

Vegetable crop	Impact of temperature on economic part, yield and quality
Tomato	There is no fruit set if night temperature goes to 13 °C and lesser or if mean maximum day temperature goes above 32 °C and mean minimum night temperature above 21 °C; below 10 °C both red and yellow colour do not develop; ideal temperature for colour development is 10–25 °C; above 30 °C red colour development stops but it redevelops when 30 °C temperature is restored; at 40 °C lycopene pigment is destroyed and thus no red colour develops further, resulting in yellow shoulders; high temperature also causes scalding of fruits
Brinjal	Adverse effect on fruit set and yield is observed above 35 °C; ovary splitting in bud stage and deformation of fruits occure below 10 °C
Chilli	High night temperature (24 °C) causes flower drop as also high day temperature (above 30 °C) coupled with low light intensity; high night temperature is responsible for higher capsaicin content of fruit
Sweet pepper	Flower drop occurs when day temperature is high (30 °C and above); low night temperature (10–15 °C) reduces pollen viability leading to parthenocarpic fruits
Cauliflower	High temperature exposure after curd formation causes loose or leafy curd along with yellowing of curds; formation of small buds over the curd (riceyness) also occurs at high temperature
Cabbage	Heads are not compact at high temperature
Knolkhol	Knob quality is poor at high temperature
Brussel's sprouts	There is no head formation at high temperature
Sprouting broccoli	At high temperature the bud clusters become loose
Chinese cabbage	If temperature drops to 15 °C or lower, seed stalks emerge before proper head is produced
Onion	Above 40 °C, bulb development is hampered
Garlic	High temperature affects bulb production adversely
Potato	Tuberization is inhibited when night temperature goes above 21 °C
Carrot	When temperature dips to lower than 15 °C, colour does nor develop properly and the roots are slender; at 24–25 °C the roots are shorter and thicker besides having less colour
Beet	When temperature is 10–15 °C seed stalks emerge prematurely; 15–18 °C is ideal temperature for roots with high sugar content and dark internal colour and when it is more than 25 °C roots show zoning, i.e., alternate white and coloured rings
Radish	At high temperature roots remain thin and become tough and pungent before reaching edible size
Turnip	Below; when temperature goes above 25 °C roots quality deteriorate and they become tough and bitter in taste; seed stalks are prematurely produced at less than 10 °C
Pea	At temperature more than 25 °C lesser pod setting occurs; quality becomes poor and maturity is hastened; peas lose tenderness and become tough at 30 °C and more; more enzymatic conversion of starch to sugar occurs at high temperature, thereby reducing quality due to less sugar content
French bean	High temperature (27–32 °C) increases flower number per plant but decreases pod set markedly

(continued)

Table 2.1 (continued)

Vegetable crop	Impact of temperature on economic part, yield and quality
Lima bean	High temperature (26–30 °C) drastically affects fruit set
Palak	Succulence of leaves is reduced drastically at high temperature and early bolting occurs when more than 25 °C thereby reducing leaf yield
Spinach	Succulence of leaves is reduced drastically; leaves turn yellow and contain reduced sugar content at high temperature (more than 25 °C)
Lettuce	Ideal temperature for compact head and crisp leaves is 18 °C night/20 °C day temperature; growth and yield are hampered at high temperature, besides quality of leaf and head is reduced when more than 22 °C
Celery	Bolting occurs at 10–15 °C; bitterness in leaves results due to high temperature
Cucurbits	More number of male flowers are produced at high temperature and thus fruit set is adversely affected
Melons	Sugar content in fruits is increased at high day temperature followed by cool night temperature thereby enhancing sweetness of fruits
Sweet potato	At both high (35–40 °C) and low (15 °C and less) temperatures, tuberization is hampered
Elephant foot yam	At high temperature, development of corms is improper and remain small
Cassava	At both high (35–40 °C) and low (15 °C and less) temperatures, tuberization is hampered
Taro	At high temperature, development of corms is improper and remain small
Yam	At high temperature, growth is hampered and tuberization is hampered
Yam bean	At high temperature, growth is hampered and bulking of tubers is hampered
Asparagus	At high temperature, production of spears is adversely affected

Source*:* Compiled from Akhtar and Hazra (2015)

intercellular spaces down the chemical potential gradient. About 90% or even more osmotically active water moves out of the cells at 10 °C. Membrane fluidity changes during temperature stresses and makes it a potential site of injury (Orvar et al. 2000; Horváth et al. 1998). Membrane fluidity basically depends on the composition of lipid molecular species, the degree of lipid saturation and the environmental temperature. Optimum functioning of the membrane at lower temperatures is influenced by the membrane lipid unsaturation. Various forms of membrane damage that may result from freeze induced cellular dehydration has been explained by Steponkus et al. (1993) encompassing cell lysis, formation of hexagonal-II phase transitions in the membrane, and resulting in “fracture jump lesions”. Hence, stabilizing membranes would be important for cold acclimatization and different mechanisms are involved in the process of which changes in lipid composition are most well documented.

In herbaceous plants like vegetables, leaf and hypocotyl wilting are the most prominent symptoms of chilling injury, besides drying of leaf tips and edges of leaf lamina, and in case of prolonged chilling, leaf necrosis is followed by even plant death (Frenkel and Erez 1996; Mitchell and Madore 1992; Hahn and Walbot 1989). Water soaked areas are also seen along with the occurrence of pits and large cavities on the plant surface, leaves and internal tissues lose their natural colour, senescence is hastened, the chilled or frozen tissues rupture, and there may be slow and uneven ripening of fruits and in many cases ripening is incomplete and the plants may become more prone to pathogen attack (Tsuda et al. 2003; Frenkel and Erez 1996; Yoshida et al. 1996; Sharom et al. 1994; McMahon et al. 1994; Cabrera et al. 1992; Dodds and Ludford 1990). The economic part may even be deformed and lose its characteristic taste and flavour (Ventura and Mendlinger 1999; Harker and Maindonald 1994). Besides, loss of moisture, shrivelling and desiccation, internal discolouration, rupture of tissue, enhanced ethylene production, lesser shelf life, change in proximate and biochemical composition, leakage of plant metabolite, stunted growth, loss of sprouting ability are other symptoms of chilling injury (Skog 1998).

There remains difficulty in seed germination of chilling sensitive plants when temperature drops to less than 10–15 °C (Ismail et al. 1997; Wolk and Herner 1982). Markowski (1988) classified vegetables into two groups based on ease of germination under cold stress. In the first group with solanaceous vegetables and pumpkin, there is no damage in seeds during imbibition at chilling temperatures, and with rising temperature, though apparent normal growth is seen, the root tip tissue remains underdeveloped leading to weak root system, and there may be necrosis of tissue of collar region and the cortex or stele may be damaged (Jennings and Saltveit 1994; Bradow 1990). On the other hand, the second group comprising of vegetables like beans, have seeds that are highly sensitive to low temperature during imbibition and may not germinate (Zemetra and Cuany 1991; Gorecki et al. 1990). The secondary infestation of pathogens may occur that enhance the plant damage (Wolk and Herner 1982).

The chilling tolerant species and varieties show marked difference in growth which is highly retarded in case of chilling-sensitive ones (Venema et al. 1999; Rab and Saltveit 1996a; Ting et al. 1991), besides late development and ultimately longer growing season (Skrudlik and Koscielniak 1996). There is delay in differentiation of the apical cone and hence the number and rate of formation of various newly formed plant parts is reduced, the root morphology is altered, flowering is decreased, fruit set hampered and seed filling is highly decreased (Rab and Saltveit 1996b; Skrudlik and Koscielniak 1996; Lejeune and Bernier 1996; Barlow and Adam 1989; Buis et al. 1988) (Table 2.2).

2.2.3 *Drought Stress*

The yield and quality of vegetables is immensely affected by availability of water and its scarcity diminishes vegetable productivity largely (Bhardwaj 2012). Drought is the most threatening environmental stress in arid and semi-arid regions, and the primary cause of crop loss worldwide, thereby reducing the mean yields for most of the crop plants by more than 50% (Sivakumar et al. 2016). The plants may show different physico-biochemical and genetic responses due to drought stress rising

Table 2.2 Chilling injury symptoms in various vegetables

Vegetable crop	Various symptoms of chilling stress injury
Tomato (mature-green stage)	Improper colour development on ripening; *Alternaria* rot
Tomato (ripe stage)	Water-soaked spots, fruit softening and decay
Brinjal	Surface scald, *Alternaria* rot, seed blackening
Pepper	Pitting, *Alternaria* rot, seed blackening
Potato	Browning, conversion of starch to reducing sugar thereby leading in sweetened tubers
Sweet potato	Internal discoloration, pitting, decay
Okra	Discoloration of fruits, water-soaked spots, pitting, decay
Cucumber	Water-soaked lesions, pitting, decay
Pumpkin	Decay, particularly *Alternaria* rot
Squash	Decay, particularly *Alternaria* rot
Snap bean	Pitting and russeting
Asparagus	Dull gray-green limp tips

Adapted from Lukatkin et al. (2012)

from scarce rainfall or lack of irrigation that hampers the yield (Vadez et al. 2012). The concentration of solute in soil is enhanced, thereby causing osmotic flow of water out of the plants which consequently results in restriction of different physio-biochemical mechanisms like photosynthesis and respiration, leading to lower vegetable yield (De la Peña and Hughes 2007; Solankey et al. 2015a). The rate of photosynthesis is reduced since the stomatal conduction is decreased (Yordanov et al. 2013), and other metabolic processes are also hampered (Dias and Brüggemann 2010). There is reduction in the activities of sucrose phosphate synthase (SPS) and invertase which is necessary for sustaining the assimilatory carbon flux from source to developing sink and resynthesis of sucrose (Isopp et al. 2008). The capability of the plant to use sucrose is hampered by the reduced activity of invertase and therefore the hexose concentration is diminished and ovary growth is adversely affected (Anderson et al. 2012), thereby leading to no fruit set (Table 2.3).

2.2.4 Waterlogging/Flooding

Vegetables are very particular about water requirement and over application or availability of water causing flood exert serious stress and hence become a chief limiting factor for growth and yield crops (Parent et al. 2008). Flooding or waterlogging impacts the vegetative and reproductive growth of plants negatively due to detrimental effects on physiological functioning (Gibbs and Greenway 2008). High sensitivity to waterlogging has been observed in most vegetables and very limited variation among the genotypes has been found specific to this trait.

Flooding condition causes slow diffusion of gases in water and consumption of the available oxygen by microbes and plant roots, thereby leading to severe oxygen

Table 2.3 Critical water application stage in vegetable crops and impact of water deficit

Name of vegetable	Critical stage of water application	Effect of drought stress
Potato	During tuberization and throughout tuber enlargement period	Tuber growth and yield is improper, besides splitting of tuber
Tomato	During flowering and stage of fruit enlargement	Flower drop is seen, there is no fertilization, fruits may be of smaller size, fruit cracking and blossom end rot may be observed
Brinjal	During flowering and fruit development phase	Yield and quality is reduced, and proper colour of fruits is not developed
Chilli and capsicum	During flowering and fruitsetting	Flower drop and immature fruit drop occur, nutrient uptake is hampered and dry matter production is reduced
Cabbag	During formation of head and its development	Tip burning and splitting of head are observed
Cauliflower	During formation of head curd and its development	Browning and buttoning are observed
Cucumber	During flowering and throughout fruit development phase	Pollen grains are deformed and non-viable, fruits are often bitter and deformed
Muskmelon	During flowering and evenly throughout fruit development	Fruit quality is poor and less sweet due to reduced TSS, reducing sugar and ascorbic acid
Watermelon	During flowering and evenly throughout fruit development	Nitrate content increases in watermelon fruit
Carrot, radish and turnip	Throughout the root enlargement phase	Roots may be distorted, thin, and tough along with strong and pungent odour, there may be harmful nitrate accumulation in roots
Onion	During bulb formation and enlargement	There may be bulb splitting and doubling, besides low storage life
Pea	During flowering and pod filling	There may be stunted plant growth, flower drop, poor seed filling, and lesser nodulation in root
Lettuce	Consistently throughout development stage	Plant growth may be improper, leaves may become tough and of poor quality, tip burning may be observed
Leafy vegetables	Throughout growth and development of plant	Foliage growth is poor, leaves may become tough, nitrate accumulation in leaves increases

Adapted from: Bahadur et al. (2011)

deficiency in the crop root zone that creates anaerobic condition in the root zone. Therefore, waterlogging in vegetables, particularly, in tomato cause accumulation of endogenous ethylene in the plants (Drew 2009), and this is expressed through epinastic growth of tomato leaves in waterlogged conditions (Kawase 2011). Waterlogging coupled with high temperature is more detrimental to vegetables and very fast wilting and plant death under such conditions has been observed in vegetables, especially tomato (Kuo et al. 2014). Waterlogging in root and bulb vegetables during development of the root or the bulb leads to serious loss in yield which may go upto 30–40%. Such stresses cause around 50% yield loss in vegetables and the stage of development, the severity and duration of the stress determine the plant response towards them (Kumar 2017).

Waterlogging in vegetables affects the plant physio-biochemical and morphological traits of the plants. Decreased stomatal conductance occurs is one of the first physiological response of plant to waterlogging (Folzer et al. 2006), causing enhanced leaf water, and also elevate the internal carbon dioxide concentration since the CO_2 exchange rate is decreased (Liao and Lin 2014), thereby leading to leaf chlorosis, decreased growth of stem and root, lesser dry matter accumulation and ultimately diminished yield (Malik et al. 2012).

The waterlogged plants are predisposed to infection by harmful pathogens and the condition is congenial for spread of the microbes particularly the fungi (Pautasso et al. 2012).

2.2.5 *Salinity*

Soil salinity is a major threat to the agricultural production system including most of the vegetables and decreases the productivity markedly. Soils that are highly saline have high concentrations of soil and hence create a water deficit in the plants and results in ion specific stresses owing to changed K^+/Na^+ ratios and there is increase in the Na^+ and Cl^- concentrations which is detrimental for the plants. There may be loss of turgidity, stunted growth, lesser photosynthesis and respiration, wilting, abscission of leaves, loss of cell integrity, necrosis of tissue and ultimately leading to plant death (Cheeseman 2008).

Among the various vegetables, tomato, brinjal, chillies and cucumber are moderately sensitive and onions are susceptible to saline soils (De la Peña and Hughes 2007). There is significant lesser germination, reduced root and shoot growth, and biomass in cabbage under saline conditions (Jamil and Rha 2014). In potato, there was significant decrease in leaf area index and canopy and improper vegetative growth due to salinity stress coupled with heat stress (Bustan et al. 2004). In pepper, the dry matter, leaf area, net assimilation and relative growth rate were severely reduced and the number of fruits per plant was adversely affected (Lopez et al. 2011). Fresh and dry weight of all cucurbits was decreased; and there was stunted growth, lesser photosynthesis, reduced stomatal conductivity stress due to decreased relative water content and total chlorophyll content thus reducing the transpiration and cell water potential under salinity stress (Kaymakanova et al. 2008). Even high salinity level of irrigation water also impacts various physiological and metabolic activities resulting in reduced growth and yield.

2.2.6 *Air Pollution and UV Radiation*

The major air pollutants are sulphur dioxide (SO_2) and ozone (O_3) along with carbon monoxide (CO), methane (CH_4), etc. other various gases. These often create acids like carbonic acid, sulphuric acid in the atmosphere and cause acid rain which directly affect the leaves and other plant tissues and make them more disposed to pathogens. Therefore, under polluted air plants show different response to foliar pathogen. Besides, host mediated impacts of the pollutants also influence the attack on roots by pathogens such as nematodes. It was observed that the germination, sporulation and infestation of plant pathogenic fungi was checked under high SO_2 and O_3 concentrations (200–300 ppb), and hence there were less severe diseases in plants growing in the stressed regions (Khan 2012). On the other hand, when the concentrations were low (50–100 ppb), the risk of the diseases increased as these concentrations created congenial situation in the plants for infection and at 50–100 ppb of SO_2 and O_3 stimulated the spore germination and spread of fungal pathogen and egg hatching, penetration and reproduction of plant parasitic nematodes, thereby increasing the severity of diseases caused by them (Khan 2012).

The ozone layer is degraded by the increasing NO_2 and SO_2 in the atmosphere resulting in penetration of the harmful ultraviolet (UV) rays on to the surface of Earth. The Ultraviolet-B (UV-B) radiation is more harmful and plant cell functions are negatively impacted by it. The advesre effects of UV radiation on vegetables like potatoes, tomatoes, cabbages and sugar beets are higher as compared to others (De La Pena and Huges 2007). The tomato plants when exposed to high UV-B shows reduced plant height, leaf area and growth through reduced photosynthetic area, and thus dry weight is also decreased (Hao et al. 1997). When exposed to UV radiation, French bean leaves show bronzing and glazing and enhanced virus susceptibility (Benda 1955). Cucumber is very sensitive to UV rays and when exposed to UV rays at early germination stages, the cotyledons are adversely affected (Zajac and Kubis 2010).

2.3 Change in Response of Pests and Diseases Due to Climatic Change

With change in climatic condition, the ecology and distribution of insect-pests is highly modified (Jat and Tetarwal 2012), their time of emergence shifts, their migration pattern changes as also their overwintering capacity, as also the host-pathogen interaction and disease incidence (Koundinya et al. 2014).

Insects are stenotherms (i.e., cold-blooded) and thus they are very sensitive to temperature. They usually respond to higher temperatures by increasing their rate of development, with less time between generations. At higher temperature conditions, the life cycle of certain insects (e.g., aphids and diamond back moth) is shortened, with enhanced fecundity for early life cycle completion, thereby leading to more generations per year or unit time (FAO 2009). The span of insect breeding season will be more under rising temperatures and increase thus the reproductive rate will increase. On the other hand, some insects require more time for completing their life cycle, particularly those living in the soil or whole life or some part of the life as these suffer more due to the fact that soil becomes an insulated medium for temperature changes more than the air (Bale et al. 2010). Gender ratios of some pests (e.g. thrips) may change due to temperature (Lewis 1997). There may be migration of insect-pests towards higher latitudes when temperature rises and some specific pest species may be adversely affected in the tropics under higher temperatures. Yukawa (2008) reported that *Nezara viridula*, a tropical and subtropical crop pest, was slowly moving northward in southwestern Japan, probably due to global warming, and the substituting the more temperate species *Nezara antennata* (FAO 2008). There has been a remarkable increase in migratory *Helicoverpa* over the period 1969–2004, which has moved inland (Cannon 2008).

There was an increment in the activity and population of sucking pests like whiteflies, aphids and thrips with increasing temperature (Awmack et al. 1997; Yamamura and Kiritani 1998). It has been predicted by Fleming and Tatchell (1995) that over the next 50 years, aphids that are the vectors of different viral diseases of vegetables, especially, potato, tomato, brinjal, peppers, okra, cucurbits and legumes, will appear at least 8 days earlier in the spring as measured by their presence in suction traps, thereby increasing chances of disease infestation and crop loss.

At high temperature, the oviposition rates, insect development, outbreak and introduction of invasive species is hastened, whereas the insect diversity in ecosystem and parasitism, reliability of economic threshold levels, effective insect biocontrol by fungi, etc. are reduced (Das et al. 2011). The development of onion maggot, cabbage maggot, European corn borer, Colorado potato beetle will be faster under higher temperatures (Newton et al. 2011). Enhanced temperatures will enable insects reach their minimum flight temperature quickly, and thus help in increased dispersal capabilities as observed in moths and aphids (Zhou et al. 2014). Higher temperatures indicate higher metabolic rates in insect, thereby depleting the stored nutrients faster leading to lesser duration of insect diapause (Hahn and Denlinger 2007). Warmer winter temperature will delay onset of diapause, whereas early summer will hasten termination of diapause and the active growth and development of insects will resume and therefore, even 1–5 °C temperature rise will result in lower winter mortality, more population build-up, faster infestation and thus higher crop damage and loss by insect-pests under the current scenario of global warming (Harrington et al. 2010).

Disease development in a plant depends on the host-pathogen-environment interaction and any alteration in the environmental components will impact the susceptibility of the host plant (Khan 2012). Plant pathogens are likely to be the first organisms to express the impacts of climate change due to their huge population size and very fast life cycle. The rate of pathogen development, the stages of its infection, its pathogenicity as well as the physiology and resistance/susceptibility of the host may be altered by the changing climate, particularly temperature and precipitation regimes (Coakley et al. 1999; Mboup et al. 2012). The growth and spread of pathogens like fungi and bacteria are greatly influenced by environmental components like temperature, sunlight, other radiation, moisture, rainfall, humidity, dew (Neumeister 2010). Limiting moisture content can check spread of several diseases since lesser root growth under moisture stress reduces the chances of infection of soil borne microbes as the chances of contact of the roots with the pathogen propagule reduces (Pertot et al. 2012). Waterlogged conditions favour spread of plant pathogens, storms may make dispersal of wind-borne pathogen propagules easier, drought and high temperature conditions can create congenial conditions for disease spread in plants (Pautasso et al. 2012). Besides, air pollution, especially UV-B radiation and ozone and nutrient availability also affect plant disease spread.

In northern latitudes, global warming may have more influence on plant pathogens, since low temperatures and long winters decrease pathogen survival, their reproduction, generations per year as well as their activity during crop growing stages (Harvell et al. 2006). The distribution of the pathogen species is influenced by the existing temperature irrespective of their huge host range and certain pathogens (i.e., *Sclerotium rolfsii* and *Macrophomina phaseolina*) do not survive in temperate climate due to their frost sensitivity (Termorshuizen 2008). Reduction in frost due to global warming allows these types of pathogens to survive and also result in very quick disease cycles (Boonekamp 2012). Viral diseases in vegetables like potatoes and sugarbeets are increasing and coming earlier due to earlier appearance of the vectors of these diseases under rising temperatures (Newton et al. 2011). Onset of several fungal diseases also earlier in cropping season, such as late blight or *Fusarium* wilt will require additional one to three sprays for the disease control in the near future (Fahim et al. 2011).

Increased temperature also increases the population of pathogen attacking the various vegetables post-harvest, particularly, aflatoxin-producing fungi thereby resulting in huge post-harvest losses (Cotty and Jaime-Garcia 2007) (Table 2.4).

2.4 Mitigation Strategies

The various mitigation strategies include development of climate resilient varieties though conventional breeding and modern approaches like marker assisted breeding, use of biotechnological tools, development of transgenics, exploitation of resistant rootstocks in grafting, besides adopting certain cultural practices that reduce greenhouse gas emission and conserve resources and also crop modelling and simulation for prediction of climate.

Table 2.4 Changing climate affecting insect pests of vegetables

Name of pest	Current distribution	Anticipated effect of climate change
Aphids (*Aphis gossypi*; *Aulacorthum solani*)	Distributed worldwide	More number of generations per year will be produced due to lower developmental zero-point, low thermal totals required for one generation, thereby leading to more infestation
Melon fly (*Bactrocera cucurbitae*)	Distributed in the southern part of Japanese archipelago	Invasion to whole Japanese archipelago
Colorado potato beetle (*Leptinotarsa decemlineata*)	Distributed in North America, southern Europe, Asia and widely over Pacific Ocean	Expansion of infestation by 400 km in its potential northern limit causing more damage and imposing threat to 99% potato growing areas
Diamond back moth (*Plutella xylostella*)	Of European origin, distributed in cosmopolitan areas of the world	Increased frequency of overwintering and drastic increase in pest status
Cabbage butter fly (*Pieris brassicae*)	Distributed in Europe, north-western part of Africa and Asia, introduced to America and Australia	Mostly increased range, abundance and diversity; decrease in diversity has been reported by Roy et al. (2001)
Fruit and pod borer (*Helicoverpa armigera*; *Spodoptera litura*)	Worldwide distribution	Increase in number of generations per year resulted in more infestation
Cabbage root fly (*Delia radicum*)	Distributed over Europe, North Africa, West Asia and North America	Earlier emergence and populations are becoming active 1 month earlier in UK with a mean temperature increase of 3 °C
Sweet potato weevil (*Cylas fprmicarius*)	Distibuted over the tropical and subtropical regions of world	Increased temperature has resulted in invasion to temperate zones

Source*:* Adapted from Yukawa (2008) and Cannon (2008)

2.4.1 Development of Climate Resilient Genotypes Through Conventional Breeding Approaches

The most feasible approach towards coping up with the ever changing climate is to breed resilient varieties. There may be conventional breeding approach or modern approaches involving molecular markers and recombinant DNA technology. Breeding in the classic way has been conducted for centuries and is still widely utilised today to produce a better crop that is more adaptable to climatic variations. Vegetable crops have immensely narrow genetic base which is a major constraint in developing tolerant genotype (Ladizinsky 1985; Paran and Van Der Knaap 2007). The biodiversity and various gene pools may be utilized through hybridization to broaden the genetic base. The classical approach makes use of existing genetic variety and hybridization for recombination of genetic material through independent chromosomal assortment and crossing-over. Induced mutations are frequently used in traditional breeding methods as well. The selection of individual plants and their progenies is emphasised in any of these approaches. For effective selection the breeding location should have similar environmental conditions of the area for which the genotype is being developed (Mickelbart et al. 2015). It is also easier to select for traits and identify genes conferring resilience among the genotypes native to the areas with varied climate since these plants can adapt to diverse range of environmental conditions (Pereira and Chaves 2007). As a result, it is critical to understand appropriate selection criteria, which vary depending on the species and stress.

2.4.1.1 Identification of Suitable Selection Criteria

Because of the unpredictability of environmental conditions, selection for developing climate resilient genotype is challenging. For example, the salinity of the salt-affected zones is often uneven, with sections of no salinity and areas of high salinity coexisting in the same field at close proximity or far apart. Presence of intense root system, stem and leaf pubescence, leaf size and orientation, photosynthetic rate, earliness, leaf rolling, biomass, dry matter accumulation, water uptake, leaf ion content, leaf necrosis, osmosis besides yield attributes form the selection criteria for developing climate resilient varieties (Akhtar et al. 2021).

The frequency of genes for temperature extremes should grow as a result of natural selection in varied populations of agricultural plants. Natural selection might be effective not only for temperature stress, but also for other stresses, if populations were produced in areas where the likelihood of a stress approaching the critical level is high. Individual plants/genotypes are separated and propagated from mixed/segregating populations using artificial selection. There must be genetic variation that can be detected and discriminated from environmental variation for selection to be successful.

2.4.1.2 Development of Improved Varieties

The best option for coping with climate change is to develop vegetable genotypes that can withstand such climate and this is an adaptation strategy (Altieri et al. 2015). However, the genes governing the abiotic stress tolerance are mostly polygenic and determining criteria for effective selection becomes tough, and hence the process complex. The process starts with characterisation of the available genotype that also makes their specific traits ready for use by the breeders. Genetic variability among the base population is the pre-requisite for any breeding programme. Selecting the resistant ones from the base population and utilizing them in breeding programmes according to their genetic diversity and distance between them allows greater chances of getting superior recombinants in the segregating population (Koundinya et al. 2018).

The cultivated varieties of vegetables often do not possess the novel traits of abiotic stress resistance and wild and related species besides some important land races have been found to be donors of these traits. In tomato, various landraces from Philippines (Divisoria-2, VC11-2-5, VC11-3-1-8) and USA (PI289309, Tamu Chico III) and lines from AVRDC CL5915 have been found to be sources of heat tolerance; PI-120256 (an old landrace from Turkey), *Solanum habrochaites* accessions LA3925, LA3921 and LA-1777 from AVRDC are sources of cold tolerance; *Solanum habrochaites* accession EC-520061 is a source of resistance to both heat and cold stress; *S. pennelli* accessions LA0716, EC-104395, EC-130042, IIHR 14-1, IIHR 146-2, IIHR 383, IIHR 553, IIHR 555, K-14, Sel-28, *S. habrochaites* accession EC- 520061, *S. pimpinellifoloium* accessions LA1579, PI-205009, EC-65992, Pan American, *S. sitiens* accessions LA1974, LA2876, LA2877, LA2878, LA2885, *S. chilense* accessions LA1958, LA1959, LA1972, besides cultivated Arka Vikas and RF-4A are sources of drought tolerance; *S. esculentum* accession PI174263, *S. pimpinellifolium* accessions LA1579 and LA1606, *S. lycopersicum* var. *cerasiforme* accession LA4133 are salinity tolerant (Rai et al. 2011; Singh 2010; Flowers 2004; Razdan and Mattoo 2007; Metwally et al. 1996; Opena et al. 1992; Foolad and Jones 1991). In brinjal, the wild species *S. sodomaeum, S. macrocarpon, S. gilo, S. macrosperma, S. integrifolium,* and cultivates genotypes Bundelkhand Deshi, SM- 1, SM- 19, SM- 30, VioletteRound, Supreme have been identified as sources of drought resistance (Rai et al. 2011; Toppino et al. 2009; Kumar and Singh 2006). In peppers, species like *C. chinense*, *C. baccatum* var. *pendulum*, *C. eximium*, variety Arka Lohit, and line IIHR – Sel.-132 have been found to be resistant to drought stress (Singh 2010); Tunisian cultivar Beldi, Demre, Ilica 250, 11-B-14, Bagci Carliston, Mini Aci Sivri, Yalova Carliston, and Yaglik 28 have been identified as sources of salinity tolerance (Yildirim and Guvenc 2006). Various drought tolerant species have been identified in potato like *Solanum chacoense, S.ajanhuiri, S.curtilobum, S.xjuzepczukii, S. acaule, S. demissum and S. stenotonum,* besides the cultivated Alpha, Bintje and Kufri Sindhuri (Arvin and Donnelly 2008; Pandey et al. 2007; Ross 1986). Several sources of drought resistance have been identified in various vegetables like, *A. caillei, A. rugosus, A. tuberosus* in okra (Charrler 1984); *Allium fistulosum, A. munzii,* Arka Kalyan, MST 42, MST 46 in onion (Singh 2010); *P. acutifolius* in French bean (Kavar et al. 2011). In tuber vegetables drought tolerance has been reported in VLS6, IGSP 10, IGSP 14, Sree Bhadra in cassava sweet potato, and CE-54, CE-534, CI-260, CI-308, CI-848, 129, 7, 16, TP White, Narukku-3, Ci-4, Ci-60, Ci-17, Ci-80 in cassava (Singh 2010). In cucurbits, *Citrullus colocynthis* (L.) Schrad renders drought tolerance in watermelon (Dane et al. 2007); *Cucurbita maxima* to winter squash (Chigumira and Grubben 2004); *Cucumis pubescens, Cucumis melo* var. *callosus, Cucumis melo* var. *chat*, INGR-98015 (AHS-10), INGR-98016 (AHS-82), CU 159, CU 196, INGR-98013 (AHK-119), AHK- 200, SKY/DR/RS-101, Arya, SC- 15 and INGR-98018 (AHC-13) are sources of drought tolerance in Cucumis sp. and cucumber (Kusvuran 2012; Pandey et al. 2011; Rai et al. 2008; Chigumira and Grubben 2004).

In vegetables, several climate resilient varieties have been developed that have resistance to different abiotic stresses and may give satisfactory yield even under stress conditions. The Table 2.5 enlists some important varieties.

Table 2.5 Vegetable varieties having resistance to different abiotic stress tolerance

Vegetable	Variety	Specific climate resilient trait
Potato	Kufri Surya	High temperature tolerant; tuberization occurs even upto 25 °C night temperature
	Gem, Jasper	Waterlogging tolerant
	Kufri Dewa, Kufri Sheetman	Frost tolerant
Tomato	Pusa Sadabahar	Ability to set fruit at both high and low night temperature
	Pusa Sheetal	Fruitset occurs at low night temperature, even at 8 °C
	Cold set	Cold tolerant
	Pusa hybrid 1	Fruitset occurs at high night temperature, even at 28 °C
	Arka Vikas	Drought tolerant
Brinjal	SM-1, SM-19, SM-30 and PKM-1	Drought tolerant
	Pusa Bindu, Pragati	Tolerance to salinity
Pepper	DLS-10-02, DLS-20-11, DLS-160-1 and DLS-152-1	High temperature tolerant
Okra	Pusa Sawani	Tolerance to salinity
Garden pea	Arka Tapas, Arka Uttam, Arka Chaitra	High temperature tolerant (upto 35 °C)
Dolichos bean	Arka Jay, Arka Vijay	Drought tolerant and photoinsensitive
French bean	Arka Garima	Heat and drought tolerant
Cucumber	Pusa Barkha	High temperature tolerance
	Pusa Uday	Can grow throughout year
Spinach beet	Jobner green	Tolerant to alkalinity (soil pH upto 10.5)
Musk melon	Jobner 96-2	Tolerant to alkalinity (high soil pH)
Bottle gourd	Pusa Santusthi	Can set fruit under high temperature as well as low temperature
Onion	Hisar-2, Punjab selection	Tolerance to salinity
Carrot	Pusa Kesar	Tolerance to high temperature
	Arka Suraj	High temeparture tolerant, flowering occurs and seeds are set under tropical condition
	Ooty-1	Drought tolerant
Radish	Pusa Chetki	High temperature tolerance
	Pusa Himani	Low temperature tolerance
Sweet potato	Sree Nandini	Tolerant to drought conditions
Cassava	H-97, Sree Sahya	Drought tolerant

Source*:* Adapted from Rai and Yadav (2005) and Koundinya et al. (2018)

2.4.2 *Modern Approaches for Development of Climate Stress Resilience*

2.4.2.1 Identification of QTLs for Abiotic Stress Resistance

Since resistance of the various abiotic stresses are polygenic in nature, the governance is complex and less comprehended (Ainsworth and Ort 2010; Collins et al. 2008; Wahid et al. 2007). Mapping populations have been developed in various vegetable crops and they provide useful information and increased knowledge on the genetic governance of tolerance towards abiotic stress through identification of quantitative trait loci (QTL). Various QTLs have been identified associated to many abiotic stress tolerances in vegetables like potato, tomato, peas, etc. by various workers. QTLs associated with heat tolerance in tomato, potato and cowpea were reported by Jha et al. (2014). Various researchers have identified QTLs for different traits for heat tolerance like canopy temperature at different stages of crop growth, higher chlorophyll fluorescence, etc. (Vijayalakshmi et al. 2010; Pinto et al. 2010; Lopes et al. 2013). QTLs for drought tolerance have been identified in tomato by Martin et al. (1989) and Foolad et al. (2010). Foolad (2014) suggested that salt tolerance in tomatoes is quantitatively inherited through QTL mapping. Molecular markers linked to different QTLs have been also developed based on mapping populations, and QTL analysis and fine mapping can pinpoint the genes and loci for resistance to different abiotic stresses in vegetables and several candidate genes for salinity, drought, waterlogging and temperature extremes (heat and cold) have been proposed (Table 2.6).

2.4.2.2 Development of Transgenics

Of the different approaches towards developing stress tolerant genotypes, development of transgenic plants is the most recent one. Introduction of genes that encode for enzymes that modify membranes or lead to formation of compounds with osmoprotective properties, or enzymes that are scavengers of reactive oxygen species or stress-induced proteins are the different transgenic approaches utilized (Ashraf 2010; Zhang and Blumwald 2001; Holmberg and Bulow 1998). In tomato, overexpression of osmotin gene has exhibited salinity and drought resistance in controlled conditions expressed through higher germination percentage, increased content of proline and higher relative water capacity (Goel et al. 2010). Tomato transformated with BADH gene from Atriplex showed tolerance to salinity (Jia et al. 2002). In tomato, drought and salinity tolerant (in controlled condition) seedlings were obtained through transformation and expression of CaXTH3, a hot pepper xyloglucan endotransglucosylase/hydrolase and even at 100 mM of NaCl, and the seedlings exhibited sufficient chlorophyll content (Choi et al. 2011). Overexpression of the strawberry D-galacturonic acid reductase gene in potato improved abiotic stress tolerance through enhanced vitamin C accumulation (Hemavathi et al. 2010).

Table 2.6 Various QTLs identified for abiotic stress tolerance in vegetables

Vegetable	Targeted abiotic stress	QTLs identified	References
Potato	Heat	9 QTLs were detected for internal heat necrosis in tubers	McCord et al. (2011)
Tomato	Heat	6 QTLs were identified for fruitsetting under high temperature conditions	Ventura et al. (2007)
Tomato	Drought	3 QTLs were identified linked to water use efficiency in *S. pennellii*	Martin et al. (1989)
Tomato	Drought	4 QTLs were identified associated with seed germination, two of which were contributed by *S. pimpinellifolium.*	Foolad et al. (2010)
Pea	Frost	161 QTLs located on seven linkage groups have been identified	Klein et al. (2014)
Pea	Drought	10 QTLs have been identified that are located on linkage groups 1, 3, and 4	Iglesias-García et al. (2015)
Common bean	Drought	14 QTLs were detected, mapped on chromosomes 1, 3, 4, 7, 8, and 9	Mukeshimana et al. (2014)
Lettuce	Heat	A major QTL, Htg6.1, for seed germination under high temperature, is linked with a temperature-sensitive gene encoding an abscisic acid biosynthesis enzyme (LsNCED4)	Argyris et al. (2008, 2011)

Transgenic potato introgressed with Nucleoside diphosphate kinase 2 (AtNDPK2) gene has been developed exhibiting salinity and drought tolerance in growth chamber (Tang et al. 2008). Δ1-pyrroline-5-carboxylate synthetase gene has been overexpressed in transgenic potato resulting in salinity tolerance (Hmida-Sayari et al. 2005). In sweet potato, the chloroplastic BADH gene from *Spinacia oleracea* (SoBADH) has been transformed and expressed exhibiting tolerance to salinity, low temperature and oxidative stress and maintained the photosynthetic activity, cell membrane integrity, and glycine betaine accumulation (Fan et al. 2012) (Table 2.7).

In vegetables, where the produce is consumed directly, the consumer acceptance of transgenic products is less and hence transgenic vegetables have not been commercialized. The transformation and expression of the transgene also differs and this has restricted the success of transgenics in vegetables. Besides, there are other issues related to market regulation, intellectual property rights, etc. transgenics may be a viable option for combating climate change and thus more studies of their influence on biodiversity, environment and human health is necessary.

2.4.2.3 Gene Silencing

Crop yields are being improved by scientists all across the world using more effective and long-lasting strategies. Gene silencing, also known as RNA interference (RNAi), is one such genetic method. RNA interference (or gene silencing) is an

Table 2.7 Transgenic Vegetables Tolerant to Abiotic Stress

Transgenic crop	Gene and gene product	Targeted stress	References
Tomato	*Arabidopsis thaliana* homeodomain-leucine zipper class I genes (ATHB-7)	Drought	Mishra et al. (2012)
Tomato	Vacuolar Na^+/H^+ antiporter (AtNHX1)	Salinity	Zhang and Blumwald (2001)
Tomato	Choline oxidase from bacterial *Arthrobacter globiformis* (codA gene)	Salinity and waterlogging	Goel et al. (2011)
Tomato, eggplant	Bacterial mannitol-1-phosphate dehydrogenase gene (MtlD)	Drought	Khare et al. (2010)
Tomato	BADH1 (betaine dehydrogenase)	Drought and salinity	Jia et al. (2002)
Tomato	C_2H_2 zinc-finger transcription factor (BcZAT12)	Drought and heat	Rai et al. (2013) and Shah et al. (2013)
Tomato	Cytosolic ascorbate peroxidase (cAPX)	Heat	Wang et al. (2006)
Tomato	Dehydrin (TAS14)	Drought and salinity	Munoz-Mayor et al. (2012)
Tomato	Glyoxalase II genes (GlyII)	Salinity	Alvarez-Viveros et al. (2013)
Tomato	H^+-Pyrophosphatase and Na^+/H^+ antiporter gene (TaNHX2)	Drought	Bhaskaran and Savithramma (2011), Yarra et al. (2012)
Tomato	Osmotin	Cold	Patade et al. (2013)
Tomato, eggplant	*Poncirus trifoliata* arginine decarboxylase (PtADC)	Drought	Wang et al. (2011), Prabhavathi and Rajam (2007)
Tomato	Polyphenol oxidases (ppo)	Drought	Thipyapong et al. (2004)
Tomato	*S*-Adenosyl-L-methionine decarboxylase (SAMDC)	Salt, cold, and drought	Alcazar et al. (2010)
Tomato	*N*-Acetylglutamate synthase (SlNAGS1)	Drought	Kalamaki et al. (2009)
Tomato	GDP-D-mannose-3′,5′-epimerase genes (SlGME1)	Drought	Zhang et al. (2011)
Tomato	*Solanum pimpinellifolium* mitogen-activated protein kinases (SpMPK1, SpMPK2, SpMPK3)	Drought	Li et al. (2013)
Tomato	*Malus domestica* subunit B of the V-ATPase (MdVHA-B)	Drought	Hu et al. (2012)
Tomato	*Lycopersicon esculentum* ethylene responsive factor (LeERF2/TERF2)	Freezing	Zhang and Huang (2010)
Potato	Pyrroline-5-carboxylate synthase (P5CS)	Salinity	Hmida-Sayari et al. (2005)
Potato	UND/PUB/ARM repeat type gene (StPUB17)	Salinity	Ni et al. (2010)
Bean	Pyrroline-5-carboxylate synthase (P5CS)	Drought, salinity and cold	Chen et al. (2009)

unique mechanism for gene down-regulation in which RNA interferes with the production of genetic information. It is initiated by endogenous or artificially introduced small interfering double-stranded RNA (siRNA) with complementary sequences to the targeted gene, causing degradation of the gene's encoded messenger RNA (mRNA) (Bosher and Labouesse 2000; Agrawal et al. 2003). RNA interference (RNAi) is a phenomena in which a high degree of targeted gene silence is achieved with little effort. It is extremely effective and can be used at various stages of growth. RNAi is being employed as a biotechnological strategy that is both alternative and advantageous. It's a sequence-based gene silencing method that's very specific and dominant. This enormous potential of RNAi has been successfully utilised for producing desirable characteristics in plants under abiotic and biotic stresses.

2.4.2.3.1 Salinity

Numerous genes, which are stimulated by this stress, have been identified by various research studies, so that the tolerance to stress can be improved by the overexpression of stress-responsive genes in plants. During stress conditions, miRNA expression, in addition to protein-coding genes, is altered in plants (Sunkar and Zhu 2004 and Zhao 2007). miR169 is one of the major miRNAs having a pivotal role in salt stress responses. Several plant species under salt stress acquire the up-regulation of miR160, miR393, and miR167. Changes in miRNA expression during stress conditions appear to be significant for the promotion of plant growth and development. enhanced accumulation of miRS1 and miR159.2 was observed in the case of Phaseolus vulgaris under salinity stress (Arenas-Huertero et al. 2009).

2.4.2.3.2 Heat Stress

For the protection of reproductive organs, a novel thermotolerance mechanism was discovered in plants. This mechanism involves the down-regulation of its target CSD genes, like CSD_1 and CSD_2, along with the gene that encodes the copper chaperone for both CSD_1 and CSD_2 genes by the induction of miR398. These results suggested that CSD_1, CSD_2, and CCS (copper chaperone for superoxide dismutase) mutants were found to tolerate the heat stress, whereas wild-type plants led to the accumulation of heat stress transcription factors and heat shock proteins, thus causing inhibition in flower damage.

2.4.2.3.3 Cold Stress

In response to cold stress, the expression of miR319 (a conserved miRNA family) was found to change in sugarcane Arabidopsis (Sunkar and Zhu 2004). It has been concluded that recently identified 21_24 nucleotide small RNAs (miRNA) and small interfering RNAs (siRNAs) regulate the expression of genes at the post-transcriptional level. They contribute to the stress-induced alterations in profiles of mRNAs or proteins and thus the RNAi technique emerged as a powerful tool to increase cold tolerance after chilling acclimation.

2.4.2.3.4 Drought Stress

Research has also shown that drought stress affects miRNA expression (Li et al. 2008; Zhao et al. 2009), and miR393, miR160, miR169 and miR167 are the most commonly affected genes in plants in response to drought stress (Sunkar 2010). Proteins of nuclear factor Y (NFY) are reported to have an important role in response of plants under stress conditions. It was also observed that the transcription factor for the Nfya5 upon drought stress is regulated by miR169, and overexpression of miR169-resistant nfya5 transgene leads to closure of stomata (Li et al. 2008) *P. vulgaris* have been studied for drought response where miRS1, miR1514a, miR2119, miR171l-n, miR1445, miR1446a-e, and miR1447 were identified for controlling water-deficient stress (Lu et al. 2008; Arenas-Huertero et al. 2009).

2.4.2.3.5 Heavy Metal Stress

The primary response of plants to heavy metal toxicity is the generation of reactive oxygen species (ROS) by interruption of their enzyme systems (Romero-Puertas et al. 2007). Gene expression at transcriptional and post-transcriptional levels is needed for plants to respond to various abiotic stresses including metal stress (Jacoby et al. 2002). Metal toxicity response in plants has been studied using different microarrays and deep sequencing of small RNA libraries, and shows the involvement of miRNAs (Hobert 2008). Different plants species were used for studying the involvement of miRNA like Brassica napus, P. vulgaris, A. thaliana, and O. sativa (Gielen et al. 2012) after exposure to different heavy metals like Cu, Zn, Mn, As, etc. In A. thaliana, miRNAs have an important role in response to heavy metal stress.

2.4.2.3.6 Nutritional Deprivation

Mineral nutrients are essential to plants and indirectly to all living organisms that are dependent on plants. Diverse factors like water content and pH change the availability of nutrients in the soil. Plants must change their structural design and metabolism to adapt to different nutrient conditions. Transgenic studies have discovered that miRNAs regulate plant adaptive responses to nutrient deprivation. The expression of Arabidopsis miR169 was reported to be down-regulated while its targets NFYA (nuclear factor Y, subunit A) family members were up-regulated by nitrogen starvation. Expression of miR169 precursors was studied and observation was made that miR169a was substantially down-regulated in both roots and shoots under nitrogen deprivation. It has also been observed that transgenic Arabidopsis plants overexpressing miRNA169a accumulated lesser amounts of N and were more sensitive to N stress as compared to the wild type (Zhao et al. 2011). Expression of miR395 is drastically up-regulated under sulfate starvation conditions and its induction is controlled by transcription factor SLIM1 (SULFUR LIMITATION 1) (Kawashima et al. 2008, 2011). at low Pi stress, expression of miR399 was up-regulated, while the target ubiquitin conjugating enzyme (UBC) was down-regulated.

2.4.3 Grafting for Stress Tolerance

Vegetable grafting was initiated in Eastern Asian countries during early twentieth century to control the soil borne diseases, particularly, *Fusarium* and other fungal and bacterial wilts hampering the production system of solanaceous and cucurbitaceous vegetables (Lee et al. 2008, 2010). At present it is a common and promising process in Korea, Japan, China as well as some European countries for modification of the plant root system for improving its tolerance to not only biotic but also abiotic stresses like salinity, waterlogging and drought (Venema et al. 2008; He et al. 2009; Martinez Rodriguez et al. 2010; Bhatt et al. 2013). Grafting could also improve high or low temperature tolerance in tomato (Abdelmageed et al. 2014). Cultivated tomato grafted onto *Solanum melongena* accession EG203 improved high temperature tolerance (Burleigh et al. 2005), while low temperature tolerance resulted when grafted onto *Solanum habrochaites* accession LA1777 rootstock (Venema et al. 2008). Tomato scions grafted on interspecific rootstocks like *Solanum lycopersicum* x *S. habrochaites* improved tolerance to low soil temperature (10–13 °C), while in brinjal, *S. integrifolium* x *S. melongena* rootstocks resulted in tolerance to low soil temperature (18–21 °C) (Okimura et al. 1986). Palada and Wu (2007) reported waterlogging tolerance in tomato grafted onto brinjal rootstock. Similarly, Anant (2016) reported that tomato grafted on brinjal rootstocks IC 354557 and IC 111056 can resist waterlogging stress for 4 days. Among cucurbits, *Cucurbita* spp. rootstocks were more tolerant to salinity than *Lagenaria siceraria* rootstocks (Matsubara 2012). Melons grafted onto hybrid squash rootstocks also showed salinity tolerance compared to non-grafted melons (Yetisir et al. 2006).

2.4.4 Cultural Practices

Good agronomic practices and modification of the existing cultural practices can combat climate change. The amount of emission of greenhouse gases may be reduced by conservation agriculture technologies, mulching and carbon sequestration by adoption of various types of cropping systems and agroforestry (Koundinya et al. 2018). For dry and humid conditions 0.39 t CO_2 equivalent/ha/year and 0.98 t

CO_2 equivalent/ha/year respectively may be mitigated through proper cultural practices (Smith et al. 2007; Milder et al. 2011).

Crop systems will be affected by both biotic (pests, infections) and abiotic (solar radiation, water, temperature) variables due to the changing climate, thereby becoming a menace to crop production and sustainability. Adaptation to this present day scenario might be better under diversified agroecosystems having broader range of characteristics and functions, which is important to combat the projected changes in biotic stress as well as and abiotic ones (Matson et al. 1997). As climate variability has grown, diverse agroecosystems have become more vital for agriculture. Temperature and precipitation, especially during the flowering and fruiting periods, is of paramount importance to crop yields. Resilient systems are the need of the day to protect the crops from the natural vagaries and extremities, especially when the crop is in critical growth phases like flowering or fruit development (Abewoy 2018). Diversified agricultural systems demonstrate how more structurally sophisticated systems might buffer the effects of changing climate on crop production in a variety of ways. Polycultures, or the cultivation of two or more crop types and wild variants in the same area, provide spatial and temporal diversity in crops as well as climate change mitigation. Agroforestry systems are an example of a high-complexity agriculture system. The multi-tier system is also beneficial for sustainable use of the available resources and also may protect the crops from temperature fluctuations (Solankey et al. 2015b). Tomato intercropped with okra, brinjal, potato and corn in high temperature conditions enhanced productivity and profitability (Kumar et al. 2012). Multi-layered cropping with Cotton + Radish + Cluster bean + Beetroot or Cotton + Radish + Beetroot + Coriander has been tested by Sankaranarayanan et al. (2012) and found to enhance productivity and proved to be profitable. A multi-tier cropping with pointed gourd + elephant foot yam + cowpea has been found to be sustainable (Solankey et al. 2015b). Short duration and bushy photo-insensitive cowpea in particular, having biotic and abiotic stress resistance can be a very important vegetable for cropping system diversification and increment of productivity as well as profitability.

Adjusting the sowing or planting dates can help in avoiding the periods of high temperature and water stress (Welbaum 2015). Precision farming is another important aspect that involves judicious fertilizer application at the proper site, at the correct time and thus does not allow excessive use of fertilizer. The water application to the crop is also very precise and applied at the critical growth stage in such a manner that it is applied at the root zone of the crop only, thereby saving water and improving quality of the vegetable besides yield. Irrigation should be applied during critical stages of growth of the crop for combating the effects of drought stress and also high temperature (Malhotra 2016). Excess water is also detrimental to vegetables, and growing them on raised beds or on furrows, particularly during rainy season, may help in combating this, and improve vegetable yield and quality (De la Peña and Hughes 2007; Welbaum 2015). In rainy season the vegetables may even be grown inside transparent polythene rain shelters and low-cost tunnels to lower the detrimental effects of waterlogging in field. Good drainage channels in the field should be prepared for drainage of rain water.

Protected structures ranging from low cost polytunnels to high cost environment-controlled glasshouses are also used in the precision agriculture, particularly, in case of high value vegetables like tomatoes, cherry tomatoes, bell peppers, cucumbers, lettuce, etc., that protects the crop from the extremities of weather (Mondal et al. 2011).

Adoption of organic farming is another way of combating climate change and it forms both adaptation and mitigation strategy for it. Chemical fertilizers and pesticides are not used in the system, and there is addition of organic sources of N that release the nutrient slowly leading to lesser N_2O release to atmosphere (Muller 2009). CO_2 emission from organic farming system is lower than conventional system, besides it reduces soil erosion, improves water infiltration and retention, thereby being more suited under moisture scarcity conditions (Sartaj et al. 2013).

Mulches, both organic and inorganic, may be used in vegetable farming system to improve quality, besides having ability to conserve soil moisture, soil temperature, reduce soil erosion and run-off and check weeds. Mulching in vegetables such as tomato, brinjal, potato, okra, melons and gourds has been done using rice straw, grasses or different coloured plastic mulches. Rice straw mulching in summer tomato has improved yield and quality and also B:C ratio (De La Pena and Hughes 2007; Pandey and Mishra 2012). The soil temperature can be increased by use of polythene mulches that are transparent, black, green and red and thereby aid in vegetable growing under cold stress conditions. On the other hand, silver poly mulch helps in reduction of temperature and aid in crop cultivation in high temperature conditions. Besides, black and green polythene mulches check weed growth, silver mulches repel white flies and aphids, red mulches suppress nematodes, yellow mulches attract white flies and aphid and blue mulches attract thrips, thereby acting as trap for them.

2.4.5 *Plant Growth Regulators and Osmoprotectants Influencing Abiotic Stress Tolerance in Plants*

Plant growth regulators and osmoprotectants have pivotal role in combating abiotic stresses, due to their direct role in plant physiology. Vegetables are particularly predisposed to quick senescence due to stress and the plant growth regulators make senescence slow due to activation of enzymes that have free radical scavenging capability. Amino vinyl glycine (AVG) and 1-methylcyclopropene (1-MCP), both having ethylene inhibiting properties, aid in extension of shelf life through suppressing ethylene formation resulting from stress (Baldwin 2003). In bell peppers and zucchini, methyl jasmonate helps in tolerating cold stress (Wang 1994). Abscisic acid also renders tolerance to chilling stress in many vegetables (Wang 1993) (Table 2.8).

Table 2.8 Influence of plant growth regulators and osmoprotectants on various abiotic stresses of vegetables

Stress	Plant growth regular/osmoprotectant used	Role
Chilling and freezing	Ammonium molybdate @ 0.15% foliar spray	Decrease negative effect
	GA3 and proline used for soaking seeds before sowing	Enhance seed germination
	Paclobutrazol	Enhance free radical scavenging enzyme activity
	Uniconzole @50 ppm	Reduction of electrolyte leakage
	Abscissic acid	Induce tolerance to freezing
Heat	Gibberellic acid @50 ppm	α-Amylase production is stimulated thus improving seed germination
	Benzyl amino purine	Senescence and lipid peroxidation are decreased
	Salicylic acid	Increases heat tolerance
	Glycine betaine	Ion leakage is decreased
	Ethylene	Improve seed germination under high temperature
Salinity	NaCl @10 mM used for seed hardening	Enhance photosynthetic activity
	Brassinolode @0.5 ppm	
	Salicylic acid @100 ppm	
	NAA @40 ppm	Check flower drop, fruit drop and bud drop
	Exogenous ABA application	Reduce toxic cl^- ion accumulation in leaves and thus release of ethylene and leaf senescence
	Jasmonic acid	Improve response to salinity in sensitive genotypes
	Ascorbic acid	Leaf number increased
	Ascorbic acid + zinc sulphate	Improved plant height and total plant biomass
	4 mM ascorbic acid +4 mM gibberellin could	Transpiration rate improved, chlorophyll *b* content, total chlorophyll content, xanthophyll content and relative water content enhanced
	Polyamines (spermine, spermidine and putrescine) GA_3, cytokinin, IAA, cycocel, thiourea	Improved salinity tolerance

(continued)

Table 2.8 (continued)

Stress	Plant growth regular/osmoprotectant used	Role
Drought	2% DAP +1% KCl as foliar application during flowering and fruit development phase	Improved drought tolerance
	Kaoline @3% as foliar spray	
	Cycocel @500 ppm, i.e., 1 ml/litre water of commercial product	
	Seed soaking in KH_2PO_4 @1% for 6–8 h	
	NAA @40 ppm	
	Cytokinin @10 ppm	
	Brassinolide @0.5 ppm	
	Salicylic acid @100 ppm	
	Ascorbic acid @100 ppm	
	Cycocel @10 ppm	
	Zinc sulphate @0.5% + boric acid @0.3% ferrous sulphate + @0.5% + urea @1% as foliar spray during critical stages of moisture application	
Waterlogging	Cycocel @500 ppm as foliar spray	Checked apical dominance and promoted lateral growth
	2% DAP +1% KCl as foliar application during flowering and fruit development phase	Improved waterlogging tolerance
	Zinc sulphate @0.5% + boric acid @0.3% ferrous sulphate + @0.5% + urea @1% as foliar spray during critical stages of stress	
	Brassinolide @0.5 ppm as foliar spray	Enhanced photosynthetic activity
	Salicylic acid @100 ppm as foliar spray	Improved use of stem reserve

2.4.6 *Crop Modelling or Simulation*

Different weather forecasting models have been developed that help in prediction of the patterns of weather based on long term available data and by using remote sensing and GIS (Vermeulen et al. 2010). Crop simulation models for different crops have also been created for prediction of the yield and other important attributes and thus help in assessing the probable effect of climate change on the particular crop and aid in determining the adaptation strategies necessary for it. For prediction of yield of potato in India, Bangladesh and Australia, Luck et al. (2010) created three global climate models (EH5OM, HadCM3Q and CCAM-Mark 3.5) and two regional climate models (RegCM3 and PRECIS). Pest-disease forecasting models in particular are of great importance to vegetable crops that can prepare growers for the upcoming pest-diseases and adoption of preventive measures. The Hyre model, Smith model, Wallin model, Blitecast, Fry model, Hartil model and Young model were used by Luck et al. (2010) for predicting late blight incidence in potato. Some important crop simulation models are Decision Support System for Agrotechnology Transfer (DSSAT), World Food Studies (WOFOST), INFOCROP, Agricultural Production Systems Simulator (APSIM), Cropping Systems Simulation Model (CropSyst) and Madhuram. DSSAT has simulation models for 42 crops including vegetables like potato, tomato and cabbage. Madhuram is specific to sweet potato for prediction of crop phenology on the basis of developmental days, both vegetative and reproductive.

2.5 Conclusion

Changing climate affects vegetable production as it does for all other crops. For feeding the ever-increasing population, the challenge is to maintain and even increase the productivity and quality. Developing climate resilient varieties, adoption of grafting, besides adoption of suitable cultural practices can be the sustainable approaches to combating the climate change and stresses. Recent techniques of forecasting and crop simulation models can be the futuristic approach to predict the climate change and be ready to face it.

References

Abdalla AA, Verkerk K (1968) Growth, flowering, and fruit set of the tomato at high temperature. Neth J Agri Sci 16:71–76. https://doi.org/10.18174/njas.v16i1.17428

Abdelmageed AH, Gruda N, Geyer B (2014) Effects of temperature and grafting on the growth and development of tomato plants under controlled conditions. In: Rural poverty reduction through research for development and transformation

Abewoy D (2018) Review on impacts of climate change on vegetable production and its management practices. Adv Crop Sci Tech 6:330. https://doi.org/10.4172/2329-8863.1000330

Agrawal N, Dasaradhi PVN, Mohmmed A, Malhotra P, Bhatnagar RK, Mukherjee SK (2003) RNA interference, biology, mechanism and applications. Microbiol Mol Biol Rev 67:657–685

Ainsworth EA, Ort DR (2010) How do we improve crop production in a warming world? Plant Physiol 154:526–530. https://doi.org/10.1104/pp.110.161349

Akhtar S, Hazra P (2015) High temperature stress on vegetable crops and breeding for resistance. In: Peter KV, Hazra P (eds) Handbook of vegetables, vol II. Studium Press, LLC, pp 453–473

Akhtar S, Naik A, Hazra P (2015) Harnessing heat stress in vegetable crops towards mitigating impacts of climate change (Chapter 10). In: Chaudhary ML, Patel VB, Siddiqui MW, Mahdi SS (eds) Climate dynamics in horticultural science, Principles and applications, vol I. Apple Academic Press, Canada, pp 173–200

Akhtar S, Naik A, Solankey SS (2021) Genotypic selection in vegetables for adaptation to climate change. In: Solankey SS, Kumari M, Kumar M (eds) Advances in research on vegetable production under a changing climate, vol 1. Springer Nature, Switzerland, pp 61–89

Alcazar R, Altabella T, Marco F, Bortolotti C, Reymond M, Koncz C, Tiburcio AF (2010) Polyamines: molecules with regulatory functions in plant abiotic stress tolerance. Planta 231:1237–1249

Altieri MA, Nicholls CI, Henao A, Lana MA (2015) Agroecology and the design of climate change-resilient farming systems. Agron Sustain Dev 35:869–890

Alvarez-Viveros MF, Inostroza-Blancheteau C, Timmermann T, Gonzálezz M, Arce-Johnson P (2013) Overexpression of GlyI and GlyII genes in transgenic tomato (*Solanum lycopersicum* mill.) plants confers salt tolerance by decreasing oxidative stress. Mol Biol Rep 4:3281–3290

Anant B, Rai N, Kumar R, Tiwari SK, Singh AK, Rai AK, Singh U, Patel PK, Tiwari V, Rai AB, Singh M, Singh B (2016) Grafting tomato on eggplant as a potential tool to improve waterlogging tolerance in hybrid tomato. Veg Sci 42:82–87

Anderson JT, Panetta AM, Mitchell-Olds T (2012) Evolutionary and ecological responses to anthropogenic climate change: update on anthropogenic climate change. Plant Physiol 160:1728–1740

Arenas-Huertero C, Pérez B, Rabanal F, Blanco-Melo D, De la Rosa C, Estrada-Navarrete G, Sanchez F, Covarrubias AA, Reyes JL (2009) Conserved and novel miRNAs in the legume *Phaseolus vulgaris* in response to stress. Plant Mol Biol 70:385–401. https://doi.org/10.1007/s11103-009-9480-3

Argyris J, Dahal P, Hayashi E, Still DW, Bradford KJ (2008) Genetic variation for lettuce seed thermoinhibition is associated with temperature-sensitive expression of abscisic acid, gibberellin, and ethylene biosynthesis, metabolism, and response genes. Plant Physiol 148:926–947. https://doi.org/10.1104/pp.108.125807

Argyris J, Truco MJ, Ochoa O, Mc Hale L, Dahal P, Van Deynze A, Michelmore RW, Bradford KJ (2011) A gene encoding an abscisic acid biosynthetic enzyme (LsNCED4) collocates with the high temperature germination locus Htg 6.1 in lettuce (*Lactuca spp.*). Theor Appl Genet 122:95–108. https://doi.org/10.1007/s00122-010-1425-3

Arvin MJ, Donnelly DJ (2008) Screening potato cultivars and wild species to abiotic, stresses using an electrolyte leakage bioassay. J Agric Sci Technol 10:33–42

Ashraf M (2010) Inducing drought tolerance in plants: recent advances. Biotechnol Adv 28:169–183

Awmack CS, Woodcock CM, Harrington R, Lether SR (1997) Host plant effects on performance of the aphid *Aulacorthum solani* (kalt) (Homoptera:Aphididae) at ambient and elevated CO_2. Glob Chang Biol 3:545–559

Bahadur A, Chatterjee A, Kumar R, Singh M, Naik PS (2011) Physiological and biochemical basis of drought tolerance in vegetables. Veg Sci 38(1):1–16

Baldwin EA (2003) Coating and other supplemental treatments to maintain vegetable quality. In: Bartz JA, Brecht JK (eds) Postharvest physiology and pathology of vegetables, 2nd edn. Marcel Dekker, New York, pp 413–456

Bale JS, Masters GJ, Hodkinson ID, Awmack C, Bezemer TM (2010) Herbivory in global climate change research: direct effects of rising temperature on insect herbivores. Glob Chang Biol 8:1–16

Barlow PW, Adam JS (1989) Anatomical disturbances in primary roots of *Zea mays* following periods of cool temperature. Environ Exp Bot 29(3):323–336

Benda GTA (1955) Some effects of ultra-violet radiation on leaves of French bean (*Phaseolus vulgaris. L.*). Ann Appl Biol 43(1):71–85

Bhardwaj ML (2012) Effect of climate change on vegetable production in India in vegetable production under changing climate scenario. In: Vegetable production under changing climate scenario, pp 1–12

Bhaskaran S, Savithramma DL (2011) Co-expression of *Pennisetum glaucum* vacuolar Na^+/H^+ antiporter and *Arabidopsis* H^+ – pyrophosphatase enhances salt tolerance in transgenic tomato. J Exp Bot 15:5561–5570

Bhatt RM, Rao NKS, Harish DM (2013) Significance of grafting in improving tolerance to abiotic stresses in vegetable crops under climate change scenario. In: Climate-resilient horticulture: adaptation and mitigation strategies, pp 159–175

Boonekamp PM (2012) Are plant diseases too much ignored in the climate change debate? Eur J Plant Pathol 133:291–294

Bosher JM, Labouesse M (2000) RNA interference, genetic wand and genetic watchdog. Nat Cell Biol 2:31–36
Bradow JM (1990) Chilling sensitivity of photosynthetic oil-seedlings. 2 Cucurbitaceae. J Exp Bot 41(233):1595–1600
Buis R, Barthou H, Roux B (1988) Effect of temporary chilling on foliar and caulinary growth and productivity in soybean (*Glycine max*). Ann Bot 61(6):705–715
Burleigh JR, Black LL, Mateo LG, Cacho D, Aganon CP, Boncato T, Arida IA, Ulrichs C, Ledesma DR (2005) Performance of grafted tomato in Central Luzon, Philippines: a case study on the introduction of a new technology among resource-limited farmers (Online). Crop Manage https://doi.org/10.1094/CM-2005-0701-01-MG
Bustan A, Sagi M, Malach YD, Pasternak D (2004) Effects of saline irrigation water and heat waves on potato production in an arid environment. Field Crop Res 90:275–285
Cabrera RM, Saltveit ME, Owens K (1992) Cucumber cultivars differ in their response to chilling temperatures. J Am Soc Hortic Sci 117(5):802–807
Cannon RJC (2008) Annexure-1 In: Climate-related transboundary pests and diseases, technical background document from the expert consultation held on 25 to 27 February 2008, FAO, Rome
Charrler A (1984) Genetic resources of the genus abelmoschus med. (Okra). International Board for Plant Genetic Resources; IBPGR Secretariat-Rome. Available online at: http://pdf.usaid.gov/pdf_docs/PNAAT275.pdf
Cheeseman JM (2008) Mechanisms of salinity tolerance in plants. Plant Physiol 87:547–550
Chen Y, Wang A, Zhao L, Shen G, Cui L, Tang K (2009) Expression of thymosin a1 concatemer in trans- genic tomato (*Solanum lycopersicum*) fruits. Biotechnol Appl Biochem 52:303–312
Chigumira NF, Grubben GJH (2004) Cucurbita maxima Duchesne. In: Grubben GJH, Denton OA (eds) PROTA2: vegetables/legumes, PROTA, Wageningen
Choi JY, Seo YS, Kim SJ, Kim WT, Shin JS (2011) Constitutive expression of CaXTH3, a hot pepper xyloglucan endotransglucosylase / hydrolase, enhanced tolerance to salt and drought stresses without phenotypic defects in tomato plants (Solanum lycopersicum cv. Dotaerang). Plant Cell Rep 30(5):879–881
Coakley SM, Scherm H, Chakraborty S (1999) Climate change and plant disease management. Annu Rev Phytopathol 37:399–426
Collins NC, Tardieu F, Tuberosa R (2008) Quantitative trait loci and crop performance under abiotic stress: where do we stand? Plant Physiol 147:469–486. https://doi.org/10.1104/pp.108.118117
Cotty PJ, Jamie-Garcia R (2007) Influences of climate on aflatoxin producing fungi and aflatoxin contamination. Int J Food Microbiol 119(1–2):109–115
Dane F, Liu J, Zhang C (2007) Phylogeography of the bitter apple, *Citrullus colocynthis*. Genet Resour Crop Ev 54:327–336
Das DK, Singh J, Vennila S (2011) Emerging crop pest scenario under the impact of climate change-a brief review. J Agric Phys 11:13–20
De la Peña R, Hughes J (2007) Improving vegetable productivity in a variable and changing climate. J SAT Agric Res 4:1–22
Dias MC, Brüggemann W (2010) Limitations of photosynthesis in *Phaseolus vulgaris* under drought stress: gas exchange, chlorophyll fluorescence and Calvin cycle enzymes. Photosynthetica 48:96–102
Dodds GT, Ludford PM (1990) Surface topology of chilling injury of tomato fruit. HortScience 25(11):1416–1419
Drew MC (2009) Plant responses to anaerobic conditions in soil and solution culture. Curr Adv Plant Sci 36:1–14
Fahim MA, Hassanein MK, Abou Hadid AF, Kadah MS (2011) Impacts of climate change on the widespread and epidemics of some tomato diseases during the last decade in Egypt. Acta Hortic 914:317–320
Fan W, Zhang M, Zhang H, Zhang P (2012) Improved tolerance to various abiotic stresses in transgenic sweet potato (*Ipomoea batatas*) expressing spinach betaine aldehyde dehydrogenase. PLoS One 7:e37344

FAO (2008) Climate-related transboundary pests and diseases, technical background document from the expert consultation held on 25 to 27 February 2008, FAO, Rome. Downloaded from ftp://ftp.fao.org/docrep/fao/meeting/013/ai785e.pdf

FAO (2009) Global agriculture towards 2050 Issues Brief. High level expert forum, Rome, pp 12–13

FAOSTAT (2021) Food and agriculture Organization of the United Nations. FAOSTAT Statistical Database, Rome, FAO

Firon N, Shaked R, Peet MM, Pharr DM, Zamski E, Rosenfeld K, Althan L, Pressman E (2006) Pollen grains of heat tolerant tomato cultivars retain higher carbohydrate concentration under heat stress conditions. Scientia Hortic 109(3):212–217. https://doi.org/10.1016/j.scienta.2006.03.007

Fleming RA, Tatchell GM (1995) Shifts in the flight periods of British aphids: a response to climate warming? In: Harrington R, Stork N (eds) Insects in a changing environment. Academic, pp 505–508

Flowers TJ (2004) Improving crop salt tolerance. J Exp Bot 55(307):319

Folzer H, Dat JF, Capelli N, Rieffel D, Badot PM (2006) Response of sessile oak seedlings (*Quercus petraea*) to flooding: an integrated study. Tree Physiol 26:759–766

Foolad MR (2014) Recent advances in genetics of salt tolerance in tomato. Plant Cell Tissue Organ Cult 76:101–119

Foolad MR, Jones RA (1991) Genetic analysis of salt tolerance during germination in *Lycopersicon*. Theor Appl Genet 81(321):326

Foolad MR, Zhang LP, Subbiah P (2010) Genetics of drought tolerance during seed germination in tomato: inheritance and QTL mapping. Genome 46:536–545

Frenkel C, Erez A (1996) Induction of chilling tolerance in cucumber (*Cucumis sativus*) seedlings by endogenous and applied ethanol. Physiol Plant 96(4):593–600

Gibbs J, Greenway H (2008) Mechanisms of anoxia tolerance in plants. I. Growth, survival and anaerobic catabolism. Funct Plant Biol 30:1–47

Gielen H, Remans T, Vangronsveld J, Cuypers A (2012) MicroRNAs in metal stress, specific roles or secondary responses. Int J Mol Sci 13:5826–15847

Goel D, Singh AK, Yadav V, Babbar SB, Bansal KC (2010) Overexpression of osmotin gene confers tolerance to salt and drought stresses in transgenic tomato (*Solanum lycopersicum* L.). Protoplasma 245:133–141

Goel D, Singh AK, Yadav V, Babbar SB, Murata N, Bansal KC (2011) Transformation of tomato with a bacterial coda gene enhances tolerance to salt and water stresses. J Plant Physiol 11:1286–1294

Gorecki RJ, Fordonski G, Bieniaszewski T, Jacunski K (1990) Comparative studies on chilling sensitivity in some legume seeds. Acta Physiol Plant 12(2):149–158

Hahn DA, Denlinger DL (2007) Meeting the energetic demands of insect diapause: nutrient storage and utilization. J Insect Physiol 53:760–773

Hahn M, Walbot V (1989) Effect of cold-treatment on protein synthesis and mRNA levels in rice leaves. Plant Physiol 91(3):930–938

Hall A (2001) Crop developmental responses to temperature, photoperiod, and light quality. In: Hall AE (ed) Crop response to environment. CRC, Boca Raton, pp 83–87

Hanif R, Iqbal Z, Iqbal M, Hanif S, Rasheed M (2006) Use of vegetables as nutritional food: role in human health. J Agric Biol Sci 1:18–22

Hao X, Hale BA, Ormrod DP (1997) The effects of ultraviolet- B radiation and carbon dioxide on growth and photosynthesis of tomato. Can J Bot 75(2):213–219

Harker FR, Maindonald JH (1994) Ripening of nectarine fruit. Changes in the cell wall, vacuole, and membranes detected using electrical impedance measurements. Plant Physiol 106(l):165–171

Harrington R, Fleming RA, Woiwod IP (2010) Climate change impacts on insect management and conservation in temperate regions: can they be predicted? Agric For Entomol 3:233–240

Harvell CD, Mitchell CE, Ward JR, Altizer S, Dobson AP (2006) Climate warming and disease risks for terrestrial and marine biota. Science 296:2158–2162

Hazra P, Som MG (2006) Vegetable Science. Kalyani Publishers, New Delhi, India

Hazra P, Samsul HA, Sikder D, Peter KV (2007) Breeding tomato (*Lycopersicon esculentum* mill.) resistant to high temperature stress. Int J Plant Breed 1(1):31–40

He Y, Zhu Z, Yang J, Ni X, Zhu B (2009) Grafting increases the salt tolerance of tomato by improvement of photosynthesis and enhancement of antioxidant enzymes activity. Environ Exp Bot 66:270–278

Hemavathi UCP, Akula N, Young KE, Chun SC, Kim DH, Park SW (2010) Enhanced ascorbic acid accumulation in transgenic potato confers tolerance to various abiotic stresses. Biotechnol Lett 32:321–330. https://doi.org/10.1007/s10529-009-0140-0

Hmida-Sayari A, Gargouri-Bouzid R, Bidani A, Jaoua L, Savouré A, Jaoua S (2005) Overexpression of Δ1- pyrroline-5-carboxylate synthetase increases proline production and confers salt tolerance in transgenic potato plants. Plant Sci 169:746–752

Hobert O (2008) Gene regulation by transcription factors and microRNAs. Science 319:1785–1786

Holmberg N, Bülow L (1998) Improving stress tolerance in plants by gene transfer. Trends Plant Sci 3(2):61–66

Horváth G, Bernáth B, Molnár G (1998) Dragonflies find crude oil visually more attractive than water: multiple-choice experiments on dragonfly polarotaxis. Naturwissenschaften 85:292–297

Hu DG, Wang SH, Luo H, Ma QJ, Yao YX, You CX, Hao and Y.J. (2012) Overexpression of MdVHA-B, V-ATPase gene from apple, confers tolerance to drought in transgenic tomato. Sci Hortic 145:94–101

Iglesias-García R, Prats E, Fondevilla S, Satovic Z, Rubiales D (2015) Quantitative trait loci associated to drought adaptation in pea (*Pisum sativum* L.). Plant Mol Biol Report 33:1768. https://doi.org/10.1007/s11105-015-0872-z

Ismail AM, Hall AE, Close TJ (1997) Chilling tolerance during emergence of cowpea associated with a dehydrin and slow electrolyte leakage. Crop Sci 37(4):1270–1277

Isopp H, Frehner M, Long SP, Nösberger J (2008) Sucrose phosphate synthase responds differently to source sink relations and to photosynthetic rates: *Lolium perenne* L. growing at elevated pCO2 in the field. Plant Cell Environ 23:597–607

Jacoby M, Weisshaar B, Vicente-Carbajosa J, Tiedemann J, Kroj T, Parcy F (2002) bZIP transcription factors in *Arabidopsis*. Trends Plant Sci 7:106–111

Jamil M, Rha ES (2014) The effect of salinity (NaCl) on the germination and seedling of sugar beet (*Beta vulgaris* L.) and cabbage (*Brassica oleracea capitata* L.). Korean J Plant Res 7:226–232

Jat MK, Tetarwal AS (2012) Effect of changing climate on the insect pest population national seminar on sustainable agriculture and food security: challenges in changing climate. Indian Hortic 3:41–49

Jennings P, Saltveit ME (1994) Temperature effects on imbibition and germination of cucumber (*Cucumis sativus*) seeds. J Amer Soc Hortic Sci 119(3):464

Jha UC, Bohra A, Singh NP (2014) Heat stress in crop plants: its nature, impacts and integrated breeding strategies to improve heat tolerance. Plant Breed 133:679–701. https://doi.org/10.1111/pbr.1221

Jia GX, Zhu ZQ, Chang FQ, Li YX (2002) Transformation of tomato with the BADH gene from Atriplex improves salt tolerance. Plant Cell Rep 21:141–146

Kalamaki MS, Alexandrou D, Lazari D, Merkouropoulos G, Fotopoulos V, Pateraki I, Aggelis A, Carrillo-Lo A, Rubio-Cabetas MJ, Kanellis AK (2009) Over-expression of a tomato N-acetyl-L-glutamate synthase gene (SlNAGS1) in *Arabidopsis thaliana* results in high ornithine levels and increased tolerance in salt and drought stresses. J Exp Bot 60(6):1859–1871

Kavar T, Maras M, Meglie V (2011) The expression profiles of selected genes in different bean species as response to water deficit. J Cent Eur Agric 12:557–576

Kawase M (2011) Anatomical and morphological adaptation of plants to water logging. Hortic Sci 16:30–34

Kawashima CG, Yoshimoto N, Maruyama-Nakashita A, Tsuchiya YN, Saito K, Takahashi H (2008) Sulphur starvation induces the expression of microRNA-395 and one of its target genes but in different cell types. Plant J 57:313–321

Kawashima CG, Matthewman CA, Huang S, Lee BR, Yoshimoto N, Koprivova A (2011) Interplay of SLIM1 and miR395 in the regulation of sulfate assimilation in Arabidopsis. Plant J 66:863–876

Kaymakanova M, Stoeva N, Mincheva T (2008) Salinity and its effects on the physiological response of bean (*Phaseolus vulgaris*. L.). J Cent Eur Agric 9:749–756

Khan MR (2012, March 27–28) Effect of elevated levels of CO_2 and other gaseous pollutants on crop productivity and plant diseases. In: National seminar on sustainable agriculture and food security: challenges in changing climate, pp 197

Khare N, Goyary D, Singh NK, Shah P, Rathore M, Anandhan S, Sharma D, Arif M, Ahmed Z (2010) Transgenic tomato cv. Pusa Uphar expressing a bacterial mannitol-1-phosphate dehydrogenase gene confers abiotic stress tolerance. Plant Cell Tissue Org Cult 103:267–277

Klein A, Houtin H, Rond C, Marget P, Jacquin F, Boucherot K, Huart M, Rivière N, Boutet G, Lejeune-Hénaut I, Burstin J (2014) QTL analysis of frost damage in pea suggests different mechanisms involved in frost tolerance. Theor Appl Genet 127:1319–1330

Koundinya AVV, Siddhya P, Pandit MK (2014) Impact of climate change on vegetable cultivation – a review. Int J Agric Environ Biotechnol 7(1):145–155

Koundinya AVV, Pradeep Kumar P, Ashadevi RK, Hegde V, Arun Kumar P (2018) Adaptation and mitigation of climate change in vegetable cultivation: a review. J Water Clim Chang 9:17–36. https://doi.org/10.2166/wcc.2017.045

Kumar SN (2017) Climate change and its impacts on food and nutritional security in India. In: Belavadi VV, Karaba NN, Gangadharappa NR (eds) Agriculture under climate change: threats, strategies and policies, p 48

Kumar R, Singh M (2006) Citation information. In: Singh RJ (ed) Genetic resources, chromosome engineering, and crop improvement vegetable crops, vol 3. CRC Press, pp 473–496

Kumar P, Nanwal RK, Dhindwal AS, Yadav SK (2012) Chapter 7: Competition relations; multistoried cropping and yield stability in Intercropping. In: Manual on cropping systems and sustainable agriculture. Department of Agronomy, CCS Haryana Agricultural University, Hisar, Haryana, India, pp 77–88

Kuo DG, Tsay JS, Chen BW, Lin PY (2014) Screening for flooding tolerance in the genus *Lycopersicon*. Hortic Sci 17:76–78

Kusvuran S (2012) Effects of drought and salt stresses on growth, stomatal conductance, leaf water and osmotic potentials of melon genotypes (*Cucumis melo* L.). Afr J Agric Res 7(5):775–781

Ladizinsky G (1985) Founder effect in crop-plant evolution. Econ Bot 39:191–199. https://doi.org/10.1007/BF02907844

Lee SG, Huh YC, Sun ZY, Miguel A, King SR (2008) Cucurbit grafting. Crit Rev Plant Sci 27:50–74

Lee JM, Kubota C, Tsao SJ, Bie Z, Echevarria PH (2010) Current status of vegetable grafting: diffusion, grafting techniques, automation. Sci Hortic 127:93–105

Lejeune P, Bernier G (1996) Effect of environment on the early steps of ear initiation in maize (*Zea mays* L.). Plant Cell Environ 19(2):217–224

Levitt J (1980) Responses of plants to environmental stresses, Chilling, freezing and high temperatures stresses, vol 1. Academic, Cambridge

Lewis T (1997) Thrips as crop pests. CAB International, University Press, Cambridge, pp 740

Li WX, Oono Y, Zhu JH, He XJ, Wu JM, Iida K (2008) The Arabidopsis NFYA5 transcription factor is regulated transcriptionally and post transcriptionally to promote drought resistance. Plant Cell 20:2238–2251

Li C, Yan JM, Li YZ, Zhang ZC, Wang QL, Liang Y (2013) Silencing the SpMPK1, SpMPK2, and SpMPK3 genes in tomato reduces abscisic acid – mediated drought tolerance. Int J Mol Sci 14:21983–21996

Liao CT, Lin CH (2014) Effect of flooding stress on photosynthetic activities of *Momordica charantia*. Plant Physiol Biochem 32:479–485

Lopes MS, Reynolds MP, McIntyre CL, Mathews KL, Jalal Kamali MR, Mossad M, Feltaous Y, Tahir ISA, Chatrath R, Ogbonnaya F et al (2013) QTL for yield and associated traits in the Seri/Babax population grown across several environments in Mexico, in the West Asia, North Africa, and South Asia regions. Theor Appl Genet 126:971–984. https://doi.org/10.1007/s00122-012-2030-4

Lopez MAH, Ulery AL, Samani Z, Picchioni G, Flynn RP (2011) Response of Chile pepper (*Capsicum annuum* L.) to salt stress and organic and inorganic nitrogen sources: i. growth and yield. Trop Subtrop Agroecosystems 14:137–147

Lu SF, Sun YH, Chiang VL (2008) Stress-responsive microRNAs in *Populus*. Plant J 55:131–151

Luck J, Asaduzzaman M, Banerjee S, Bhattacharya I, Coughlan K, Debnath GC, De Boer D, Dutta S, Forbes G, Griffiths W, Hossain D, Huda S, Jagannathan R, Khan S, O'Leary G, Miah G, Saha A, Spooner-Hart R (2010) The effects of climate change on pest and diseases major food crops in the Asia Pacific region. Asia–Pacific Network for Global Change Research. https://www.apn-gcr.org/resources/files/original/1534fe7a80b1be6e9d00d2cd6934fae0.pdf. Accessed 14 Apr 2012

Lukatkin AS, Brazaitytė A, Bobinas C, Duchovskis P (2012) Chilling injury in chilling-sensitive plants: a review. Žemdirbystė=Agriculture 99(2):111–124

Malhotra SK (2016) Recent advances in seed spices research-a review. Ann Plant Soil Res 18:300–308

Malik AI, Colmer TD, Lambers H, Setter TL, Schortemeyer M (2012) Short term waterlogging has long term effects on the growth and physiology of wheat. New Phytol 153:225–236

Markowski A (1988) Sensitivity of different species of field crops to chilling temperature. II. Germination, growth and injuries of seedlings. Acta Physiol Plant 10(3):275–283

Martin B, Nienhuis J, King G, Schaefer A (1989) Restriction fragment length polymorphisms associated with water use efficiency in tomato. Science 243:1725–1728

Martinez Rodriguez MM, Estan MT, Moyano E, Garcia Abellan JO, Flores FB (2010) The effectiveness of grafting to improve salt tolerance in tomato when an 'excluder'genotype is used as scion. Environ Exp Bot 63:392–401

Matson PA, Parton WJ, Power AG, Swift MJ (1997) Agricultural intensification and ecosystem properties. Science 277:504–509

Matsubara S (2012) Studies on salt tolerance of vegetables, 3: Salt tolerance of rootstocks. Scientific reports of the faculty of agriculture Okayama University

Mboup M, Bahri B, Leconte M, Vallavieille P, D. and Kaltz, O. (2012) Genetic structure and local adaptation of European wheat yellow rust populations: the role of temperature specific adaptation. Evol Appl 5:341–352

McCord PH, Sosinski BR, Haynes KG, Clough ME, Yencho GC (2011) QTL mapping of internal heat necrosis in tetraploid potato. Theor Appl Genet 122:129–142. https://doi.org/10.1007/s00122-010-1429-z

McMahon MJ, Permit AJ, Arnold JE (1994) Effects of chilling on *Episcia* and *dieffenbachia*. J Am Soc Hortic Sci 119(1):80–83

Metwally E, El-Zawily A, Hassan N, Zanata O (1996) Inheritance of fruit set and yields of tomato under high temperature conditions in Egypt. In: First Egypt-Hungary horticulture conference, vol I, pp 112–122

Mickelbart MV, Hasegawa PM, Bailey-Serres J (2015) Genetic mechanisms of abiotic stress tolerance that translate to crop yield stability. Nat Rev Genet 16(4):237–251. https://doi.org/10.1038/nrg3901

Milder JC, Majanen T, Scherr SJ (2011) Performance and potential of conservation agriculture for climate change adaptation and mitigation in sub-Saharan Africa: final report on an assessment of WWF and CARE projects in support of the WWF–CARE Alliance's rural futures initiative. Eco Agriculture Partners, Washington, DC, USA

Mishra KB, Iannacone R, Petrozza A, Mishra A, Armentano N, La Vecchia G, Trtilek M, Cellini F, Nedbal L (2012) Engineering drought tolerance in tomato plants is reflected in chlorophyll fluorescence emission. Plant Sci 182:79–86

Mitchell DE, Madore MA (1992) Patterns of assimilate production and translocation in musk-melon (*Cucumis melo* L.): low temperature effects. Plant Physiol 99(3):966–971

Mondal P, Basuand PB, Bhadoria S (2011) Critical review of precision agriculture technologies and its scope of adoption in India. Am J Exp Agric 1(3):49–68

Mukeshimana G, Butare L, Cregan PB, Blair MW, Kelly JD (2014) Quantitative trait loci associated with drought tolerance in common bean. Crop Sci 54:923–938. https://doi.org/10.2135/cropsci2013.06.0427

Muller A (2009) Benefits of organic agriculture as a climate change adaptation and mitigation strategy for developing countries. Discussion paper series–EfD DP 09–09. Environment for development. http://www.ifr.ac.uk/waste/ Reports/BenefitsOfOrganicAgriculture.pdf. Accessed 8 Nov 2013

Muñoz-Mayor A, Pineda B, Garcia-Abellán JO, Antón T, Garcia-Sogo B, Sanchez-Bel P, Flores FB, Atarés A, Angosto T, Pintor-Toro JA, Moreno V, Bolarin MC (2012) Overexpression of dehydrin tas14 gene improves the osmotic stress imposed by drought and salinity in tomato. J Plant Physiol 169:459–468

Neumeister L (2010) Climate change and crop protection -anything can happen. In: Pesticide action network asia and the pacific, Penang, Malaysia, pp 4–41

Newton AC, Johnson SN, Gregory PJ (2011) Implications of climate change for diseases, crop yields and food security. Euphytica 179:3–18

Ni X, Tian Z, Liu J, Song B, Li J, Shi X, Xie C (2010) StPUB17, a novel potato UND/PUB/ARM repeat type gene, is associated with late blight resistance and NaCl stress. Plant Sci 178:158–169

Nullis C (2018) Communications and public affairs WMO IPCC issues special report on global warming of 1.5 °C. Bulletin 67(2)

Okimura M, Matsuo S, Arai K, Okitsu S (1986) Influence of soil temperature on the growth of fruit vegetable grafted on different stocks. Bull Veg Ornam Crops Res Stn, Japan, Ser. C.9: 43–58. (in Japanese with English summary)

Opena RT, Chen JT, Kuo CG, Chen HM (1992) Genetic and physiological aspects of tropical adaptation in tomato. In: Adaptation of food crops to temperature and water stress, AVRDC, Shanhua, pp 321–334

Orvar BL, Sangwan V, Omann F, Dhindsa RS (2000) Early steps in cold sensing by plant cells: the role of actin cytoskeleton and membrane fluidity. The Plant J 23(6):785–794

Palada MC, Wu DL (2007) Increasing off-season tomato production using grafting technology for peri-urban agriculture in Southeast Asia. Acta Hortic 742:125–131. https://doi.org/10.17660/ActaHortic.2007.742.17

Pandey VK, Mishra AC (2012) Effect of mulches on soil moisture and fruit yield in summer tomato. Agric Eng Today 36(1):15–17

Pandey SK, Naik PS, Sud KC, Chakrabarti SK (2007) CPRI – perspective plan vision 2025. Central Potato Research Institute, Shimla, HP, India, pp 1–70

Pandey S, Ansari WA, Jha A, Bhatt KV, Singh B (2011) Evaluations of melons and indigenous *Cucumis spp*. genotypes for drought tolerance. In: 2nd International symposium on underutilized plant species, 27th June – 1st July, The Royal Chaulan Kuala Lumpur, Malaysis, (A-61), pp 95

Paran I, van der Knaap E (2007) Genetic and molecular regulation of fruit and plant domestication traits in tomato and pepper. J Exp Bot 58(14):3841–3852

Parent C, Capelli N, Berger A, Crèvecoeur M, Dat JF (2008) An overview of plant responses to soil waterlogging. Plant Stress 2:20–27

Paťade VY, Khatri D, Kumari M, Grover A, Gupta SM, Ahmed Z (2013) Cold tolerance in Osmotin transgenic tomato (*Solanum lycopersicum* L.) is associated with modulation in transcript abundance of stress responsive genes. Springerplus 2:117

Pautasso M, Doring TF, Garbelotto M, Pellis L, Jeger MJ (2012) Impacts of climate change on plant diseases-opinions and trends. Eur J Plant Pathol 133:295–313
Peet MM, Willits DH, Gardner R (1997) Response of ovule development and post-pollen production processes in male-sterile tomatoes to chronic, sub-acute high temperature stress. J Exp Bot 48:101–112
Pereira JS, Chaves MM (2007) Plant responses to drought under climate change in Mediterranean-type ecosystems. In: Global change and Mediterranean-type ecosystems, pp 140–160
Pertot I, Mach FE, Elad Y (2012) Climate change impact on plant pathogens and plant diseases. In: *Envirochange project booklet*, pp 4
Pinto RS, Reynolds MP, Mathews KL, McIntyre CL, Olivares-Villegas JJ, Chapman SC (2010) Heat and drought adaptive QTL in a wheat population designed to minimize confounding agronomic effects. Theor Appl Genet 121:1001–1021. https://doi.org/10.1007/s00122-010-1351-4
Prabhavathi V, Rajam MV (2007) Polyamine accumulation in transgenic eggplant enhances tolerance to multiple abiotic stresses and fungal resistance. Plant Biotechnol 24(3):273–282
Rab A, Saltveit ME (1996a) Differential chilling sensitivity in cucumber (*Cucumis sativus*) seedlings. Physiol Plant 96(3):375–382
Rab A, Saltveit ME (1996b) Sensitivity of seedling radicles to chilling and heat shock-induced chilling tolerance. J Am Soc Hortic Sci 121(4):711–715
Rai N, Yadav DS (2005) Advances. In: Vegetable production. Researcho Book Centre, New Delhi
Rai M, Pandey S, Kumar S (2008, May 21–24) Cucurbitaceae. In: Pitrat M (ed) Proceedings of the IXth EUCARPIA meeting on genetics and breeding of Cucurbitaceae, INRA, Avignon (France), pp 285–293
Rai N, Tiwari SK, Kumar R, Singh M, Bharadwaj DR (2011, April 4–5) Genetic Resources of Solanaceous Vegetables in India. In: National symposium on vegetable biodiversity. Jawaharlal Nehru Krishi Vishwa Vidyalaya, Jabalpur, M.P, pp 91–103
Rai AC, Singh M, Shah K (2013) Engineering drought tolerant tomato plants over-expressing BcZAT12 gene encoding a C2H2 zinc finger transcription factor. Phytochemistry 85:44–50
Raison JK, Lyons JM (1986) Chilling injury: a plea for uniform terminology. Plant Cell Environ 9:685–686
Razdan MK, Mattoo AK (2007) Genetic improvement of solanaceous crops: tomato, 2nd edn. Science Publishers, p 47
Robinson DS (1990) Food biochemistry and nutritional value. Longman scientific and technical publisher, New York, USA
Romero-Puertas MC, Corpas FJ, Rodriguez-Serrano M, Gomez M, del Rio LA, Sandalio LM (2007) Differential expression and regulation of antioxidative enzymes by cadmium in pea plants. J Plant Physiol 164:1346–1357
Ross H (1986) Potato breeding: problems and perspectives. J Plant Breed 13:1–132
Roy DB, Rothery P, Moss D (2001) Butterfly numbers and weather: predicting historical trends in abundance and the future effects of climate change. J Animal Ecol 70(2):201–217
Sankaranarayanan K, Nalayini P, Praharaj CS (2012) Multi-tier cropping system to enhance resource utilization, profitablity and sustainablity of Btcotton (Gossypium hirsutum) production system. Indian J Agric Sci 82(12):1044–1050
Sartaj SA, Chand S, Najarand GR, Teli MA (2013) Organic farming: as a climate change adaptation and mitigation strategy. Curr Agric Res J 1(1):45–50
Shah K, Singh M, Rai AC (2013) Effect of heat-shock induced oxidative stress is suppressed in BcZAT12 expressing drought tolerant tomato. Phytochemistry 95:109–117
Sharom M, Willemot C, Thompson JE (1994) Chilling injury induces lipid phase changes in membranes of tomato fruit. Plant Physiol 105(1):305–308
Siddique KHM, Loss SP, Regan KL, Jettner RL (1999) Adaptation and seed yield of cool season grain legumes in Mediterranean environments of South-Western Australia. Aust J Agric Res 50:375–388. https://doi.org/10.1071/A98096
Singh HP (2010, October 29–30) Ongoing research in abiotic stress due to climate change in horticulture. In: Curtain raiser meet on research needs arising due to abiotic stresses in agriculture

management in India under global climate change scenario, Baramati, Maharashtra, pp 1–23. http://www.niam.res.in/pdfs/DDG_Hort_lecture.pdf

Sivakumar R, Nandhitha GK, Boominathan P (2016) Impact of drought on growth characters and yield of contrasting tomato genotypes. Madras Agric J 103:78–82

Skog LJ (1998) Chilling injury of horticultural crops. Ontario Ministry of Agriculture, Food and Rural Affairs Factsheet. http://www.omafra.gov.on.ca/english/crops/facts/98–021.htm. Accessed 11 May 2011

Skrudlik G, Koscielniak J (1996) Effects of low-temperature treatment at seedling stage on soybean growth, development and final yield. J Agron Crop Sci 176(2):111–117

Smith P, Martino D, Cai Z, Gwary D, Janzen H, Kumar P, McCarl B, Ogle S, O'Mara F, Rice C, Scholes B, Sirotenko O (2007) Agriculture. In: Metz B, Davidson OR, Bosch PR, Dave R, Meyer LA (eds) Climate change 2007: mitigation. Contribution of working group III to the fourth assessment report of the intergovernmental panel on climate change. Cambridge University Press, Cambridge, UK, pp 492–540

Solankey SS, Singh RK, Baranwal DK, Singh DK (2015a) Genetic expression of tomato for heat and drought stress tolerance: an overview. Int J Veg Sci 21(5):496–515. https://doi.org/10.1080/19315260.2014.902414

Solankey SS, Akhtar S, Aditya (2015b) Multitier cropping system for profitability and scalability in vegetable production. In: Choudhary ML, Aditya (ed) Family farming and rural economic development. New India Publishing Agency, New Delhi, pp. 331–338

Steponkus PL (1984) Role of the plasma membrane in freezing injury and cold acclimation. Annu Rev Plant Physiol 35:543–584

Steponkus PL, Uemura M, Webb MS (1993) A contrast of the cryostability of the plasma membrane of winter rye and spring oat-two species that widely differ in their freezing tolerance and plasma membrane lipid composition. In: Steponkus PL (ed) Advances in low-temperature biology, vol 2. JAI Press, London, pp 211–312

Sunkar R (2010) MicroRNAs with macro-effects on plant stress responses. Semin Cell Dev Biol 21:805

Sunkar R, Zhu JK (2004) Novel and stress-regulated micro RNAs and other small RNAs from *Arabidopsis*. Plant Cell 16:2001–2019

Tang L, Kim MD, Yang KS, Kwon SY, Kim SH, Kim JS, Yun DJ, Kwak SS, Lee HS (2008) Enhanced tolerance of transgenic potato plants overexpressing nucleoside diphosphate kinase 2 against multiple environmental stresses. Transgenic Res 17:705–715

Termorshuizen AJ (2008) Climate change and bio invasiveness of plant pathogens: comparing pathogens from wild and cultivated hosts in the past and the present. Pests Clim Chang:6–9

Thipyapong P, Melkonian J, Wolfe DW, Steffens JC (2004) Suppression of polyphenol oxidase increases stress tolerance in tomato. Plant Sci 167:693–703

Ting CS, Owens TG, Wolfe DW (1991) Seedling growth and chilling stress effect on photosynthesis in chilling-sensitive and chilling tolerant cultivars of *Zea mays*. J Plant Physiol 137(5):559–564

Toppino L, Acciarri N, Mennella G, Lo Scalzo R, Rotino GL (2009, September 16/19) Introgression breeding in eggplant (*Solanum melongena* L.) by combining biotechnological and conventional approaches. In: Proceedings of the 53rd Italian society of agricultural genetics annual congress Torino, Italy

Tsuda H, Niimura Y, Katoh T (2003) Chill injury in *Saintpaulia* leaf with special reference to leaf spot formation. J Agric Sci Tokyo Nogyo Daigaku 47(4):283–289

Vadez V, Berger JD, Warkentin T, Asseng S, Ratnakumar P (2012) Adaptation of grain legumes to climate change: a review. Agron Sustain Dev 32:31–44

Venema JH, Posthumus F, de Vries M, van Hasselt PR (1999) Differential response of domestic and wild *Lycopersicon* species to chilling under low light: growth, carbohydrate content, photosynthesis and the xanthophyll cycle. Physiol Plant 105(1):81–88

Venema JH, Dijk BE, Bax JM, Van Hasselt PR, Elzenga JTM (2008) Grafting tomato (*Solanum lycopersicum*) onto the rootstock of a high altitude accession of *Solanum habrochaites* improves suboptimal temperature tolerance. Environ Exp Bot 63:359–367

Ventura Y, Mendlinger S (1999) Effects of suboptimal low temperature on yield, fruit appearance and quality in muskmelon (*Cucumis melo* L.) cultivars. J Hortic Sci Biotechnol 74(5):602–607

Ventura G, Grilli G, Braz LT, Gertrudes E, Lemos M (2007) QTL identification for tolerance to fruit set in tomato by fAFLP markers. Crop Breed Appl Biotechnol 7:234–241

Vermeulen SJ, Aggarwal PK, Ainslie A, Angelone C, Campbell BM, Challinor AJ, Hansen J, Ingram JSI, Jarvis A, Kristjanson P, Lau C, Thornton PK, Wollenberg E (2010) Agriculture, food security and climate change: outlook for knowledge, tools and action: CCAFS Report 3. CGIAR–ESSP program on climate change. Agriculture and Food Security, Copenhagen, Denmark. https://cgspace.cgiar.org/rest/bitstreams/15414/retrieve. Accessed 4 Jan 2013

Vijayalakshmi K, Fritz AK, Paulsen GM, Bai G, Pandravada S, Gill BS (2010) Modeling and mapping QTL for senescence-related traits in winter wheat under high temperature. Mol Breed 26:163–175. https://doi.org/10.1007/s11032-009-9366-8

Wahid A, Gelani S, Ashraf M, Foolad M (2007) Heat tolerance in plants: an overview. Environ Exp Bot 61:199–223. https://doi.org/10.1016/j.envexpbot.2007.05.011

Wang CY (1993) Approaches to reducing chilling injury of fruits and vegetables. Hortic Rev 15:63–95

Wang CY (1994) Reduction of chilling by Methyl Jasmonate. Acta Hortic 368:901–907

Wang Y, Wisniewski M, Meilan R, Cui M, Fuchigami L (2006) Transgenic tomato (*Lycopersicon esculentum*) overexpressing cAPX exhibits enhanced tolerance to UV-B and heat stress. J Appl Hortic 2:87–90

Wang BQ, Zhang QF, Liu JH, Li GH (2011) Overexpression of PtADC confers enhanced dehydratation and drought tolerance in transgenic tobacco and tomato: effect on ROS elimination. Biochem Biophys Res Commun 413:10–16

Welbaum GE (2015) Vegetable production and practices. Centre for Agriculture and Bioscience International, p 476

WMO (2012) WMO statement on the status of the global climate in 2011. WMO-No. 1085, World Meteorological Organization

Wolk WD, Herner RC (1982) Chilling injury of germinating seeds and seedlings. Hort Science 17(2):169–173

Xin Z, Browse J (2000) Cold comfort farm: the acclimation of plants to freezing temperatures. Plant Cell Environ 23(9):893–902

Yamamura K, Kiritani K (1998) A simple method to estimate the potential increase in the number of generations under global warming in temperate zones. Appl Entomol Zool 33:289–298

Yarra R, He SJ, Abbagani S, Ma B, Bulle M, Zhang WK (2012) Overexpression of wheat Na^+/ H^+ anti- porter gene (TaNHX2) enhances tolerance to salt stress in transgenic tomato plants (*Solanum lycopersicum* L). Plant Cell Tissue Org Cult 111:49–57

Yetisir H, Caliskan ME, Soylu S, Sakar M (2006) Some physiological and growth responses of watermelon [*Citrullus lanatus* (Thunb.) Matsum. and Nakai] grafted onto *Lagenaria siceraria* to flooding. Environ Exp Bot 58:1–8

Yildirim E, Guvenc I (2006) Salt tolerance of pepper cultivars during germination and seedling growth. Turk J Agric For 30(347):353

Yordanov I, Velikova V, Tsonev T (2013) Plant responses to drought, acclimation, and stress tolerance. Photosynthetica 38:171–186

Yoshida R, Kanno A, Sato T, Kатeya T (1996) Cool temperature-induced chlorosis in rice plants: relationship between the induction and a disturbance of etioplast development. Plant Physiol 110(3):997–1005

Yukawa J (2008) Annexure-3: Northward distribution range extensions of plant pests, possibly due to climate change: examples in Japan. In: Climate-related transboundary pests and diseases, technical background document from the expert consultation held on 25 to 27 February 2008, FAO, Rome

Zajac MR, Kubis J (2010) Effect of UV-B radiation on antioxidative enzyme activity in cucumber cotyledons. Acta Biol Cracov Ser Bot 52(2):97–102

Zemetra RS, Cuany RL (1991) Variation among inbreds for seed response to low temperatures in maize (*Zea mays* L.). Maydica 36(1):17–23

Zhang HX, Blumwald E (2001) Transgenic salt tolerant tomato plants accumulate salt in foliage but not in fruit. Nat Biotechnol 19:765–768

Zhang Z, Huang R (2010) Enhanced tolerance to freezing in tobacco and tomato overexpressing transcription factor TERF2/LeERF2 is modulated by ethylene biosynthesis. Plant Mol Biol 73(3):241–249

Zhang C, Liu J, Zhang Y, Cai X, Gong P, Zhang J, Wang T, Li H, Ye Z (2011) Overexpression of SlGMEs leads to ascorbate accumulation with enhanced oxidative stress, cold, and salt tolerance in tomato. Plant Cell Rep 30:389–398

Zhao B (2007) Identification of drought-induced microRNAs in rice. Biochem Biophys Res Commun 354:585–590

Zhao BT, Ge LF, Liang RQ, Li W, Ruan KC, Lin HX (2009) Members of miR-169 family are induced by high salinity and transiently inhibit the NF-YA transcription factor. BMC Mol Cell Biol 10:29

Zhao M, Ding H, Zhu JK, Zhang F, Li WX (2011) Involvement of miR169 in the nitrogen-starvation responses in *Arabidopsis*. New Phytol 190:906–915

Zhou X, Harrington R, Woiwod IP, Perry JN, Bale JS (2014) Effects of temperature on aphid phenology. Glob Chang Biol 1:303–313

Chapter 3
Impact of Climate Change on Nutraceutical Properties of Vegetables

Meenakshi Kumari, Shashank Shekhar Solankey, D. P. Singh, and Rajiv

Abstract Due to the fact that vegetables are the only cheapest source of nutrients, vitamins, and minerals, hence they are a crucial part of the human diet. They provide good remunerative to the growers as they fetch more money from market. The effects of climate change, such as global warming, modifications to seasonal and monsoon patterns, and biotic and abiotic variables, are also having an impact on these crops, just like they do on other crops. Crop failures, low yields, declining quality, and an increase in pest and disease issues are frequent under climate change-related conditions, which make unprofitable to cultivate vegetables. Because of many physiological and enzymatic processes depend on temperature, they will be significantly impacted. The two most significant effects of temperature rise on vegetable cultivation are drought and salt. Crop yields may improve as a result of increased CO_2 fertilisation; however, this positive effect decreases after certain point. Greenhouse gases produced by human activity, such as CO_2, CH_4, and CFCs, are a major factor in global warming, while sulphate and nitrogen dioxides weaken the ozone layer and allow dangerous UV rays to enter the atmosphere. These climate change effects also have severe impact on the prevalence of pests and diseases, as well as on the nutritional value (vitamins, minerals, proteins, etc.) and aesthetics of vegetable crops. Iron and zinc levels, as well as the amount of protein in vegetable crops, were dramatically lowered by higher CO_2 levels. In the end, the quality and volume of global vegetable output are falling due to climate change.

Keywords Vegetables · Climate change · Greenhouse gasses · Nutritional value

---

M. Kumari (✉)
Department of Vegetable Science, Shree Guru Gobind Singh Tricentenary University, Gurugram, Haryana, India

S. S. Solankey
Department of Vegetables & Floriculture, Agricultural Research Institute, Patna, Bihar, India

D. P. Singh · Rajiv
Department of Vegetable Science, Chandra Shekhar Azad University of Agriculture and Technology, Kanpur, Uttar Pradesh, India

S. S. Solankey, M. Kumari (eds.), *Advances in Research on Vegetable Production Under a Changing Climate Vol. 2*, Advances in Olericulture,
https://doi.org/10.1007/978-3-031-20840-9_3

3.1 Introduction

The main cause of abiotic stress in today's vegetable cultivation is climate change. The cropping systems of vegetable should be climate change resistant, so that the vegetable production technology can achieve both economically successful and environmentally sustainable. The productivity of vegetable crop is severely threatened by environmental stresses including flooding, drought, and excessive temperatures which finally results in complete crop failures (Singh et al. 2020). The head of the Foundation for Innovation in Medicine (FIM) and executive Director Dr. Stephen L. De Felice, invented the term "nutraceutical" in 1979 as a combination of the phrases "nutrition" and "pharmaceutical" (Crawford, New Jersey) (Kalra 2003). It is a food item or food-related product that provides both medicinal and health advantages, such as illness prevention and therapy. They are products that have been separated or purified from foods, usually offered in therapeutic forms unrelated to food, and have been shown to provide physiological benefits or offer protection from chronic illness (Singh and Devi 2015). Due to an unbalanced diet, approximately worldwide 3 billion people suffer from malnutrition. A balanced diet must include vegetables because they are an excellent source of nutraceutical substances and phytonutrients. Climate change has a significant negative impact on the production, quality, and productivity of vegetable crops (Kumari et al. 2021). Under adverse environmental stress, such as heat, cold, drought, flood, and salinity, many plant species' genes are activated, enabling them to withstand a variety of stress conditions (Solankey et al. 2021a). Vegetables are recognised as protective foods since they are a great supply of vitamins, minerals, carbs, and proteins. They also offer health protection because of the presence of secondary metabolites with medicinal value. The most prominent phytonutriceuticals in vegetables with biological activity against chronic diseases include vitamins, minerals, dietary fibre, organosulfur compounds (glucosinolates and thiosulfides), and flavonoids. Among all crops, potatoes are particularly sensitive to climate change since they need a certain environment for a number of physiological processes (Singh and Devi 2015). Fighting against hunger and malnutrition are now two main priorities of developing nations. In India, around 43.5% children under the age of 5-year-old are chronically undernourished. Vegetable consumption is usually thought to have a number of beneficial impacts on health. Risk of cancer and other cardiovascular disease are directly linked with low consumption of fruit and vegetables Martinez-Gonzalez et al. (2011); Krebs and Kantor (2001); Lock et al. (2005); and Mosby et al. (2011). Vegetables with a wide variety and high nutritional value are crucial for reducing malnutrition. As mentioned in below Table 3.1, each vegetable includes a distinct combination of phytonutrients.

Vegetables with high amount of anthocyanin (broccoli, black/ purple carrot, purple brinjal and purple cauliflower) are becoming more or more popular day-by-day due to their increased activity of antioxidant (Table 3.2). The colour features of radish and potato extracts are quite comparable to those of Allura red (Shipp and Abdel 2010).

Table 3.1 Biochemical compound that have nutritional importance found in vegetables

Nutraceuticals	Vegetables
Glucosinolates, Sulforaphane	Cole crops
Lycopene	*Solanum lycopersicum* & various Nightshade family crops, *Citrullus lanatus*
Silymarin	*Cynara cardunculus* var. *scolymus*
Ascorbic acid	*Brassica oleracea* var. *capitata*, *Brassica oleracea* var. *italica*, green leafy vegetables
Tocopherol	Green leafy vegetables
Allyl sulphides	Alliaceae family crops
Retinol	*Daucus carota*, *Cucurbita moschata*, *Cucumis melo* var. *cantalupensis*
Ascorbic acid	*Momordica charantia*, *Capsicum annuum* var. *grossum*
Folates	Green leafy vegetables
Alliin, Methiin	*Allium species*
Quercetin	*Allium cepa* & *Allium sativum*
Kaempferol, Myricetin, Fisetin	*Allium cepa*, *Lactuca sativa*, endive, *Armoracia rusticana*
Luteolin	*Apium graveolens*, *Brassica oleracea* var. *italica*
Apigenin	*Apium graveolens*, *Brassica oleracea* var. *capitata* and *Lactuca sativa*
Isoflavonoids	Legume vegetables, *Brassica oleracea* var. *italica* and *Abelmoschus esculentus*
Genistein and Daidzein	*Glycine max*
Glucoraphanin	*Brassica oleracea* var. *capitata* f. *rubra* and *Brassica oleracea* var. *italica*
Glucobrassicin, Progoitrin, Gluconasturtiin	*Brassica oleracea* var. *italica*
Glucoerucin, glucoraphanin	*Brassica rapa* and *Brassica napus* var. *napobrassica*
Lysine, Chlorgenic acid	*Solanum tuberosum*
Caffeic acid, Chlorgenic acid	*Solanum melongena*
Nasunin	*Solanum melongena*
Angelicin, Xanthotoxin	*Pastinaca sativa*
Ferulic acid, Betanin	*Beta vulgaris*
Anthocyanin and Chlorgenic acid	*Ipomoea batatas*
Rutin	*Asparagus officinalis, Capsicum annuum* (green color)
Patuletin, Spinacetin	*Spinacea oleracea*
2″-xyloside vitexin and 6″-malonyl-2″-xyloside vitexin	Swiss Chard
Betanin	*Beta vulgaris*
Capsaicin	*Capsicum annuum* (red color)
Carnitine	*Asparagus officinalis*
Curcumin	*Curcuma longa*
Hesperitin	Green leafy vegetables
Lignan	*Glycine max* and *Brassica oleracea* var. *italica*
Resveratol	*Allium cepa* (red)

Adapted from Singh and Devi (2015)

Table 3.2 Anthocyanin concentration in different vegetables

Vegetable	Anthocyanin (mg/100 g)	References
Brassica oleracea var. *capitata* f. *rubra*	322	Wu et al. (2006)
Raphanus sativus (red)	100–154	Wu et al. (2006)
Allium cepa (red)	23.3–48.5	Ferreres et al. (1996)
Solanum melongena	8–85	Koponen et al. (2007)

Adapted from Singh and Devi (2015)

3.2 Improvement of Nutrition in Vegetables

As we go into the twenty-first century, improving the nutritional quality of horticulture products, particularly the nutraceutical importance of green veggies will make plant breeder's efforts profitable. It is becoming increasingly clear that eating healthful meals may help maintain a healthy lifestyle and that eating is not only for body growth and subsistence in industrialised nations when the majority of the population has access to enough food. People are starting to eat more nutritious foods that can help with "diseases of excess" and chronic diseases linked to diet, such some forms of obesity, heart disease, and some types of cancer. Along with other agricultural professionals and extension agents, plant breeder's services, are primarily responsible for the world's population's access to an abundance of food, better health and nutrition, and stunning landscapes. Breeding plants to increase their mineral and vitamin content has a number of beneficial advantages. The majority of breeding and genetic work has been focused on crops like *Daucus carota, Ipomoea batatas, Capsicum annuum, Solanum lycopersicon, Cucurbita moschata, and Cucumis sp.* that are already reasonably rich suppliers of vitamins (Singh and Devi 2015). In Tomato, dominant gene (Aft) Anthocyanin fruit responsible for purple colour, which generates restricted pigmentation upon stimulation by high light intensity was introduced into tomato by crossing domestic tomato plants with *S. chilense* (Mes et al. 2008; Jones et al. 2003). A robust and varied pigmentation may also be induced in the tomato peel from *Solanum lycopersicoides* Dunal by the gene Aubergine (Abg). Red cabbage's anthocyanin production and accumulation are mediated by the transcriptional activation of the anthocyanin structural genes by the bHLH and MYB transcription factors (Yuan et al. 2009).

An intriguing genetic mutation known as spontaneous reported in Cauliflower (*Brassica oleracea* var. *botrytis*) which is responsible for semi dominant Orange (Or) mutant that causes carotenoid deposition in typically unpigmented tissues (Dickson et al. 1998). The Or gene causes the plant's tissues to accumulate large quantities of β-carotene, which is responsible for orange colour. This is especially noticeable in the plants with white edible curd and shoot apical meristem. Mano et al. (2007) reported the finding of a new R2R3-type MYB gene, IbMYB1, and its predominant expression in the root of tuberous vegetable of purple-fleshed cultivars using a purple-fleshed sweet potato cDNA library. Purple colour of *Ipomoea batatas* tuberous roots in the flesh is caused by the gene IbMYB1. A ripening-inducible E8 promoter and a yeast S-adenosylmethionine decarboxylase gene (ySAMdc;

Spe2) in tomato fruit were coupled to raise the polyamines spermidine and spermine levels. By increasing the conversion of putrescine into higher polyamines, the ySAMdc gene promoted spermidine and spermine, which are ripening-specific compound. According to these results, the overall quality of fruit juice was improved due to increase in lycopene and shelf life of vine. Since cultivated tomatoes contain generally low amounts of lycopene and raising lycopene levels will improve the fruit's nutritional value. Romer et al. (2000) boost the carotenoid content and profile of tomato fruit, trans-genic lines have been created that express the bacterial carotenoid gene (crtI), which makes the enzyme phytoene desaturase, which converts phytoene into lycopene. However, the amount of β -carotene more than tripled, reaching 45% of the total carotenoid content. Chromosome fragments of *Brassica villosa*, a wild progenitor has been introgressed to increase the amount of glucosinolates. Depending on the *B. villosa* allele, indole-3-carbinol or sulphoraphane is produced during hydrolysis. *Brassica rapa* is a root vegetable that is most often consumed in Asia. As compared to *B. oleracea, B. rapa* contain different types of isothiocynates and a new research shows that it also gives protective benefits to people who lack GSTM1 (Gasper et al. 2005). Table 3.3 list of gene which is responsible for enhancing various nutraceutical components in various vegetables.

3.3 Quality of Vegetables/Fruits and Elevated CO_2

Additionally, it has been noted that in various vegetables, increased CO_2 raises the concentrations of some bioactive substances. The impact of elevated CO_2 on physiology of vegetables/fruits have been summarised by Moretti et al. 2010. In their research, it was found that several vegetables/fruits had reduced alkaloids and organic acids while increased ascorbic acid, flavonoids, sugars, phenols, starch anthocyanin and also firmness and colour (Shivashankara et al. 2013). According to Zhang et al. (2014), tomato fruits with increased CO_2 had significantly higher concentrations of compounds like Vitamin A, lycopene, Vitamin C, which are essential for development of our health and also had high amount of chemicals like titrable acidity, total soluble solids and sugar/acid ratio which is known as flavor-enhancing chemicals. The tomato fruit firmness, colour, fragrance, and sensory qualities were also markedly improved by CO_2 enrichment. Yield contributing characters like fruits per plant and average weight of fruits were main contributing characters for yield n tomato under heat stress (Solankey et al. 2017).

3.4 Vitamin C, Sugars and Acidity

Protective antioxidant substances like ascorbate and phenolics are created by plants by extra carbon which is fixed during enrichment of CO_2. In Tomato fruits, when enhanced CO_2 was provided at various degrees of maturity, some quality metrics

Table 3.3 Vegetable gene list responsible for nutraceutical enhancement

Vegetable crop	Gene	Nutrient enhancement
Solanum tuberosum	*Or*	β-carotene
Brassica oleracea var. *botrytis*	*Or*	β-carotene
Solanum tuberosum	*AmA1*	Protein
Solanum tuberosum	*Crt B*	β-carotene
Solanum lycopersicum	*B*	β-carotene
Ipomoea batatas	*asp-1*	High protein
Solanum lycopersicum	*Phytoene synthase – 1(Psy-1)*	Carotenoids
Solanum lycopersicum	*chi-a*	High flavonols
Solanum lycopersicum	*LC and C1*	Kaempferol
Solanum lycopersicum	*Aft, Abg*	Anthocyanin
Cucumis sativus	*Ore*	β-carotene
Brassica oleracea var. *capitata* f. *rubra*	*MYB*	Anthocyanin
Brassica oleracea var. *botrytis* (purple)	*Pr*	Anthocyanin
Ipomoea batatas	*IbMYB1*	Anthocyanin
Solanum lycopersicum	*Cry-2*	Lutein
Solanum lycopersicum	*ySAMdc; Spe2*	Lycopene
Solanum tuberosum	*Dxs*	Phytoene
Solanum lycopersicum	*GCH1*	Folate
Lactuca sativa	*Gch1*	Folate
Lactuca sativa	*Pfe*	Iron
Lactuca sativa	*Gul oxidase*	Ascorbate
Solanum lycopersicum	*hmgr-1*	Tocopherols

Adapted from Singh and Devi (2015)

like organic acids were lower whereas, ascorbic acid and sugars were maximum (Islam et al. 1996). The increased CO_2 improved fruit colour and development. In tomato at the pink stage, acidity and ascorbic acid levels are maximum whereas, slightly going down during ripening stage. In bean sprouts, ascorbic acid levels were also found to increase by two folds even with a CO_2 concentration that was doubled for 1 h each day for 7 days (Tajiri 1985). High CO_2 increased total sugars and acidity in grapes, although the impact was only noticeable during middle stage of ripening (Kurooka et al. 1990 and Bindi et al. 2001).

3.5 Total Phenols, Anthocyanins and Flavonoids

Under high CO_2 concentrations, tomato antioxidant levels increased at very slow rate (Barbale 1970; Madsen 1971, 1975; Kimball and Mitchell 1981).

3.6 Volatile Aroma Compounds

In field-grown strawberries (*Fragaria ananassa* Duch), Wang and Bunce (2004) examined how elevated CO_2 affected the volatile aroma composition and fruit quality. Under high CO_2 levels, ethyl hexanoate, ethyl butanoate, methyl hexanoate, methyl butanoate, hexyl acetate, hexyl hexanoate, furaneol, linalool, and methyl octanoate content of these key strawberry scent esters increased significantly (Shivashankara et al. 2013).

3.7 Mineral Nutrients

It has been suggested that increased CO_2 has an impact on nutrients supply of vegetable and fruits. Lettuce produced in high CO_2 environments reported lower ash content (McKeehen et al. 1996). In a number of types of woody and herbaceous plants, significant reductions in minerals including Iron, nitogen, sulphur, Magnesium, Calcium, Zinc (15–25%) were observed when CO_2 levels were high (Loladze 2002). Chronic exposure to high CO_2 levels may also have an impact on product quality (Gruda 2005), with total soluble solids, Vitamin C and capacity of antioxidant being enriched, while other macronutrients and micronutrients in green vegetables, such magnesium, iron, and zinc, may be depleted (Dong et al. 2018).

3.8 Effect of High Temperature on Quality

Heat waves may significantly affect plant growth, production, and product quality in horticulture. Open field crops are extensively exposed to sunlight and high temperatures during heat waves. Physiological problems associated with calcium (Ca) uptake are frequently brought on by a lack of protection against stress. Plants transpire a lot of water when it's hot outdoors, which causes all the calcium in the transpiration stream to flow directly to the leaves (Bisbis et al. 2019). Since Ca doesn't reach the developing tip and the enclosing leaves in lettuce, this commonly results in tip burn, which results in necrosis on the margins of new leaves (Collier and Tibbitts 1982). Low transpiration results in insufficient Ca being allocated to the fruits, which results in blossom end rot in fruiting plants like tomato or pepper. Under the influence of global warming, winter dormancy may be hindered, which may have an impact on the output of perennial vegetables. For instance, during the cold season, asparagus becomes dormant and accumulates frigid temperatures of 0–7 °C (Nie et al. 2016). Since cauliflower starts curd formation only at a temperature of 7–10 °C, therefore, higher temperatures might delay the process. When the temperature was increased by 2.9 °C above ambient the head development took place 49 days later in cauliflower (Wurr et al. 1996). Heat stress alters the physical characteristics of biomolecules directly (Solankey et al. 2021b).

In an experiment with broccoli, Kaluzewicz et al. (2009) found that the higher the temperature was maintained for longer periods of time throughout the initial phase of growth that occurs after sowing as well as the period just before harvest, the greater the yield. If broccoli heads were exposed over 20 °C temperature for a longer period then the resulting yields were lower. The proportion of broccoli heads with an uneven surface decrease with the amount of time spent at temperatures between 5 and 15 °C during harvest and between 20 and 25 °C during the growth period before harvest. At the time of harvest, if broccoli heads exposed over 20 °C temperatures for a longer time results in development of loose heads.

3.9 Vitamin C, Sugars and Acidity

The healthy growth and development of plants, as well as the determination of the phonological phases, depend greatly on temperature. Among various fruit crops *Vitis venifera* are the most significant which are impacted by high temperatures for prolonged periods of time. In Tomato, due to high temperature biochemical compounds like sugar, acidity and dry weight are reduced (Bikash Khanal 2012).

3.10 Phenols, Flavonoids and Anthocyanins

Increases in temperature hinder the colour development. The anthocyanin concentration is more sensitive to night temperatures than to day temperatures (Mori et al. 2005). The polyphenol content of several tomato varieties ranged from 104 to 400 mg kg^{-1}, according to George et al. (2004). This behaviour may be viewed as the plant's acclimatisation to heat stress (Rivero et al. 2001). Fruits lose some of their antioxidant capacity at lower temperatures (Wang and Zheng 2001). Even at 25 °C temperatures, heat in broccoli can result in diseases and deformities such as uneven heads and oversized flower buds (Kaluzewicz et al. 2009). Heat triggered bracting in sensitive cultivars during this stage of head development (Wiebe 1972), and harvest temperatures exceeding 25 °C caused loose heads and early ripening (Kaluzewicz et al. 2009). High temperatures also resulted in uneven and less sweet heads, but they also raised the flavonol content and changed the composition of the glucosinolates in the florets (Molmann et al. 2015).

3.11 Lycopene and Carotenoids Content

High temperatures typically result in smaller size of tomato fruit and higher dry matter content. Lycopene content is also impacted by the high day/night temperature treatments (30°/25 °C) in tomato (*Lycopersicon esculentum* Mill., cv. "Laura") compared to the control temperature (28°/23 °C) (Fleisher et al. 2006). The amount of lycopene and other nutritional value components in tomatoes are further diminished by high sun radiation and temperature (Dumas et al. 2003; Helyes et al. 2003; Rosales et al. 2006). It has been shown that beyond 40 °C, ß-carotene content and synthesis decreases (Gautier et al. 2005). The quality of the fruit is greatly influenced by the high temperature caused by direct solar exposure rather than plant temperature due to lycopene degradation. Instead of plant temperature, direct solar radiation-induced high temperature on the fruit surface has a significant impact on the fruit's quality because lycopene is degradable (Dumas et al. 2003). As a result, tomato fruits produced in greenhouses have 40% more lycopene than tomatoes grown in fields as a result (Helyes et al. 2007).

3.12 Terpenoids

High temperatures change the volatile fragrance molecules in many vegetables in addition to the bioactive components. The effect of high temperature on the soybean isoflavone quality was observed to fluctuate with increased CO_2 and water stress (Caldwell et al. 2005).

3.13 Stress from Water's Impact

The disruptions in Ca allocation brought on by heat are not the sole factor contributing to lettuce tip burn; Ca absorption is also a factor. Poor Ca absorption may also result from insufficient soil water uptake (Bisbis et al. 2019). Additionally, too much water may result in buttoning, nitrate leaching, and a consequent decrease in output because the soil is depleted of nutrients (Kaiser et al. 2011). The flowering stalk's elongation from the core causes bolting, which happens in lettuce right before bloom induction (Kumar et al. 2012; Chatterjee and Solankey 2015). Bolting in lettuce causes bitterer leaves and worse head development, which are undesirable in all except stem lettuce. High temperatures might also cause premature bolting (Simko and Hayes 2015).

3.14 Sugars, Ascorbic Acid and Acidity

Stress of water makes fruits less juicy, which raises their sugar content (Chartzoulakis et al. 1999). The consequences, however, depends mostly on the phenological stage of water stress and may render fruits utterly non-commercial (Romero et al. 2006). By applying the treatment at later phases of fruit maturity, deficit irrigation therapy is utilised in some fruit crops to boost the sugar content. Under deficit irrigation circumstances, tomato showed higher total soluble solids and sugars (Mitchell et al. 1991; Birhanu and Tilahun 2010). However, water stress causes a drop in tomato marketable yield.

3.15 Phenols, Flavonoids and Anthocyanins

Strawberries accumulated more proanthocyanidins and anthocyanins due to inadequate watering. The increased activity of phenylalanine ammonia-lyase under water scarcity condition is primarily responsible for the increase in anthocyanins and phenolic compounds (Tovar et al. 2002). Mineral movement like nitrogen, potassium and phosphorus within the tree is going to be affected by the water stress. However, lack of water has a negative impact on the quality during fruit set and the early stages of fruit growth. The water stress affects the impact of transport of minerals like nitrogen, potassium, and phosphorus inside the tree (Kirnak et al. 2001). If this happens during the active period of fruit growth will have an impact on the quality of the fruit.

3.16 Lycopene and Carotenoids

The primary pigments in many fruits and vegetables are due to carotenoids, which is isoprenoids that are naturally occurring and have antioxidant characteristics.

3.17 Salinity Stress

One of the key environmental conditions that inhibits plant development, yield, and output is salt stress. It has been discovered that salt stress during fruit development stage limits the vegetative growth and fruit quality (Shivashankara et al. 2013).

3.18 Phenols, Flavonoids and Anthocyanins

Reactive oxygen species (ROS) and their scavengers, enzymes, or nonenzymatic low molecular mass antioxidants are known to arise in response to salt stress.

3.19 Lycopene and Carotenoids

Antioxidants like lycopene, carotenoids and ascorbic acid accumulated in tomato fruits during salt stress. Reactive oxygen species (ROS) and their scavengers, enzymes, or nonenzymatic low molecular mass antioxidants are known to arise in response to salt stress (D'Amico et al. 2003). However, under salt stress, the leaves of *Solanum lycopersicum* plant exhibit a reduced expression of carotenoid biosynthesis genes, which significantly slows down photosynthesis and lowers plant productivity and yield (Merlene et al. 2011). Abiotic stressors have a significant impact on the antioxidant capacity of fruits.

3.20 Conclusion

Regular eating of a diet high in vegetables has undeniably beneficial benefits on health since the phytonutrients in vegetables can shield the body against many chronic illnesses. Cruciferous vegetables, bulb crops, tomatoes, cucurbits, soybeans, carrots, okra, and underutilised vegetables including lettuce, coleus, sweet potatoes, yams, moringa, winged beans, basella, horse purslane, and cluster beans are rich sources of bioactive chemicals. Abiotic stress effects on antioxidant quality are further amplified by climate change. Adaptation tactics, including as the cultivation of robust crop varieties, effective irrigation systems, unique pollination techniques, and agricultural technologies, will be needed to adjust to changing environmental circumstances and preserve the supply of foods that are crucial for human sustenance. As a result, efforts must be made to comprehend the impact of various abiotic stresses on various fruit crops as well as the critical stages of fruit growth at which the overall quality of the vegetables are adversely affected. Additionally, strategies must be developed to counteract the negative effects of abiotic stresses.

References

Barbale D (1970) The influence of the carbon dioxide on the yield and quality of cucumber and tomato in the covered areas. Augsneun Raza (Riga) 16:66–73

Bikash Khanal (2012) Effect of day and night temperature on pollen characteristics, fruit quality and storability of tomato. Master thesis, Department of Plant and Environmental Science, Norwegian University of Life Sciences. http://hdl.handle.net/11250/189462

Bindi M, Fibbi L, Miglietta F (2001) Free Air CO2 Enrichment (FACE) of grapevine (*Vitis vinifera* L.): II. Growth and quality of grape and wine in response to elevated CO2 concentrations. Eur J Agro 14:145–155

Birhanu K, Tilahun K (2010) Fruit yield and quality of drip-irrigated tomato under de fi cit irrigation. Afr J Food Agric Nutr Dev 10(2):2131–2151

Bisbis MB, Gruda NS, Blanke MM (2019) Securing horticulture in a changing climate – a mini review. Horticulturae 2019(5):56. https://doi.org/10.3390/horticulturae5030056

Caldwell CR, Britz SJ, Mirecki RM (2005) Effect of temperature, elevated carbon dioxide and drought during seed development on the iso flavone content of dwarf soybean [*Glycine max* (L.) Merrill] grown in controlled environments. J Agric Food Chem 53:1125–1129
Chartzoulakis K, Michelakis N, Stefanoudaki E (1999) Water use, growth, yield and fruit quality of 'Bonanza' oranges under different soil water regimes. Adv Hortic Sci 13(1):6–11
Chatterjee A, Solankey SS (2015) Functional physiology in drought tolerance of vegetable crops-an approach to mitigate climate change impact. Clim Dyn Hortic Sci 1:149–171
Collier GF, Tibbitts TW (1982) Tipburn of lettuce. Hortic Rev 4:49–65
D'Amico ML, Izzo R, Tognoni F, Pardossi A, Navari F (2003) Application of diluted seawater to soilless culture of tomato (*Lycopersicon esculentum* Mill.): effect on plant growth, yield, and fruit quality and antioxidant capacity. J Food Agric Environ 1:112–116
Dickson MH, Lee CY, Blamble AE (1998) Orange-curd high carotene cauliflower inbreds, NY 156, NY 163, and NY 165. HortScience 23:778–779
Dong J, Gruda N, Lam SK, Li X, Duan Z (2018) Effects of elevated CO_2 on nutritional quality of vegetables: a review. Front Plant Sci 9:924
Dumas Y, Dadomo M, Lucca GD, Grolier P (2003) Effects of environmental factors and agricultural techniques on antioxidant content of tomatoes. J Sci Food Agric 83:369–382
Ferreres F, Gil MJ, Tomas, Barberan FA (1996) Anthocyanins and flavonoids from shredded red onion and changes during storage in perforated films. Food Res Int 29:389–395
Fleisher DH, Logendra LS, Moraru C, Both A, Cavazzoni J, Gianfagna T, Lee T, Janes HW (2006) Effect of temperature perturbations on tomato (*Lycopersicon esculentum* Mill.) quality and production scheduling. J Hortic Sci Biotechnol 81(1):125–131
Gasper AV, Al-Janobi A, Smith A, Bacon JR, Fortun P, Atherton C, Taylor MA, Hawkey CJ, Barrett AD, Mithen RF (2005) Glutathione-*S*-transferase M1 polymorphism and metabolism of Sulforaphane from standard and high-Glucosinolate broccoli. Am J Clin Nutri 82:1283–1291
Gautier H, Rocci A, Buret M, Grasselly D, Dumas Y, Causse M (2005) Effect of photo selective filters on the physical and chemical traits of vine-ripened tomato fruits. Can J Plant Sci 85(2):439–446
George B, Kaur CH, Khurdiya DS, Kapoor HC (2004) Antioxidants in tomato (*Lycopersicon esculentum*) as a function of genotype. Food Chem 84:45–51
Gruda N (2005) Impact of environmental factors on product quality of greenhouse vegetables for fresh consumption. Crit Rev Plant Sci 24:227–247
Helyes L, Pék Z, Brandt S, Varga GY, Barna É, Hóvári J, Lugasi A (2003) Influence of harvest date on fruit technological traits of five processing tomato varieties. Acta Hortic 613:213–216
Helyes L, Lugasi A, Pék Z (2007) Effect of natural light on surface temperature and lycopene content of vine ripened tomato fruit. Can J Plant Sci 87:927–929
Islam MS, Matsui T, Yoshida Y (1996) Effect of carbon dioxide enrichment on physico-chemical and enzymatic changes in tomato fruits at various stages of maturity. Sci Hortic 65(2):137–149
Jones CM, Mes P, Myers JR (2003) Characterization and identification of the anthocyanin fruit (Aft) tomato. J Hered 94:449–456
Kaiser E, Stützel H, Heuvelink E (2011) Effect of cooling irrigation on the vernalisation of cauliflower (Brassica oleracea L. Botrytis). Master's thesis, Wageningen University and Research/ Leibniz Universität Hannover, Wageningen/Hannover
Kalra EK (2003) Nutraceutical: definition and introduction. AAPS PharmSci 5(3):1–2
Kaluzewicz A, Krzesinski W, Knaflewski M (2009) Effect of temperature on the yield and quality of broccoli heads. Veg Crops Res Bull 71:51–58. https://doi.org/10.2478/v10032-009-0026-7
Kimball BA, Mitchell ST (1981) Effects of CO2 enrichment, ventilation, and nutrient concentration on the flavour and vitamin C content of tomato fruit. Hortic Sci 16:665–666
Kirnak H, Kaya C, Ismail TAS, Higgs D (2001) The influence of water deficit on vegetative growth, physiology, fruit yield and quality in eggplants. Bulg J Plant Physiol 27(3–4):34–46
Koponen MJ, Happonen MA, Mattila HP, Torranen (2007) Contents of anthocyanins and ellagitannins in selected foods consumed in Finland. J Agric Food Chem 55:1612–1619

Krebs Smith SM, Kantor LS (2001) Choose a variety of fruits and vegetables daily: understanding the complexities. J Nutr:4875–5015
Kumar R, Solankey SS, Singh M (2012) Breeding for drought tolerance in vegetables. Veg Sci 39:1–15
Kumari M, Kumar M, Solankey SS, Tomar S (2021) Effect of greenhouse gases on vegetable production. In: Solankey SS et al (eds) Advances in research on vegetable production under a changing climate, vol. 1. Springer, Cham. pp 211–219. ISBN 978-3-030-63497-1
Kurooka H, Fukunaga S, Yuda E, Nakagawa S, Horiuchi S (1990) Effect of carbon-dioxide enrichment on vine growth and berry quality of kyoho grapes. J Jpn Soc Hortic Sci 59(3):463–470
Lock K, Pomerleau J, Causer L, Altmann DR, McKee M (2005) The global burden of disease attributable to low consumption of fruit and vegetables: implications for the global strategy on diet. Bull World Health Organ 83:100–108
Loladze I (2002) Rising atmospheric CO2 and human nutrition: toward globally imbalanced plant stoichiometry? Trends Ecol Evol 17:457–461
Madsen E (1971) The influence of CO2 -concentration on the content of ascorbic acid in tomato leaves. Ugeskr Agron 116:592–594
Madsen E (1975) Effect of CO2 environment on growth, development, fruit production and fruit quality of tomato from a physiological viewpoint. In: Chouard P, de Bilderling N (eds) Phytotronics in agricultural and horticultural research. Bordas, Paris, pp 318–330
Mano H, Ogasawara F, Sato K, Higo H, Minobe Y (2007) Isolation of a regulatory gene of anthocyanin biosynthesis in tuberous roots of purple-fleshed sweet potato. Plant Physiol 143:1252–1268
Martinez-Gonzalez MA, de la Fuente-Arrilaga C, Lopez-del-Burgo C, Vazquez-Ruiz Z, Benito S, Ruiz-Canela M (2011) Low consumption of fruit and vegetables and risk of chronic disease: a review of the epidemiological evidence and temporal trends among Spanish graduates. Public Health Nutr 14:2309–2315
McKeehen JD, Smart DJ, Mackowiak CL, Wheeler RM, Nielsen SS (1996) Effect of CO2 levels on nutrient content of lettuce and radish. Adv Space Res 18(4–5):85–92
Merlene AB, Devesh S, Gothandam KM (2011) Effect of salt stress on expression of carotenoid pathway genes in tomato. J Stress Physiol Biochem 7(3):87–94
Mes PJ, Boches P, Myers JR (2008) Characterization of tomatoes expressing anthocyanin in the fruit. J Amer Soc Hortic Sci 133(2):262–269
Mitchell JP, Shennan C, Grattan SR, May DM (1991) Tomato fruit yields and quality underwater deficit and salinity. J Am Soc Hortic Sci 116(2):215–221
Molmann JA, Steindal ALH, Bengtsson GB, Seljåsen R, Lea P, Skaret J, Johansen TJ (2015) Effects of temperature and photoperiod on sensory quality and contents of glucosinolates, flavonols and vitamin C in broccoli florets. Food Chem 172:47–55
Moretti CL, Mattos LM, Calbo AG, Sargent SA (2010) Climate changes and potential impacts on post-harvest quality of fruit and vegetable crops: a review. Food Res Int 43:1824–1832
Mori K, Sugaya S, Gemma H (2005) Decreased anthocyanin biosynthesis in grape berries grown under elevated night temperature condition. Sci Hortic 105:319–330
Mosby TT, Cosgrove M, Sarkardei S, Platt KL, Kaina B (2011) Nutrition in adult and childhood cancer: role of carcinogens and anti-carcinogens. Anticancer Res 32:4171–4192
Nie LC, Chen YH, Liu M (2016) Effects of low temperature and chilling duration on bud break and changes of endogenous hormones of asparagus. Eur J Hortic Sci 81:22–26
Rivero RM, Ruiz JM, Garcıa PC, Lopez-Lefebre LR, Sanchez E, Romero L (2001) Resistance to cold and heat stress: accumulation of phenolic compounds in tomato and watermelon plants. Plant Sci 160:315–321
Romer S, Fraser PD, Kiano JW, Shipton CA, Misawa N, Schuch W, Bramley PM (2000) Elevation of the provitamin a content of transgenic tomato plants. Nat Biotechnol 18:666–669
Romero P, Navarro JM, Pérez PJG, Garcia SF, Gomez GA, Porras I, Martinez V, Botia P (2006) Deficit irrigation and rootstock: their effects on water relations, vegetative development, yield, fruit quality and mineral nutrition of Clemenules mandarin. Tree Physiol 26(12):1537–1548

Rosales MA, Ruiz JM, Hernández J, Soriano T, Castilla N, Romero L (2006) Antioxidant content and ascorbate metabolism in cherry tomato exocarp in relation to temperature and solar radiation. J Sci Food Agric 86:1545–1551

Shipp J, Abdel Aal EISM (2010) Food applications and physiological effects of anthocyanins as functional food ingredients. Open Food Sci J 4:7–22

Shivashankara KS, Srinivasa Rao NK, Geetha GA (2013) Impact of climate change on fruit and vegetable quality. In: Singh HP et al (eds) Climate-resilient horticulture: adaptation and mitigation strategies. Springer, New Delhi, pp 237–244. https://doi.org/10.1007/978-81-322-0974-4_21

Simko I, Hayes RJ (2015) Breeding lettuce for improved fresh-cut processing. In: Proceedings of the III international conference on fresh-cut produce: maintaining quality and safety, Davis, CA, USA, 13 September 2015, Acta horticulturae 1141, pp 65–76

Singh S, Devi MB (2015) Vegetables as a potential source of nutraceuticals and phytochemicals: a review. Int J Med Pharm Sci 5(2):1–14

Singh H, Sethi S, Kaushik P, Fulford A (2020) Grafting vegetables for mitigating environmental stresses under climate change: a review. J Water Clim Change 11(4):1784–1797

Solankey SS, Shirin A, Pallavi N, Meenakshi K (2017) Effect of high temperature stress on morpho-biochemical traits of tomato genotypes under polyhouse condition. Indian J Ecol 44(Special issue 5):497–502

Solankey SS, Kumari M, Akhtar S, Singh HK, Ray PK (2021a) Challenges and opportunities in vegetable production in changing climate: mitigation and adaptation strategies In: Solankey SS et al (eds) Advances in research on vegetable production under a changing climate, vol. 1. Springer, Cham, pp 13–59. ISBN 978-3-030-63497-1.

Solankey SS, Kumari M, Kumar M, Silvana N (2021b) The role of research for vegetable production under a changing climate future trends and goals. In: Solankey SS et al (eds) Advances in research on vegetable production under a changing climate, vol. 1. Springer, Cham, pp 1–12. ISBN 978-3-030-63497-1

Tajiri T (1985) Improvement of bean sprouts production by intermittent treatment with carbon dioxide. Nippon Shokuhin Kogyo Gakkai Shi 32(3):159–169

Tovar MJ, Romero MP, Girona J, Motilva MJ (2002) L-Phenylalanine ammonia lyase activity and concentration of phenolics in developing olive (*Olea europaea* L. cv. Arbequina) fruit grown under different irrigation regimes. J Sci Food Agric 82:892–898

Wang SY, Bunce JA (2004) Elevated carbon dioxide affects fruit fl avor in fi eld-grown strawberries (*Fragaria* × *ananassa* Duch). J Sci Food Agric 84:1464–1468

Wang SY, Zheng W (2001) Effect of plant growth temperature on antioxidant capacity in strawberry. J Agric Food Chem 49(10):4977–4982

Wiebe HJ (1972) Effeect of temperature and light on growth and development of cauliflower. Gartenbauwissenschaft 37:165–178

Wu X, Beecher GR, Holden JM, Haytowitz DB, Gebhardt SE, Prior RL (2006) Concentrations of anthocyanins in common foods in the United States and estimation of normal consumption. J Agric Food Chem 54:4069–4075

Wurr DCE, Fellows JR, Phelps K (1996) Investigating trends in vegetable crop response to increasing temperature associated with climate change. Sci Hortic 66:255–263

Yuan Y, Chiu LW, Li L (2009) Transcriptional regulation of anthocyanin biosynthesis in red cabbage. Planta 230:1141–1153

Zhang Z, Liu L, Zhang M, Zhang Y, Wang Q (2014) Effect of carbon dioxide enrichment on health-promoting compounds and organoleptic properties of tomato fruits grown in greenhouse. Food Chem 153(2014):157–163

Chapter 4
Nutritional Stress Management in Vegetable Crops Under Changing Climate Scenario

Bholanath Saha, K. Madhusudhan Reddy, Sushanta Saha, Ayesha Fatima, and Shashank Shekhar Solankey

Abstract Climate change impacts the yield reduction, decreased quality and sometimes entire crop failure and thus render reduced interest of farmers. Deficiency of nutrients along with impact of climate change in vegetable seedlings are also common in vegetable based production systems. Large scale occurrence of multi-nutrient deficiencies leads to reduction in yield. Judicious nutrient application through soil test based fertilizer recommendation in vegetables could reduce the possibility of yield loss caused by climatic aberrations as some nutrient elements have established effects on drought tolerance. The importance of maintaining soil fertility and nutrient management is becoming more widely understood, especially in developing countries where there is a high population pressure on the land. The hopes of humanity to successfully overcome the challenges of food shortage have undoubtedly been brightened by the intensive fertilizer use, intensive cropping and high-yielding varieties, but at the same time they have also given rise to numerous problems associated with soil fertility, fertilizer use, and soil and water management. The food and nutritional security should warrant availability of adequate and

B. Saha (✉)
Department of Soil Science, Dr. Kalam Agricultural College (Bihar Agricultural University, Sabour, Bhagalpur), Kishanganj, Bihar, India

K. M. Reddy
Department of Horticulture (Vegetable and Floriculture), Bihar Agricultural College, Sabour (Bihar Agricultural University, Sabour), Bhagalpur, Bihar, India

S. Saha
Directorate of Research, Bidhan Chandra Krishi Viswavidyalaya, Kalyani, West Bengal, India

A. Fatima
Department of Agronomy, Dr. Kalam Agricultural College (Bihar Agricultural University, Sabour, Bhagalpur), Kishanganj, Bihar, India

S. S. Solankey
Department of Vegetables & Floriculture, Agricultural Research Institute, Patna, Bihar, India

S. S. Solankey, M. Kumari (eds.), *Advances in Research on Vegetable Production Under a Changing Climate Vol. 2*, Advances in Olericulture, https://doi.org/10.1007/978-3-031-20840-9_4

quality food to people to meet their dietary and nutritional needs for a healthy and productive life. Vegetable crops being short term grown and potential high yields, have high demand for added nutrients to maintain the quality of the produce. Managing nutrients for these short term grown vegetable crops under changing climatic conditions needs special cares to be taken for efficient crop production. For instance, nitrogen being a highly mobile nutrient needs its judicious application during unprecedented high rainfall. Another example is potassium, a costly fertilizer element which could enhance the drought tolerance in crops under current days climatic changes, needs its judicious application through soil test based fertilizer application. The ability of a soil to supply nutrients is a measurement of its fertility, and how fully it is expressed, depends on the impact of the plants that are cultivated there, the surrounding environment, and soil management.

Keywords Climate change · Soil · Nutrient management · Vegetable crops

4.1 Introduction

Soil is an essential resource for the foundation of agricultural development and sustained quality of human life. The changes in climate due to 'greenhouse effect', the changes in atmospheric temperature and associated global water cycle, changing rainfall pattern have major impact on vegetation, soil process and ecosystem functioning including changes in nutrient availability which have profound implications of agricultural production. The probable effect of anthropogenic greenhouse gases induced climate change on different soil processes, nutrient dynamics and soil health. Since the industrial revolution era, human activity has increased the amount of greenhouse gases (GHGs) emission in the atmosphere, leading to increased radiative forcing from carbon di-oxide (CO_2), methane, CFCs, tropospheric ozone and nitrous oxide. Since 1750, there has been an increase in CO_2 and methane concentrations by around 36% and 148%, respectively. These levels are significantly greater than they have ever been in the previous 650,000 years. The source of GHG emissions from agriculture are primarily from methane emission from paddy fields (23%), enteric fermentation in ruminants (59%) and nitrous oxides released from manures and fertilizers application. The FAO estimated that land use and agriculture accounted for 20% of all GHG emissions (Ullah et al. 2021).

A majority of the climatic studies under controlled condition have found significant variation in soil processes including net nitrogen (N) mineralization and soil respiration, which may have a large impact on how well ecosystems function by changing the availability of nutrients (Emmett et al. 2004). Synthesis of findings from numerous research conducted across a number of ecosystem types, global warming could result in an average 20% increase in soil respiration and an average 46% increase in soil net N mineralization rates (Rustad et al. 2001).

A mineral element is considered to be essential, when a plant is unable to complete its reproductive stage of life cycle owing to its deficiency. Regarding postharvest disorders of various vegetables, the importance of calcium nutrition during growth has been well-documented (Sams and Conway 2003). Calcium has also been suggested as a putative signaling molecule which is actively involved in the development of cross tolerance to abiotic stresses (Bowler and Fluhr 2000). Therefore, the role of pre-harvest calcium nutrition in post-harvest stress resistance may be complex, and depends on whether the vegetable is also exposed to abiotic stresses. Relatively high pre-harvest nitrogen is often associated with poor post-harvest vegetables quality. In cabbage, excessive nitrogen fertilization leads to higher accumulation of zinc and aluminum and nitrate-induced Mn deficiency (Berard et al. 1990). In potatoes, susceptibility to black spot (Stevens and Davelaar 1997) is influenced by nitrogen fertilization. On the other hand, vegetables will typically contain more vitamin C when there is a nitrogen deficiency or when nitrogen application rates are lower than recommended. Vitamin C content has been closely correlated with storage life potential (Hodges et al. 2001), which is likely a consequence of the importance of this antioxidant nutrient in forestalling oxidative injury that leads to quality losses in storage. Potassium deficiency in carrots is linked to increased weight loss during storage (Shibairo et al. 1998).

Raising vegetables and fruit trees requires the use of nutrient components, particularly micronutrients, not only to prevent yield loss but also to enhance product quality. Particularly for fruits and vegetables, the reduced quality of the produced product owing to micronutrient deficiency frequently results in greater economic losses than a reduction in yield. For instance, peanut prices in US markets were reduced by 2% or more in kernels with hollow hearts (Morrill et al. 1977). Low B levels may also compromise the ability to store fruits and vegetables (Smith et al. 1997).

4.2 Effect of Nutrient Deficiency Stress in Vegetable Crops

It takes the coordinated efforts of soil scientists, ecologists, physiologists, biochemists, agronomists, and molecular biologists to understand the complicated phenomenon of nutrient stress. The presence of high concentrations or insufficient levels of element availability can both cause nutrient stress. A deficiency of one element can sometimes be due to the presence of another element in excess amounts. In this regard, efforts were made to illustrate the availability, functional characteristics, and toxic and deficient symptoms of 14 elements necessary for plant survival as well as other elements that cause phytotoxicity. Visual deficiency signs offer a fundamental basis for determining the plant‘s nutritional condition.

The traditional methods of fertilising vegetable crops with copious amounts of organic manures—like FYM, compost, and green manures—maintain the soil's necessary levels of micronutrients. Due to limited usage of organic manures, the utilization of short-duration and high yielding cultivars, and the application of macro elements at higher dosages of, output has stalled and it is possible that there will be an increase in the lack of micronutrients. Thirty five different elements are needed by plants in total, 16 of which are crucial to their growth and development. The two categories of macro and micro nutrients can further be used to separate the important plant nutrients. The basic nutrients—nitrogen, phosphorous, and potassium—require larger amounts than micronutrients, which are substances that are essential for growth of plants but are needed in lesser amounts. Zinc (Zn), Iron (Fe), Boron (B), Copper (Cu), Manganese (Mn), Molybdenum (Mo), and Chlorine (Cl) are the necessary micronutrients. These substances are extremely effective and deliver the best results in small amounts. Micronutrient deficiencies have a significant impact on plant development, metabolism, and the reproductive stage. On the other hand, plants are harmed by even a tiny excess or shortage. A crucial element in controlling nutrient supply is soil pH. All macro and micronutrients are most readily available in soils that range in pH 6.5–7.0 (slightly acidic to neutral).

It is increasingly recognized throughout the world that soil fertility and fertilizer management is of considerable importance, particularly in developing countries where there is a significant population strain on the land. High-yielding varieties, intensive cropping, and intensive fertiliser use have undoubtedly increased humanity's chances of successfully overcoming the challenges of food scarcity, but they have also resulted in a number of issues related to soil fertility, fertiliser use, and soil and water management. In an ideal scenario, the soil would ensure adequate nutrient distribution and retention, support and sustain root growth, maintain the biotic environment of the soil, respond to management, and prevent deterioration. The nutrient-supply capability of soil is an important indicator of its fertility, whose full expression depends on the plants grown on it, the environment, and soil management.

4.3 The Essential Nutrient Elements

One of the most important factors affecting plant productivity is soil nutrition. Nutrient availability is highly vulnerable to climate change. The availability of Carbon, Nitrogen and Phosphorus has serious impacts on plants since they are essential nutrients for plant growth and development (Elbasiouny et al. 2022). It is believed that crop plants cannot grow properly and reproduce without seventeen (17) specific elements. On the basis of their concentration in plants, the macronutrients C, H, O, N, P, S, Ca, Mg, K are generally at concentrations greater than 1000 mg/kg (plant dry matter basis); and the micronutrients *viz*., Fe, Mn, Zn, Cu, B, Ni, Cl and Mo are generally less than 100 mg/kg in concentration (Table 4.1). Carbon (C), Hydrogen (H) and Oxygen (O) are only of brief significance, as they are supplied by CO_2 and H_2O, which are available in abundance in the both atmosphere and hydrosphere respectively. Other than N, which makes up 78% of the atmosphere, parent material and weathering minerals in the soil are the main sources of the other important elements. Other elements are also absorbed by plants, some in significant amounts, such Sodium (Na), which is beneficial for many other species since it can partially replace K and is required for halophytes like saltbush (*Atriplex vesicaria*) (Marschner 1995).

Table 4.1 List of essential and beneficial elements according to their concentration in plant tissues

Macronutrient (>0.5 g/kg plant dry weight)		Micronutrient (<0.5 g/kg plant dry weight)	
C	Carbon	Fe	Iron
H	Hydrogen	Mn	Manganese
O	Oxygen	Cu	Copper
N	Nitrogen	Zn	Zinc
P	Phosphorus	Mo	Molybdenum
K	Potassium	B	Boron
Ca	Calcium	Cl	Chlorine
Mg	Magnesium	Ni	Nickel
S	Sulphur	(Na)	Sodium
(Si)	Silicon	(Co)	Cobalt

Both humans and animals require sodium. In addition to being advantageous for several species of commercial significance, such as paddy, sugarcane, and sugar-beet, silicon (Si) is necessary for some marsh grasses. Both legumes and non-legumes require cobalt (Co) for symbiotic N_2 fixation, and ruminant mammals also require cobalt. Iodine (I) and selenium (Se) are essential for human and animal health, while not being beneficial to plants other than a few algae. Se deficiency in animals' results in white muscle disease, while Iodine (I) deficiency in humans' results in goitre. Vanadium (Va) can partially replace Mo in the process by which bacteria fix N_2. Lanthanum (La) and cerium (Ce), rare earth elements, have allegedly been claimed to be helpful for plant development in China, although this claim has not been supported elsewhere.

Understanding how mobile micronutrients are in plants might help identify nutritional deficiencies. In the event of a deficiency, a mobile micronutrient in the plant moves to the growth points, causing deficiency symptoms to manifest on the lower leaves, as opposed to immobile nutrients, which do not migrate to the growing points, causing deficiency symptoms to manifest on the younger plant parts (Table 4.2 and Fig. 4.1). The availability of nutrients to plants and the technique of delivery are influenced significantly by the mobility of micronutrients in soil.

4.4 Movement in Soil

(i) Mobile: The micronutrients that are mobile are highly soluble and they are not adsorbable to clay complexes, for instance, $B0_3^{=}$, $Mn^{++}$and Cl^-
(ii) Less mobile: They are likewise soluble, but because they are adsorbed on clay complexes, their mobility is diminished. *e.g.* Cu^{++}.
(iii) Immobile: Because these ions are so reactive, they become stuck in the soil. *e.g.* Zn^{++}

Table 4.2 Typical deficiency symptoms/disorders in common vegetables

Vegetable crop	Disorder
Beans	Hypocotyl necrosis
Brussels sprouts	Internal browning
Cabbage	Internal tip burn
Cauliflower	Whiptail, buttoning, browning
Carrots	Cavity spot, cracking
Celery	Black heart
Lettuce	Tip burn
Parsnip	Cavity spot
Tomatoes	Blossom end rot
Radish	Akashin

Fig. 4.1 Visual deficiency symptoms in vegetables. (Source: www.commons.m.wikimedia.org)

4.5 Nitrogen (N)

The availability of nitrogen (N) to plants affects how they use carbohydrates. Vegetative cells will accumulate carbohydrates, which will cause them to thicken when N supplies are limited. Proteins are created from the manufactured carbohydrates when N supplies are sufficient and growth circumstances are favourable.

Less glucose is deposited in the vegetative section, which causes more protoplasm to develop. Because protoplasm is highly hydrated, this results in a more succulent plant. In plants, excessive succulence weakens the plant, and lodging may happen in grain crops. In certain circumstances, leafy vegetables extreme succulence makes them more susceptible to illness and insects. In vegetables like beet leaves, palak, Amaranthus and other leafy vegetables, excessive succulence enhances susceptibility to diseases and insects. On the other hand, adequate amounts of N in cucurbitaceous and cruciferous vegetables promotes the fruit and curd quality. Changing environmental factors like inadequate and untimely rainfall, severe drought may reduce the utilization efficiencies of added nitrogenous fertilizers.

4.6 Visual Deficiency Symptoms of N

Vegetables will show obvious signs of nitrogen deficit if the soil is lacking in it or if the plants cannot access it. Since they have an impact on the entire plant, nitrogen deficits are rather simple to spot: older leaves in the Brassica family (the cabbage family) frequently take on a red, orange, or purple hue.

Yellowish leaves are a sign of N deficiency in plants. Chlorosis, a sign of N insufficiency, is produced when older leaves lose protein N from their chloroplasts. When there is a significant N shortage, lower leaves become brown and eventually die while the higher leaves stay green. Chlorosis first manifests on the lower leaves. When the entire leaf is dead, the necrosis spread from the leaf tip to the midrib. The mobility of N in the plant is indicated by the tendency of the younger growth to stay green when the lower leaves turn yellow or die. The mobility of N in the plant is indicated by the tendency of the newer growth to stay green when the lower leaves turn yellow or die. Vegetables, with particular reference to cole crops like cabbage, cauliflower and broccoli, inadequate nitrogen may lead to lower curd/head size, nitrogen induced molybdenum deficiency.

4.7 Phosphorus (P)

Numerous physiological processes in plants involve phosphorus. In plants, phosphorus is referred to as the "energy currency" since it is involved in energy transfer and storage. Within plants, ADP and ATP serve as a form of currency. Numerous physiological processes in plants involve phosphorus. In plants, phosphorus is referred to as the "energy currency" since it is involved in energy transfer and storage. Within plants, ADP and ATP serve as a form of currency. P is a crucial component of ribonucleic acid and deoxyribonucleic acid (DNA) (RNA). Increased root growth is related to enough P. It is necessary to boost the strength of cereal straw and the ability of leguminous crops to fix nitrogen. Adequate P is associated with increases root growth in carrot, beets, turnip *etc*. Adequate content of P enhances the strength of stems in vegetable crops and boosts the N-fixing capacity of leguminous vegetables like faba bean and cowpea.

4.8 Visual Deficiency Symptoms of P

The plant will generally become shorter overall, and the colour of the leaves will darken. The prominent foliar symptoms that are visible with N or K deficit are not frequently detected because of the influence of P deficiency on slowing overall growth. The dark green tint shifts to a greying-green to bluish-green metallic sheen as the P shortage increases.

In some crops, like sugar beet, young seedlings have dark green leaves that eventually turn into older leaves with brown, netted veining as the plant matures. In maize and other grasses, purple leaf colouring is frequently correlated with P shortage. P is translocated from ageing to growing tissues in plants and is mobile. As a result, early growth stage reactions to P are typical. P moves to fruit and seeds during the reproductive stage.

4.9 Potassium (K)

In plant cells, potassium plays a role that is notably related to solution ionic strength. K's great mobility in plants is due to its involvement in charge balance, water relations, and osmotic pressure inside cells and across membranes. K is significant in many crop quality parameters due to its role in the synthesis and transportation of photosynthates to reproductive and storage organs in plants and their subsequent transformation into carbs, oils, lipids, protein and other products. A sufficient K level improves fruit size, colour, flavour, and peel thickness in vegetables, all of which are crucial for storage and shipping quality. Plants need K for the photosynthetic conversion of solar energy into chemical energy through synthesis of ATP. In vegetables, an adequate level of K significantly improves fruit size, colour, taste and peel thickness, which is highly crucial for storage and shipping quality.

4.10 Visual Deficiency Symptoms of K

K is mobile in plants, therefore symptoms of a visual deficiency usually first appear in older or lower leaves, similar to what happens with N or P, and proceed to the upper leaves as the severity of the deficit rises. Young leaves of high-yielding, quickly growing crops can also have a K shortage. Indicators of a K deficit brought on by dry conditions include chlorosis at the leaf tips and eventually the entire leaf.

The severity of crop losses caused by fungal and bacterial diseases; mites and insect infestation and nematode and viral infection can all be exacerbated by K stress. Weakened straw is another sign of a K deficit, which causes crops to lodge. K insufficiency manifests in citrus fruits and vegetables as decreased size, reduced size, decreased peel or skin thickness, increased skin breaking, and skin discolouration. Lack of K also results in poor shelf-life and storage properties.

4.11 Calcium (Ca)

Ca's coordination ability, which it uses to create stable but reversible intermolecular connections, primarily in cell walls and in the plasma membrane, is correlated with its activity. Even at high concentrations, calcium is a non-toxic mineral nutrient and is particularly good at detoxifying large amounts of other mineral elements in plants. The metal component of the metalloenzyme amylase is calcium. The main cation in the middle lamella of cell walls is calcium, and calcium pectate is its main component. Calcium therefore gives tissues mechanical strength while promoting cell division, plant growth, protein synthesis, glucose transport, and acid-base balance in cells.

Indian soils are abundant in calcium and hardly ever display any visual signs of a calcium deficit. A decrease in meristematic tissue is a sign of a Ca deficiency. The youngest leaves and developing tips are first affected by deficiency, and as the condition progresses, the leaf margins begin to necrotize. Apple discolouration is one sign of a Ca deficit in the fruit. The bitter apple pit is a common sign of a deficit. The typical Ca deficiency symptom in tomato is blossom end rot (BER). Calcium deficiency in vegetable crops also induces the deficiency of boron as their functions are inter-related.

4.12 Magnesium (Mg)

Magnesium is a crucial component of plant chlorophyll. Additionally, magnesium is the most frequent activator of enzymes involved in energy metabolism. The mobility of Mg^{2+} inside cells and its ability to interact with ligands through ionic bonding are both important for its roles in plants. The modulation of cellular pH and the cation-anion balance accounts for a large fraction of the total Mg^{2+} ions. By removing calcium from the functional groups, it can occasionally prevent magnesium from having its activating effect.

4.13 Visual Deficiency Symptoms of Mg

Plants growing in acidic soils with a coarse texture frequently have magnesium deficiencies. Many fruits and vegetables have leaves with chlorotic interveinal patterns. While leaf veins are still green, lower leaves with a magnesium shortage first turn a golden bronze or reddish colour. Plants have a reduced ability to absorb Ca and Mg in very acidic soils.

4.14 Sulphur (S)

Sulphur is found in igneous and sedimentary rocks as sulphides. Sulphur can also be found in seawater, industrial waste, organic molecules in soil, and gaseous emissions in the environment. Sulphur shortage typically affects old, heavily weathered land surfaces, soils that have been heavily leached, and soils that are far from the sea and industrial areas. The production of vitamins requires sulphur. Three amino acids that are present in both plants and animals must be produced in order to survive. The majority of the sulphur found in plant tissues is incorporated into proteins.

4.15 Visual Deficiency Symptoms of S

Protein synthesis is inhibited by a sulphur shortage. It might be challenging to discern between sulphur and nitrogen deficiency in field crops. A common sign of S shortage in wheat is yellowing of the leaves. Tomato plants have restricted development and are smaller and paler in colour. S-deficiency is typically characterised by mottled yellow-green leaves with yellowish veins in younger leaves.

4.16 Iron (Fe)

The porphyrin molecules cytochromes, hemes, haematin, ferrichrome, and leghaemoglobin all contain iron as a structural component. These elements participate in the oxidation-reduction processes that occur during respiration and photosynthesis. Almost 90% of the Fe in leaves is found in lipoprotein of the chloroplast and in membranes of mitochondria, and up to 75% of the Fe in cells is linked to the chloroplasts. Ferrodoxin, a Fe-S protein, serves as an electron acceptor in photosynthetic reduction processes where Fe-containing cytochromes in chloroplasts are involved. Ferrodoxins are the first stable component of the photosynthetic electron transport chain. Fe compounds frequently perform the respiration-related O_2 to H_2O reduction. Fe is a component of nitrogenase, an enzyme used by N-fixing bacteria to fix N_2. Additionally, Fe could be able to partially replace the molybdenum (Mo) necessary for the nitrate reductase activity in soybeans.

4.17 Visual Deficiency Symptoms of Fe

The Fe sufficiency range in plant tissue is from 50 to 250 ppm. Generally speaking, deficiency is likely to happen when tissue Fe level is less than 50 ppm. Since Fe is not mobile in the plant, its deficiency symptoms typically manifest in young leaves. Young leaves first develop interveinal chlorosis, which quickly spreads to the entire leaf. In extreme circumstances, leaves become completely dead and white. Fe poisoning can be seen in specific situations. For instance, when rice leaves contain more than 300 ppm Fe, leaf bronzing symptoms manifest themselves in poorly drained or buried soils.

Mn^{2+} and organically complexed, low-molecular-weight Mn are absorbed by plants. Plants normally have 20–500 ppm of manganese, but Mn-deficient plants only have 15–20 ppm. Prior to being absorbed by plant roots, Mn must first be converted to Mn^{2+}. Photosynthesis reactions, enzyme activation, and root growth all depend on Mn. Several electron transport processes also involve Mn. Similar to Cu, Mn activates many enzymes which produce many essential amino acids and phenols for the synthesis of lignin. These substances are used to create lignin as well as phenolic acids and alcohols that protect against pathogen infection.

4.18 Visual Deficiency Symptoms of Mn

Because of its crucial function in photosynthesis, Mn deficient plants have much slower root and shoot development rates. N and P build up as a result, which raises the risk of root and leaf illnesses. A lack of Mn also limits the production of lignin and phenolic acids, which both contribute to the development of illnesses. Disease in Mn-deficient plants can be brought on by soil fungi that normally do not affect plant roots. Grasses having low Mn are more frequently prone to diseases like root rot. Since Mn is stationary in the plant, the deficiency symptoms are first visible on the younger leaves. In most crops, interveinal chlorosis results from a Mn deficit. Younger leaf chlorosis caused by low Mn in some crops can be mistaken for Fe insufficiency. Such phrases as grey speck of oat, marsh spot of pea, and speckled yellows of sugar beet have been used to denote Mn deficit in a variety of crops.

4.19 Zinc (Zn)

Plant accessible Zn is primarily controlled by the solubility of Zn minerals in soil, soil organic matter, and Zn adsorbed on clay and organic matter soil surfaces. Zn is absorbed by plant roots both as a cation and in synthetic and organic complexes. Zn is involved in a variety of enzymatic processes; however, it is unclear whether it serves as a cofactor that is functional, structural, or regulatory. Tryptophane, a substance found in certain proteins and is necessary for the synthesis of growth promoting hormones (auxins) such indole acetic acid (IAA), is synthesised with the help of zinc. Plants lacking in zinc produce fewer growth hormones, which results in shorter internodes and smaller-than-average leaves. Additionally, the creation of chlorophyll, enzyme activation, and cell membrane integrity all depend on zinc.

4.20 Visual Deficiency Symptoms of Zn

Zinc shortage is noticeable visually in the leaves, but it can also affect overall plant growth including fruit or storage organs. The typical symptoms of its deficiency are light green, yellow, or white patches in between the leaf veins which are especially found in older leaves. Bushy, rosetted leaves on account of shortened stem or stalk internodes, is another symptom of Zn deficiency. A distinctive rosetting or clusters of little leaves at the top of the stem in plants indicate a zinc shortage. Rosetting is a common occurrence on citrus and fruit trees. Zinc is sensitive to corn, cotton, potatoes, and other vegetable crops. Lack of zinc inhibits tillering and causes midrib chlorosis near the base of new leaves. Brown patches or yellow-orange leaf tips gradually develop in older leaves that eventually spread to the entire leaf.

4.21 Copper (Cu)

Cu is taken up by plants as cuprous (Cu^{2+}) and as a constituent of either natural or artificial chemical complexes. The concentration of copper in plant tissue typical ranges from 5 to 20 ppm. Copper may participate in a variety of oxidation-reduction reactions in the plant thanks to how easily it can collect and donate electrons. Copper is necessary for the transmission of electrons involved in photosynthesis and respiration. Although Fe and Mn are also involved in the transport of electrons, they cannot take the place of Cu. The primary energy source for the formation of proteins, lipids, cell wall membranes, and active food uptake is adenosine triphosphate (ATP), which is produced via electron transport processes involved in photosynthesis and respiration. Lignin is a component of cell walls that adds stiffness and strength, which is necessary for plants to stand upright. Several enzymes necessary for the formation of lignin, including polyphenol oxidase and diamine oxidase, contain copper. Deformed leaves and stems from a copper shortage enhance the likelihood of lodging. Lignin helps plants naturally resist illnesses. Plants lacking in copper are more prone to illness.

4.22 Visual Deficiency Symptoms of Cu

Cu deficiencies do happen in sensitive crops cultivated on low Cu-soils, albeit being less common than other micronutrient deficiencies. Because of the activity of the Cu-containing enzymes in the chloroplasts, chlorosis of younger leaves is a common indication pertaining to Cu shortage, however the symptoms vary depending on the crop. Young leaves of maize and small grains begin yellowing and become dwarf; and as the shortage worsen, younger leaves turn pale and older mature leaves begin to wither. Cu deficient plants can have stem melanosis, take-all root rot, and ergot infection, ultimately resulting in smaller grains or fruits. The leaves of numerous vegetable crops lack turgor, take on a bluish tinge, get chlorotic, and cur, and flower formation is unsuccessful. Reduced lignifications with low copper levels in vegetable crops are known to cause lodging, wilting, and an increase in disease incidence.

4.23 Boron (B)

B is not required by animals, fungi, or microbes, while being necessary for higher plants and some types of algae. B plays a major role in maintaining the structural integrity of plant cell walls. In order for cells to expand, regulate H+ transport, retain cellular Ca^{2+}, and control the generation of lignin after cell growth, B provides cross connections between the polysaccharides that make up the cell wall. Because boron connections are adaptable, cell wall growth is possible. Normal cell wall growth is hampered by a B shortage. Among the dicots and monocots, the grasses are less reliant on B for cell wall construction, it nevertheless plays a significant role, however, these roles are different. When pollen tubes are growing, which is necessary for seed formation, cell wall stability is particularly crucial. B is necessary for the delivery of photosynthates (sugars) to meristematic (growing) tissues that are quickly developing, like root tips, leaves, buds and conductive tissues. Boron is also important in the regular transportation of water, nutrients, and organic substances to the plants' new development organs. A sufficient B level promotes the growth of seeds and fruits as well as flower creation and retention.

4.24 Visual Deficiency Symptoms of B

In areas that are actively expanding, boron shortage frequently manifests as a structural malformation. B deficiency has a greater impact on reproductive growth than on vegetative growth. A sufficient B level boosts floral formation, retention, seed, and fruit development. Since B cannot easily go from older to actively expanding tissues, the first sign of a visible shortage is the cessation of growth of terminal buds, which is then accompanied by the senescence of the young leaves. The youngest leaves of plants with a B deficiency turn a pale green tint, losing more pigment at the base than the tip. If growth persists, the basal tissues disintegrate, giving leaves a twisted appearance. A swollen, cracked, or water-soaked petioles and stems, as well as discolouration, cracking or decaying of fruit, tubers, or roots, are common manifestations of B deficient symptoms. B insufficiency is the root cause of apple internal cork. Citrus fruits with low B levels include inconsistent peel thickness, lumpy fruit and sticky deposition inside the fruit.

The symptoms of micronutrient deficiencies can be brought on by a variety of biotic and abiotic sources. Because the oxygen-depleted waterlogged soils are anoxic, plants growing in them frequently produce tiny, yellow leaves and have dieback signs that are similar to those of iron deficiency. Instead of applying micronutrients under these circumstances, good irrigation management would produce more beneficial outcomes. When there is a drought, plants become stunted and their leaf margins may burn, especially in the summer. The same symptoms result from having too much or too little of certain substances. Foliage yellowing brought on by root infections can resemble nutritional shortages. Some herbicides, when misused, can cause symptoms that are similar to those of iron or zinc deficiency. When the underlying reason cannot be determined, micronutrient deficiencies are frequently held responsible. In order to properly identify micronutrient deficiency symptoms in vegetable crops, one should have the necessary understanding of damage caused by insect infestation, symptoms of disease, herbicide injury, moisture stress conditions, *etc.* when diagnosing micronutrient deficiency symptoms. The following micronutrient statuses of soils and vegetable crops can be monitored via soil tests and plant analyses.

Micronutrient availability is influenced by soil pH, texture, and organic matter content. Due to the mineralization of organic matter, mineral soils with high organic matter may have an adequate supply of micronutrients. However, availability of B, Cu, Fe, Mn, and Zn often declines with an increase in soil pH, although Mo availability increases with an increase in pH. For growing vegetables, soil pH should be between 6.5 and 7.0. Vegetable production is not at all acceptable for soil pH levels below 5.5. The majority of Mn and Fe deficiency occurrences develop in mineral soils with pH levels higher than 7.5.

Soil testing is crucial for developing a sound vegetable production programme based on soil fertility. To promote wise micronutrient recommendations, soil testing is required. The interpretation of soil testing is based on the soil samples. As a result, it's crucial that the soil samples are carefully taken. After clearing the surface litter, each sampled field should have 15–20 sites where a uniform core of a thin slice of soil from the surface to plough depth (15–22 cm) can be obtained. Before conducting a chemical analysis, a little amount of soil should be sampled and air dried on a plastic sheet.

4.25 Fertilizer Nutrient Management for Enhancing Productivity and Nutrient Use Efficiency

The primary elements of integrated nutrient management are fertilizer, organic manures, agricultural residues, industrial by-products, and biofertilizers (INM). There is a dearth of comprehensive databases on topics besides fertilizers, which must be generated.

4.26 Components Used for Enhancing Fertilizer Use Efficiency

4.26.1 *Chemical Fertilizers*

The most significant INM component is fertiliser. Because extensive farming requires supplying significant amounts of nutrients, our reliance on fertiliser has been steadily growing. However, fertiliser consumption is both insufficient and seriously out of balance. Improper application of fertiliser, especially major nutrients (N, P, and K) and occasionally S-based fertilisers for greater crop productivity, when micronutrients are frequently disregarded. Only around 30–40% of the applied fertiliser nutrients are actually used by the crops; the remainder is lost via a variety of processes, including leaching, surface runoff, volatilization, denitrification, soil erosion, and fixation in the soil. As a result, every effort should be made to improve fertiliser use efficiency.

Farmyard manure (FYM), composts, and other organic manures have historically been crucial inputs for preserving soil fertility and assuring yield stability. Due to the poor nutritional content, bulky nature, and limited availability of these nutrient sources, they gradually lost importance in crop production in favour of easily accessible chemical fertilisers. However, the rising price of fertilisers and their finite supply forced plant nutrition to be found in alternative and renewable sources, which greatly stoked interest in organic recycling. In India, the enormous potential of both rural and urban compost is underutilised. Less than half of livestock's manurial potential is now untapped since a significant amount is lost as fuel and waste in non-agricultural areas. A sizeable fraction of the nutrients required for agricultural production are lost due to inappropriate handling, storage, and assimilation in soil. Other potential organic nutrition sources include food processing waste, press mud, non-edible oil cakes, and urban compost.

Macro and macronutrients are directly provided by organic manures, and the soils' physical, chemical, and biological qualities are improved indirectly. Apart from providing nutrients to the existing crop, these manures also have a significant long-term impact on the system's subsequent crops. Numerous studies have found that FYM can partially replace the fertiliser N requirements of monsoon crops without having an unfavourable effect on the overall productivity of cereal-based cropping systems. Furthermore, it was discovered that in some areas, by substituting 25% of the N requirements of the preceding monsoon crop through FYM, the winter wheat's fertiliser requirements might be decreased by up to 25%. The productivity gained with the combined application of chemical fertilisers and FYM was significantly higher than that obtained with fertilizer alone. Nutrient composition of some important organic sources/manures is listed in Table 4.3.

4.27 Inclusion of Legumes in the Cropping System

The symbiotic interaction between rhizobium and legumes is crucial for agriculture. Certain *Rhizobia* can infect plant roots and cause nodules to develop there. The ambient nitrogen is converted to ammonia in these nodules. L-ketoglutamic acid is next produced from the ammonia, and eventually glutamic acid is produced. The *Rhizobium* species and the legume species do, however, have a particular relationship. There are 20 different species of *Rhizobium*, which may be relevant to mention at this point. Legumes can make a significant contribution to INM when grown as a green manuring crop or as fodder in a cropping system. Mann and Garrity 1994 reported that legumes increased productivity and restored soil fertility in rice-wheat system. As much as 50–60 kg N ha may be made accessible, according to an evaluation of several leguminous crops for partially covering the N needs of the next crop. There may be a carryover of N for the subsequent cereal crop to the extent of 75 kg in Indian clover, 35–60 kg in fodder cowpea, 55 kg in black gram, 60 kg in peanut, 68 kg in Bengal gram, and 50 kg in lathyrus. When a legume was the prior crop as opposed to a cereal, the subsequent crop's grain output increased significantly. In general, pulse crops do leave 30–50 kg of leftover N per ha for the next crop to use. By providing the right species of *Rhizobia* to legume crops, this benefit could be increased. Legumes undoubtedly play a significant role in the residual fertility build-up that occurs in a nation where the average national consumption of plant nutrients from chemical fertilisers is quite low. This contribution must be fully utilised. The plants themselves use the majority of the nitrogen fixed by the nodules. Four significant N_2-fixing connections, including the cyanobacterial lichen symbiosis with *Rhizobium*-leguminous plants, Frankia-actinorhizal plants, *Anabaena-Azolla* symbiosis, and *Rhizobium*-leguminous plants, have been thoroughly researched. Because of these symbiotic relationships, plants can typically flourish in soils that are deficient in nitrogen (Saha et al. 2017).

4.28 Legumes as Green Manures

Green manuring was regarded as essential for increasing agricultural output in the pre-Green Revolution era. However, with the more widespread use of intensive cropping systems and the easier availability of fertilisers, the practise of green

Table 4.3 Major nutrient contents in some manures

Material	N	P_2O_5	K_2O
Bulky organic manures			
Farmyard manure	0.5	0.3	0.5
Compost (urban)	1.0	0.5	1.5
Compost (rural)	0.5	0.5	1.0
Cow dung	1.5	0.5	0.5
Cow dung slurry from biogas plant	1.8	1.0	1.0
Horse dung	2.5	1.5	1.5
Sheep and goat dung	3.0	1.0	2.0
Buffalo dung	1.5	1.0	1.0
Pig dung	3.0	2.5	2.0
Camel dung	1.3	0.5	0.5
Poultry litter	3.0	2.5	1.5
Night soil	5.0	3.0	2.0
Sludge (activated)	2.5	2.0	0.5
Tea leaves (wastes)	3.0	0.3	1.0
Water hyacinth	2.0	1.0	2.0
Green manure	0.7	0.2	0.6
Concentrated organic manures non-edible oil cakes			
Castor cake	5.5	2.0	1.5
Cotton seed cake (undecorticated)	4.0	2.0	1.5
Mahua (ippi) cake	2.5	1.0	1.5
Karanj (honge) cake	4.0	1.0	1.0
Neem cake	5.0	1.0	1.5
Safflower cake	5.0	1.5	1.0
Undi cake	3.5	1.5	2.0
Edible cakes			
Groundnut cake	7.0	1.5	1.5
Cottonseed cake (decorticated)	6.0	3.0	2.0
Linseed cake	5.0	1.5	1.0
Mustard cake	4.5	1.5	1.0
Niger cake	4.5	2.0	1.0
Rapeseed cake	5.0	2.0	1.0
Safflower cake (decorticated)	8.0	2.0	2.0
Sesame (*til*) cake	6.0	2.0	1.0
Meals from animal wastes			
Bone meal	3.0	20.0	–
Meat meal	10.5	2.5	0.5
Blood meal	10.0	1.5	1.0
Fish meal	7.0	6.0	1.0

Source: Biswas and Mukherjee (2008)

manure farming was all but abandoned. There has been a resurgence of interest in green manuring in recent years due to signs of productivity decline brought on by the widespread use of just chemical fertilisers 30 (NPK). Due to atmospheric N-fixation, green manuring with legumes enriches the soil N content. The soil's micronutrients, N, P, and K all become soluble on decomposition of green manure. It also lessens the gaseous and leaching losses of nitrogen. Additionally, green manures enhance the soil physical, chemical and biological characteristics. The two most significant green manure crops are sun hemp (*Crotolaria juncea*) and Dhaincha (*Sesbania aculeata*) however, cluster bean, berseem, Indian clover, *etc*. are also infrequently utilised.

4.29 Crop Residues

The available crop residues in various regions of the nation and the technologies that can be used to utilise such residues influence the possible uses of crop residues. Crop wastes have historically been used for a variety of competing purposes, including livestock feed, fodder, fuel, roof thatching, packaging, and composting. Cereal crop leftovers are mostly used as cow fodder. Straw and husk from paddy are sometimes used as boiler fuel and for heating homes. Crop residues are either used by farmers directly or sold to middlemen or landless people who then sell them to industry. The residual waste is either burned on-site or left unused.

A significant quantity of rice crop residue is burned on farms in places like Punjab and Haryana when it is not used as animal feed (Saha et al. 2019). In the majority of the country, sugarcane tops are either burned on-farm to grow a ratoon crop or utilised as feed for dairy cows. Groundnut leftovers are used as fuel in lime and brick kilns. Cotton, chilli, pulse, and oilseed crop leftovers are primarily used as fuel for domestic purposes. Domestic fuel is made from coconut shells, rapeseed and mustard stalks, pigeon pea, jute, mesta, and sunflower. Because of the multiple nutrient movements in an agricultural system, adding crop waste helps maintain the balance of nutrients in the soil.

Although there is a huge potential for crop residues in India, many crop wastes with feed value are required to support a large animal population and may not be available as an element in INM. However, where mechanical harvesting is used, as in north-west India, a sizeable amount of crop residues are left in the field, which might serve as a potential nutrient source. There are a lot of leftovers from crops such as potatoes, sugarcane, vegetables, *etc*., which are ultimately squandered.

Even though cereal crop wastes provide excellent bovine fodder, they can be used to complement chemical fertilisers in areas where their supply exceeds local demand and there is no accessible market. Even with conventional harvesting techniques, the amount of stubbles left in the field varies by crop and ranges from 0.5 to 1.5 t ha^{-1}. This amount is significantly more when harvesting is done mechanically. During land preparation, it is typical to collect and burn the tough-to-decompose residue from coarse grains like sorghum, maize, pearl millet, etc., which results in a large loss of plant nutrients. It is necessary to develop suitable management techniques to utilise these stubbles, either by adding cellulose-decomposing microbial cultures or providing a portion of the chemical N required for the next crop during land preparation (Singh et al. 2009).

4.30 Bio-fertilizers

Microbial inoculants, often known as bio-fertilizers, are substances that contain microorganisms that are helpful to agriculture. The most popular bio-fertilizer is *Rhizobium*, which colonises the roots of some legumes to produce root nodules. These nodules serve as ammonia manufacturing plants. In a single crop season, the N can be fixed up to100–300 kg/ha by a *Rhizobium*-legume association, and in some cases, leaves significant N for the crop that follows. More than 80% of the N needs of the leguminous crop can be satisfied by this symbiosis. *Azotobacter* is a different type of free-living N-fixing bacteria. The mechanisms by which the plants inoculated with *Azotobacter* gain advantages in terms of increased grain yield and N uptake include the formation and branching of roots, synthesis of plant growth hormones, improvement in uptake of NO_3, NH_4, H_2PO_4, K and Fe, improved water status of the plants, increased nitrate-reductase activity, and formation of anti-fungal compounds.

In 342 out of 411 on-farm trials conducted under the AICRP-CS, considerable response to *Azotobacter* inoculation in irrigated wheat was noted (Hegde and Dwivedi 1994). *Azospirillum* being associative symbiotic bacteria fixes N in loose association with plants by colonising the root mass. *Azotobacter* has been reported to fix on an average 20–25 kg N/ha. With an average response similar to 15–20 kg/ha of applied N, it has demonstrated favourable interaction with applied nitrogen in a number of field crops.

P-solubilizing bacteria (PSB) and fungal strains have been discovered in recent years. Increased dissolution of insoluble and sparingly soluble soil P has been observed following inoculation with P-solubilizing microbial cultures. About 30–35 kg P_2O_5/ha might be added if the microbial cultures and low-grade rock phosphate were used together. *Pseudomonas striata* soil inoculation increased wheat grain output while also having a lasting effect on subsequent crops of maize in Delhi's alluvial soil (Hegde and Dwivedi 1993).

4.31 Industrial By-Products and Municipal Wastes

There are other additional organic sources with good nutritious potential. To determine their fertiliser equivalents, these nutrient transporters have not yet undergone a thorough evaluation. These sources must be integrated based on their availability in suitable crops and cropping systems. The waste from various food processing sectors, as well as by-products from companies such spent-wash from distilleries, molasses, pressmud, etc., have good manorial value. Sulphitation pressmud (SPM) has positive effects on soil characteristics as well as a significant capacity to deliver nutrients. In the past three decades, SPM has grown significantly in significance as a nutrient supplement in intensive cropping systems of sugarcane-ratoon-wheat and other crops in regions where sugarcane is predominated. The other significant nutrient sources that can be combined with fertiliser inputs are municipal solid wastes (MSW) and sewage sludge, however these must be utilised with caution due to the risk of pathogens and heavy metal load.

4.32 Enhancing Crop Nutrient Use Efficiency

The following may be regarded as the most significant of the several factors that affect fertilizer consumption and, consequently, how well crops respond to fertilizers.

A. Crop characteristics

   i. Kind of plant and root system
   ii. Plant population
   iii. Varieties
   iv. Crop rotation and crop residues

B. Soil characteristics

   i. Soil nutrient status
   ii. Soil reaction
   iii. Soil moisture
   iv. Soil temperature
   v. Physical condition of the soil
   vi. Soil chemical properties
   vii. Effect of soil amendments

C. Fertilizer characteristics and fertilizer manipulations

   i. Type of fertilizer
   ii. Time of fertilizer application
   iii. Method of fertilizer application
   iv. Use of nitrification inhibitors
   v. Use of chelating substances

4.33 Crop Characteristics

Because of a variety of factors, including yield level, which directly affects the amount of nutrients removed by various crops, the responses to fertiliser application can vary greatly from crop to crop. The information provided in the preceding chapters makes this point quite evident. Although the ability of different crops to absorb nutrients is a good indicator of how they will react to fertilisers, the root properties of various plant species can frequently alter how they will react to various fertilisers. For instance, the rate and scope of root system growth varied significantly among various plant species.

4.34 Soil Characteristics

The soil's chemical makeup in relation to the plant nutrients available directly affects how crops react to fertiliser application. It has been demonstrated that the amount of exchangeable potash in the soil is inversely correlated with the size of the reaction to added potash. It has been demonstrated through potash tests that as the amount of exchangeable potash in the soil increases, less potash is needed to provide the highest yield. In the case of other nutrients, a similar relationship also applies. According to the soil testing values obtained, soils are classified as having "low," "medium," or "high" levels of plant nutrients, and the use of appropriate fertiliser is advised. Crops on soils with low phosphorus content should respond quickly to phosphate application. If the rating is high, there may be little to no reaction to phosphorus fertilizer application however, if the rating is medium, then response is likely. A soil's fertiliser requirements can be calculated on the basis of results obtained from the soil test in order to produce the desired yields. In general, pot cultures produce greater correlations than outdoor settings. This is due to the fact that there are numerous combinations of environmental and managerial factors that affect crop development and yield in field circumstances. In addition to enhancing the effectiveness and economy of fertiliser application, soil testing is useful in identifying the areas that respond to various plant nutrients and in designing the fertilisers that should be applied there.

Due to diverse variations in the parent material or kind of weathering, soils can have significantly different chemical compositions. For instance, because sandy soils are typically formed from grains that lack potassium, they are typically poor in potassium. On such soils, fertilisation with potash is likely to have a positive reaction. On the other hand, potassium aluminium feldspars, which are abundant in potassium, are typically the source of clay soils. The majority of the literature on the connection between soil-water use and fertiliser use has focused on how fertilisers might encourage the wise use of water. The reaction of a crop to fertiliser application and the ongoing availability of soil moisture have been discovered to be closely related. The application of fertilizer under these circumstances may even negatively influence the yield since higher early vigorous growth may cause the scarce water supply to be consumed more quickly, if soil moisture becomes a limiting factor at any point throughout a crop's growth.

4.35 Fertilizer Characteristics and Fertilizer Manipulations

(i) **Fertilizer types:** Different fertilisers may be distinct from one another in various ways. They could differ in terms of the amount or type of nutrients they contain. Thus, nutrients can be provided in the forms of nitrate, ammonium or amide in case of nitrogenous fertilisers. Similar to this, phosphorus in phosphatic fertilisers can either be water-soluble, citrate-soluble, or citrate-insoluble. Additionally, the fertilisers can differ substantially in terms of analysis. Because of this, a single superphosphate will only provide 16% P.O., whereas a triple superphosphate will provide 3x as much. The water solubility of the fertilizers as well as their ability to become fixed in the soil or revert back into insoluble forms may also vary, causing them to move very little after being added to the soil. The nitrate and amide fertilizers, which are susceptible to leaching during heavy rains, are among the first group of fertilizers with a high water solubility. Phosphatic fertilizers fall in the second group and they do not move in the soil at all and, therefore, special care has to be taken in their case so that they could be available the roots of the crop plants. Fertilizers may also affect the salt concentration of the soil solution. The salt index of a fertilizer is a measure of this phenomenon. Fertilizer salts differ greatly in their effect on the concentration of the soil solution because of the differences in their salt index.

(ii) **Time of fertilizer application:** The time of fertilizer application has been found important only in the case of nitrogenous fertilizers which have a tendency to leach with irrigation or rains. In the case of phosphatic and potassic fertilizers, application of entire quantity at the time of sowing has given the best results with most of the crops. The early vegetative stage and the panicle-initiation stage have been recognised as the most critical stage for nitrogen application in the case of rice. An adequate supply of nitrogen at the early vegetative stage should be ensured to encourage satisfactory plant development and for the proper development of the grains, it is crucial to apply nitrogen at panicle initiation stage. Split applications of nitrogen to rice are more important in the case of light-textured soils where nitrogen application done during transplanting may get leached by the time the plant reaches the panicle-initiation stage. Thus, split applications may be helpful in augmenting the nitrogen supply of the plant during its reproductive phase. Similar responses to nitrogen application have been obtained in most of the rice-growing areas of the world. In many other crops also, the split application of nitrogen has been found to be superior over single dose. The split application of nitrogen has also been found to be more beneficial than the conventional method of applying the entire quantity of fertilizer at the time of sowing in the case of hybrid maize. The general recommendation is 120 kg/ha which should be applied in three equal splits at sowing, at the knee-high stage and at the silking stage of the crop.

(iii) **Method of fertilizer application.** The band/point placement of phosphate-containing fertilizers has been found to be beneficial almost universally. The response to phosphatic fertilizers has been reported to be much higher when the fertilizer was placed in a band 5 cm wide to the side and 5 cm below the seed than that from the broadcast fertilizer application.

4.36 Conclusion

The adverse impacts of climate change are already being felt across the world. Climate change is perceived to bring about significant changes in the soil processes particularly net nitrogen mineralisation and soil respiration. This in turn would exert profound influence on the availability of native nutrients in the soil and soil health. Vegetables constitute an important part of human diet and climate change can bring about nutritional stress in vegetable crops thus challenging food and nutritional security in an era when there is a rampant increase in population all over the world. Hence, there is a need for development of nutrient recommendation and application modules that are specific to vegetable crops which can help in counteracting the deleterious effects of nutrient stress on these crops as well as enhance the nutrient use efficiency of applied nutrients. It is high time when a sea-change in package of practices for vegetables production is required along with the adoption of natural methods of farming (inclusion of legumes, biofertilizers, crop residues, *etc.*) which has the ability to cope up with the changing climate scenario.

References

Berard LS, Senecal M, Vigier B (1990) Effects of nitrogen fertilization on stored cabbage. II. Mineral composition in midrib and head tissues of two cultivars. J Hortic Sci 65(4):289–296

Biswas TD, Mukherjee SK (2008) Soil fertility and fertilizer use. In: Textbook of soil science. Tata McGraw Hill Publishing Copany Limited, New Delhi

Bowler C, Fluhr R (2000) The role of calcium and activated oxygens as signals for controlling cross-tolerance. Trends Plant Sci 5(6):241–246

Elbasiouny H, El-Ramady H, Elbehiry F, Rajput VD, Minkina T, Mandzhieva S (2022) Plant nutrition under climate change and soil carbon sequestration. Sustainability 14(2):9–14

Emmett BA, Beier C, Estiarte M, Tietema A, Kristensen HL, Williams D, Penuelas J, Schmidt I, Sowerby A (2004) The response of soil processes to climate change: results from manipulation studies of shrub-lands across an environmental gradient. Ecosystems 7:625–637

Hegde DM, Dwivedi BS (1993) Integrated nutrient supply and management as a strategy to meet nutrient demand. Fert News 38:49–59

Hegde DM, Dwivedi BS (1994) Crop response to bio-fertilizers in irrigated areas. Fert News 39:19–26

Hodges DM, Wismer WV, Forney CF (2001) Antioxidant responses in leaves of two cultivars of spinach (*Spinacia oleracea* L.) differing in their senescence rates. J Amer Soc Hortic Sci 126(5):611–617

Mann RA, Garrity DP (1994) Green manures in rice-wheat cropping systems in Asia. In: Ladha JK, Garrity DP (eds) Green manure production systems for Asian Rice lands. International Rice Research Institute, Los Banos, pp 27–42

Marschner H (1995) Mineral nutrition of higher plants. Academic, Orlando

Morrill LG, Hill WE, Chrudinsky WA, Ashlock L, Trip LD, Tucker BB, Weatherly L (1977) Boron requirements of Spanish peanuts in Oklahoma: effects on yield and quality and interaction with other nutrients. Agricultural Experimental Station, Oklahoma State University, MP-99

Rustad LE, Campbell JL, Marion GM, Norby RJ, Mitchell MJ, Hartley AE, Cornelissen JHC, Gurevitch J (2001) GCTE-news. A meta-analysis of the response of soil respiration, net N mineralization, and above-ground plant growth to experimental ecosystem warming. Oecologia 126:543–562

Saha B, Saha S, Das A, Bhattacharyya PK, Basak N, Sinha AK, Poddar P (2017) Biological nitrogen fixation for sustainable agriculture. In: Meena VS et al (eds) Agriculturally important microbes for sustainable agriculture. Springer Nature Singapore Pte Ltd. https://doi.org/10.1007/978-981-10-5343-64

Saha B, Verma DK, Shamim M, Das T, Saha S, Om H, Jha VB (2019) Crop residue management in small holding farming sector of Bihar—its importance and challenges. Curr J Appl Sci Tech 37(6):1–4

Sams CE, Conway WS (2003) Pre-harvest nutritional factors affecting postharvest physiology. In: Bartz JA, Brecht JK (eds) Postharvest physiology and pathology of vegetables, 2nd edn. Marcel Dekker, Inc, New York, pp 161–176

Shibairo SI, Upadhyaya MK, Toivonen PMA (1998) Potassium nutrition and postharvest moisture loss in carrots (*Daucus carota* L.). J Hortic Sci Biotech 73(6):862–866

Singh M, Dwivedi BS, Datta SP (2009) Integrated nutrient management for enhancing productivity, nutrient use efficiency and environmental quality. In: Proceedings of the Platinum Jubilee symposium on soil science in meeting the challenges to food security and environmental quality, pp 55–67

Smith TE, Stephenson RA, Asher CJ, Hetherington SE (1997). Boron deficiency in avocado. 2. Effects on fruit size and ripening. In: Bell RW, Rerkasem B (eds) Boron in Soils and Plants. Proceedings of the international symposium on boron in soils and plants, Chiang Mai, 7–11 September, 1997. Developments in plant and soil sciences, vol 76. Kluwer Academic Publishers, Dordrecht, pp 135–138

Stevens LH, Davelaar E (1997) Biochemical potential of potato tubers to synthesize blackspot pigments in relation to their actual blackspot susceptibility. J Agric Food Chem l45(11):4221–4226

Ullah A, Bano A, Khan N (2021) Climate change and salinity effects on crops and chemical communication between plants and plant growth-promoting microorganisms under stress. Front Sustain Food Syst 5:618092

Chapter 5
Impact of Climate Change on Leafy and Salad Vegetables Production

Menka Pathak, Satyaprakash Barik, Sunil Kumar Dash, and Durga Prasad Moharana

Abstract Due to their significant nutrient, vitamin, and mineral content, leafy vegetables are an essential part of the human diet. In addition to other vegetable crops, these are also being impacted by the effects of climate change, such as global warming, changes to the seasonal and monsoon patterns, and biotic and abiotic factors. Crop failures, low yields, declining quality, and an increase in pest and disease issues are frequent in climatically changing regions, making it unprofitable to grow leafy and salad vegetables. They will be significantly impacted because many physiological processes and enzymatic activities depend on temperature. The two main effects of temperature rise that make vegetable cultivation more difficult are drought and salinity. Crop yields may increase as a result of increased CO_2 fertilization, but this effect only lasts to a certain point. These effects of climate change also have an impact on the occurrence of pests and diseases, host-pathogen interactions, insect distribution and ecology, timing of appearance, migration to new locations, and their capacity for overwintering, which is a significant hindrance to the cultivation of leafy vegetables. Therefore, appropriate preventive measures must be taken as soon as possible to lessen the aforementioned challenges.

Keywords Global warming · Leafy vegetables · Amaranthus · Spinach · Salad vegetable · Lettuce

---

M. Pathak (✉) · S. K. Dash
Department of Vegetable Science, Odisha University of Agriculture and Technology, Bhubaneswar, Odisha, India

S. Barik
Department of Agriculture and Allied Sciences, C.V. Raman Global University, Bhubaneswar, Odisha, India

D. P. Moharana
Department of Horticulture, Institute of Agricultural Sciences, Banaras Hindu University, Varanasi, Uttar Pradesh, India

S. S. Solankey, M. Kumari (eds.), *Advances in Research on Vegetable Production Under a Changing Climate Vol. 2*, Advances in Olericulture,
https://doi.org/10.1007/978-3-031-20840-9_5

5.1 Introduction

Due to the alarming rate of increase in industrialization and deforestation, catastrophic changes have occurred in the global climatic conditions (Rakshit et al. 2009). The changes in climate refer to fluctuations in temperature, increases in soil salinity, water logging, increased CO_2 content, and UV radiation. High temperature is the result of increased greenhouse gases such as carbon dioxide and methane. This rise in temperature leads to extreme weather conditions including droughts, heat waves, floods, or storms, changes in ocean currents, and hastens the rate of ozone depletion (Minaxi et al. 2011; Usman and Balsalobre-Lorente 2022; Murshed et al. 2022; Kumar 2012). An increase in temperature also leads to the melting of glaciers, ultimately leading torising sea levels causing flooding and an increased level of salinity in small Pacific states and low-lying countries. Agriculture in India mostly depends upon the climatic condition of a region. The Indira Gandhi Institute of Development Research has found that global warming may lead to a decline in India's GDP by upto 9% (Priyadarshini and Abhilash 2019). Changes in climate raise the risk of particular food, water, and vector-borne diseases; a coronavirus pandemic is a visible example (Abbass et al. 2022). Increased rates of malnutrition, poor health, hunger, starvation, and food and water insecurity are all indirectly correlated with climate change. Thus, a solution is one suggestion to resolve global issues, and be taken into account that gives equal weight to human nutrition, health, (Sarker et al. 2022).

Leafy vegetables embedded with mineral matter, phytochemicals, and provitamin must be in a substantial proportion for a balanced diet (Adenipekun and Oyetunji 2010). Dieticians recommend daily consumption of at least 116 g of leafy vegetables for a balanced diet. Leafy vegetables are also good sources of fibre, which helps in the functioning of the digestive system. Consuming plenty of leafy vegetables assists in protection against bowl cancer which is one of the most common cancers. There is a wide variation in the consumption of leafy vegetables in different parts of the world. Due to wide variation in the climate of the country, various types of wild herbaceous/perennials are consumed by local people. In this chapter, relationships between the production of leafy vegetables and environmental factors namely temperature, light, soil, and water have been discussed.

5.2 Importance of Leafy Vegetables and Salad Crops

Leafy vegetables serve as essential sources of protective foods (Nnamani et al. 2009). The bioactive compounds present in the leafy vegetables possess numerous antimicrobial and antioxidant properties (Kim et al. 2013) and are recommended for controlling and managing age-related ailments and oxidative strains (Gacch et al. 2010). Leafy vegetables are rich sources of minerals (Fe, P, and Ca), vitamin C, Vitamin E, carotenoids, and flavonoids (Fasuyi 2006). Owing to the high accumulation of photosynthates, a significant amount of Potassium can be availed in leafy vegetables as Potassium is directly involved in the photosynthesis process. Besides, leafy salad vegetables synthesize a number of secondary metabolites, which act as antimicrobial agents. They also possess antidiabetic (Kesari et al. 2005), antihistaminic (Yamamura et al. 1998), anticarcinogenic (Rajeshkumar et al. 2002), and hypolipidemic (Khanna et al. 2002) properties.

Besides the major constituents of food (carbohydrate, protein, and fat) and micronutrients (minerals, vitamins, and trace elements), many health-promoting substances such as flavonoids, carotenoids, and other polyphenols, allylic sulfides, monoterpenes dietary fibers, phytosterols, and phenolic acids (Kris-Etherton et al. 2002) are also present in the leafy vegetables. Gupta et al. (2005) found the leafy vegetables to possess some antinutritional compounds such as saponins having the potential to reduce diseases like high blood pressure, heart disease, stroke, and other cardio vascular diseases in human beings (Williamson et al. 1997). The fibre present in the leafy vegetables assists to reduce the intake of starchy food, constipation prevention, and reducing the incidence of metabolic diseases such as hypercholesterolemia and *Diabetes mellitus*. Leafy Vegetables as soup are reported to enhance fertility in females (Mensah et al. 2008). Also, growing these alternative vegetables will improve both human health and the income of farmers in poor nations. (Sarker et al. 2022).

5.3 Factors Responsible for Climate Change on Leafy and Salad Vegetable Production

5.3.1 Temperature

Temperature variation within seasons is a sign of global warming as a result of climate change (Solankey et al. 2021). Among climatic factors influencing vegetable production, the temperature is the most important component. Temperature influences the yield, quality, and shelf life of produce, seed production, bud and seed dormancy, viability and longevity of seed and the occurrence of insects-pests and diseases.

5.3.1.1 Effects of Temperature on Seed Germination

Optimum growth of most of vegetables occurs between 10 °C and 30 °C, but when the temperature goes above 30 °C or falls below 10 °C, a decrease in plant vigor and growth is observed, due to the discrepancy in the rate of metabolic pathways (Downs and Hellermers 1975). Seed germination in leafy vegetables like amaranthus is significantly affected by temperature. The optimum temperature for germination in amaranthus is observed to be between 15 °C to 40 °C, while the optimum temperature is 25 °C. However, when the temperature falls below 10 °C, no seed germination is observed (Tiryaki 2009). Hence, in cold areas, where the soil temperature is the major hindrance, increasing the media temperature and the provision of dark conditions during sowing has been successful in enhancing the germination percentage in amaranthus (Aufhammer et al. 1998; Loonat et al. 2003). Similarly, in crops like lettuce exhibiting thermo-dormancy, high temperature (more than 30 °C) due to climate change has been a major challenge (Berrie 1966; Reynolds and Thompson 1971). Besides cool season leafy vegetables, tropical leafy vegetable such as drumstick (*Moringa oleifera*) is vastly affected by high temperature. 20/30 °C temperature regime (TR) leads to a significant enhancement in germination rate as well as uniformity (Muhl et al. 2011). However, as the temperature increases, a gradual decline can be observed in germination in drumsticks.

Germination and initial establishment of seedlings of Spinach (*Spinacia oleracea* L.) are highly susceptible to chilling and freezing injury, heat, and moisture stress (Ashraf and Foolad 2005). For optimal germination success, pericarp removal through mechanical means is recommended, however, when the temperature exceeds 25 °C, a significant reduction in germination percentage was observed (Atherton and Farooque 1983). The germination of seeds of spinach is found to be restricted when the temperature 35 °C (Leskovar et al. 1999). In a germination study, carried out by Katzman et al. (2001) on spinach, the results showed that the maximum germination in spinach can be achieved at 18 °C.

5.3.1.2 Effect of Temperature on Growth and Development

The optimum temperature is that at which the photosynthetic rate is highest with normal respiration rate or which the net assimilation rate (NAR) and yield realization are highest. During the daytime, the plants synthesize carbohydrates and simultaneously utilize in respiration. At high day temperatures, rate of respiration will be very high resulting in low net photosynthesis. Net photosynthesis refers to the amount of carbohydrates synthesized by the plant minus the amount of carbohydrates utilized in respiration in a specific period. Net photosynthesis, thus, determines the amount of carbohydrates available for growth of plant during the following night.

Warm season leafy vegetables like amaranthus requires a temperature in the range of 25–30 °C for the maximum accumulation of biomass (Khandaker et al. 2010). Similarly, in intensive care requiring crops like lettuce both ambient temperature as well as root zone temperature play key role in obtaining maximum leaf yield. An ambient temperature of 19–24 °C leads to the maximum plant growth (Lafta et al. 2017; Chen et al. 2021). However, low root zone temperature leads to the reduction of leaf area, fresh weight, stem size, and water content of lettuce (Sakamoto and Suzuki 2015). Yield and quality of lettuce are drastically affected by higher temperature due to abnormal acceleration in the elongation of lettuce stems (Zhao et al. 2003; Rader and Karlsson 2006). High temperature orchestrated stem elongation leads to the formation of loose heads, ultimately becoming unmarketable (Jie and Kong 1998).

5.3.1.3 On Nutritional Properties of Leaves

Leafy vegetables are mostly consumed owing to their nutritional properties, which are significantly affected by fluctuating temperatures. The major pigment with high nutritional value in amaranthus is anthocyanin, especially betacyanin in the case of red amaranthus. The mean optimum temperature for the highest accumulation of betacyanin content has been recorded to be 28–29 °C (Khandaker et al. 2010). In the presence of high temperature, red coloration with enhanced color index can be achieved in red amaranth. Beta-carotene and lutein are among the most important pigments in vegetable crops broadly available in kale and spinach. Research shows the maximum accumulation of beta-carotene and lutein occurs at 30 °C and 10 °C for kale and spinach, respectively. Lutein and beta carotene concentration are positively associated with an increase in temperature for kale and negatively associated with increasing temperature in the case of spinach (Lefsrud et al. 2005). As evidenced by numerous researchers, spinach is one of the best sources of ascorbic acid, oxalic acid, and sugar. These compounds are also significantly affected by the temperature regime of the growing environment. The low temperature of 5–10 °C during growing conditions is found to be the optimum temperature for ascorbic acid, oxalic acid, and sugar accumulation in spinach leaves (Tamura 2004; Proietti et al. 2009). Sugar and ascorbic acid content are found to increase rapidly when the minimum air temperature decreases below 5 °C (Tamura 2004). In coriander, besides ambient temperature, root zone temperature plays a key role in enhancing the secondary metabolite content. The secondary metabolites such as ascorbic acid, carotenoids, phenolic compounds, and chlorogenic acid content of the coriander leaves increase substantially the root zone temperature by 15–35 °C (Nguyen et al. 2020). In lettuce and Japanese parsley (*Oenanthe stolonifera*), the accumulation of the health-promoting pigments such as anthocyanin and chlorophyll b in leaves is the highest at the low temperature of 20 °C (Hasegawa et al. 2001; Gazula et al. 2005).

5.3.1.4 Effect of Temperature on Quality of Produce

In most of the leafy and salad vegetables, the quality of produce is usually influenced by temperature. It affects the quality of produces not only by increasing respiration and transpiration but also by damaging the tissues or altering morphology following exposure to excessive heat or chilling by bringing changes in relative amounts of sugar and starch of edible parts.

Temperature can be regarded as the only factor governing the post-harvest quality of leafy vegetables. Generally, a combination of high humidity and low temperature significantly enhances the self-life of most of leafy vegetables (Cantwell and Kasmire 2002), due to delayed chlorophyll degradation (Pogson and Morris 1997) owing to the high transpiration rate, as leafy vegetables have more exposed surface area. Hence, accelerated loss of color and senescence is a major issue if plants get exposed to high temperatures.

5.3.1.5 On Disease Development

High temperature and High humidity both favor the development of foliage diseases and pests' population. Pathogens like *Rhizoctonia solani* and *Sclerotini asclerotiorum* in lettuce have more devastating effects on plants in the presence of high temperatures (Grosch and Kofoet 2003; Clarkson et al. 2014). Similarly, one of the major soil-borne pathogens *Fusarium oxysporumf.sp. lactucae* infestation in lettuce increases rapidly with an increase in temperature (Ferrocino et al. 2013). Similarly, in spinach, rust and anthracnose are the two major diseases diminishing the market value significantly. A high temperature of 15–18 °C leads to a significant increase in rust infection, while more than 20 °C results in severe anthracnose infestation in spinach (Sullivan et al. 2002; Uysal and Kurt 2017). A similar negative impact of high temperature can also be observed in parsley for septoria blight diseases. When ambient temperature increases from 20 °C to 30 °C, an increase in the mean lesion number is observed (Kurt and Tok 2005).

5.3.1.6 Effect of Temperature on Physiological Disorders

Temperature is the most important component actively influencing the growth and development of leafy vegetables and salad crops up to varying extents (Table 5.1). For example, high temperature for a short period of time leads to accelerated growth, however prolonged exposure causes tip burn and bolting in lettuce (Cox et al. 1976;

Table 5.1 Classification of leafy vegetables and salad crops based on reaction to high temperature

Degree of sensitivity	Vegetables	Effect due to prolong period
Sensitive	Parsnip	Inferior root quality
	Lettuce	Bolting, loose heads, bitterness, tip burn, small & light heads, thermo-dormancy od seeds, Ca deficiency
	Celery	Less quality stems
Moderate sensitive	Spinach	Reduced yields
Tolerant	Shallot	Tolerant
	Basella	Tolerant
	Parsley	Tolerant
	Amaranthus	Tolerant
Highly tolerant	Malabar Spinach	Highly tolerant
	New Zealand Spinach	Highly tolerant

Fukuda et al. 2009). High temperature during growing conditions even for 2 weeks can lead to discoloration of ribs in lettuce (Jenni 2005).

The bitterness of the leaves is observed in lettuce upon prolonged exposure to high temperatures. Spinach beet can tolerate frost better than other vegetables. It can also tolerate warm weather but high temperature leads to premature bolting without giving economic yield. Palak can tolerate high temperatures, unlike spinach which is purely a cool-season crop. It fits well in different crop rotations due to its adaptation to high temperatures condition, short cropping duration, and appreciable biomass yield per unit area. Amaranthus although being considered a warm season crop can also be grown successfully in temperate climate during summer. *Amaranthus* species that grow under varying climatic conditions differ in their day length requirements and respond differently to changes in photo and thermoperiodism.

However, in temperature fluctuation area, some varieties of the mentioned leafy vegetable and salad crops can be grown successfully (Table 5.2).

5.3.2 Light

The biochemical processes that take place in plants in the presence of light produces carbohydrates and chemical energy. Plants require a greater amount of light, which furnishes energy for the combination of carbon dioxide and water to form first manufactured compound i.e., glucose ($C_6H_{12}O_6$) is necessary for survival, growth, and development of plant. Therefore, greater the amount of light, the greater the photosynthesis and accumulation of carbohydrates in plant body. Several components of light requirement such as light intensity, light source, color of light plays active role in plant growth of leafy vegetables. Light intensity in a range of 200–600 μmol m^{-2} s^{-1} is found to be the most suitable for high light use efficiency and yield in lettuce (Fu et al. 2012). High light intensity leads to low light use efficiency and yield. Similar results such as reduced vigor, leaf area as well as shoot: root ratio and increased amount of oxalate and nitrate are observed in spinach when plants are exposed to relatively low light intensities (Proietti et al. 2004). Besides, yield and yield

Table 5.2 Recommended cultivars to grow in high and low temperature areas

Vegetable	Recommended cultivars
Lettuce	Skyphos, Forlina, Green Butter, Red Butter, Salanova (Butterhead type), Starfighter (Leaf type), Arroyo and Dove (Romain type), Cultivars 9547 and 9542, Salma (a new summer-autumn lettuce cultivar), Elisa (lettuce cultivar for summer cultivation), Florida Butter crisp, Glacier and Misty Dat, Great Lakes (Slow bolting type)
Palak	**Ooty-1 (Drought and frost), Punjab Green, Pusa Harit, New Zealand spinach**
Amaranthus	Chhoti Chaulai (early summer and rainy), Badi Chaulai (summer season), Pusa Kiran (rainy season), Pusa Lal Chaulai: (summer and rainy seasons)
Celery	Sac yuquin

attributing traits, nutritional qualities of leafy vegetables are also affected by light intensity. Optimum light intensity provisions lead to of a greater amount of simple sugars and carotenoid pigments, while in the absence of light, ascorbic acid content reduces at a significant rate in leafy vegetables like chinese kale (Noichinda et al. 2007).

Besides light intensity, supplemental light sources have a significant effect on growth and nutritional properties of leafy vegetables. Supplemental UV-A and blue light results in significant increase in the concentration of anthocyanin pigments. Supplemental red light results in enhancement in phenolics content, while supplemental far red light enhances the fresh weight, dry weight, stem length, leaf length, and leaf width in lettuce (Li et al. 2009). Similarly, white and/or red fluorescent lamps with a photosynthetic photon flux of 300 $\mu mol\ m^{-2}\ s^{-1}$ leads to the accumulation of biomass, β-carotene, and lutein content in spinach. Another source of light such as light-emitting diodes (LEDs) can be used to decrease the bitter component 'gluconapin' and increase in 'glucoraphanin' content in roots, total phenolic, and anthocyanins, as well as the strongest antioxidant capacity of leafy vegetable crops like chinese kale (Qian et al. 2016). Similarly, RGB LED light exhibits the maximum amount of chlorophylls, β-carotene, lutein, neoxanthin and violaxanthin, while plants grown in incandescent light results in the maximum protein content and total thiocyanates content in kale (Fiutak and Michalczyk 2020).

5.3.3 *Effect of CO_2*

CO_2 plays a major role in the global warming which is predicted to keep rising at an alarming rate to dangerous levels in the coming decades. Several experiments have been carried out to figure out the possible potential of elevated CO_2 to affect crop growth and production potential (Zhang et al. 2015). The elevated CO_2 was found to enhance yield and accumulation of biomass due to the increased rate of photosynthesis (Ainsworth et al. 2002; Prior et al. 2011). The elevated CO_2 also possess the potential to reduce the stomatal conductance of plant which ultimately leads to reduction of transpiration, enhancement in water use efficiency which help plant to combat drought stress (Radoglou and Jarvis 1992). Nonetheless, increase in CO_2 content leads to the reduction in nutritional quality and growth characteristics of lettuce and spinach (Giri et al. 2016). CO_2 concentration has also been found to inhibit disease infestation in leafy vegetable crops. Fusarium wilt of lettuce was found to be inhibited at 800 ppm CO_2 concentration as put forth by Ferrocino et al. (2013).

5.3.4 Rainfall

Rainfall is one of the most important factors, especially when vegetables are grown under dry land conditions, however, high rainfall may cause flood damage, partial drowning on certain soil types, and will often favour disease development. The pattern of rainfall in the region should be studied before taking any decision concerning the types of crop to be cultivated. Local variations in rainfall can influence the methods of cultivation to be practiced and the types of crop, which can be grown successfully. Heavy rainfall helps in leaching down of salts and reduces the salinity level in the soil but seed germination is deteriorated with an increase in the number of rainy days.

5.3.5 Humidity

Humidity, or air moisture content, plays a key role since it influences varieties phases of vegetable production, like seed germination, vegetative growth, flowering, fruit set, quality of vegetables, etc. Besides seed production and seed viability in storage, humidity plays most important role in the occurrence of insect-pests and disease. Certain diseases (*e.g.*, powdery mildew) are associated with dry weather but humid conditions are known to favour diseases, like leaf rust, downy mildew and fungal blights affecting foliage. Humidity and temperature in combinations are well known to influence the growth and production of commercial vegetable crops. High humidity leads to the enhanced disease incidence, such as anthracnose in spinach and *Septoria petroselini* in parsley (Kurt and Tok 2005; Uysal and Kurt 2017).

Excessively high humidity favors the growth of microbes during storage and makes the fruits and vegetables susceptible to disease infection. Besides low temperature, relative humidity is equally important factor in cold storage for maintaining firmness of fresh produce as water loss from vegetables having high water content is rapid at low storage humidity. High humidity prevents wilting of leafy vegetables particularly after harvest. The moisture content positively influences the total plant, stem, and leaf and root weight (Ejieji and Adeniran 2010).

5.3.6 Frost

Most vegetables are injured at or slightly below freezing temperature but tropical and subtropical vegetables may be damaged or permanently killed by temperature below 10 °C but above freezing point. The extent of the damage caused by frost fluctuate species and even with varieties of the same species or to some degree with the stage of plant development.

Frost has a devastating impact on lettuce which is easily identifiable, the extent of which vary according to the duration of exposure to frost and growth stage of lettuce. Frost damage manages itself by the separation of the outer leaf cuticle from the underlying cell tissue, which leads to bronzing of leaves. This occurs due to the damage to the epidermal cells. Severe frost conditions also lead to necrotic spotting as well as interveinal lesions. Sometimes, young leaves tips are dried up and upon the continuation of the leaf growth, curling of the leaves occurs. Shorter exposure of frost causes roughening and thickening of the leaf tissue.

In case of celery, when temperature falls below – 0.5 °C, freezing injury can generally be seen. Water soaked spots appear on thawing and wilted leaves of lettuce. Freezing injury can appear in the field condition in Romain and Crisphead lettuce, resulting in epidermis separation from the leaves. This ultimately leads to attack of bacterial pathogens during storage condition, especially if lettuce is stored at less than 2 °C.

In spinach, freezing injury appears when exposed to −0.3 °C ambient temperature, which leads to water soaking spots accompanied by bacterial soft rot. Palak crop can withstand frost and tolerate warm weather but high temperature leads to early bolting without giving sufficient cuttings. During hot weather leaves pass edible stage quickly.

5.3.7 Hail

Hail storms can beat down a standing crop nearing harvest; do considerable damage to the harvested sheaves left in the field for drying and cause necrotic spots on fruits like tomato, which happened to be hit by hail. Hails pose a serious threat through their mechanical damages to the branches and development of corky tissues in fruits in and around the hail hit spots. Hail damage to leafy tissues can reduce photosynthetic surface and depress yield.

5.3.8 Soil Factors

Vegetables crops generally grow well in fertile soils with regular supply of essential nutrients and moisture. Light textured soils are suitable for lettuce. But, in general loam soils are suitable for almost all vegetable crops. Cool season vegetables are shallow rooted crops, which have short growing season so these crops should be grown in soils warm up quickly but such soils are often low in nutrients and poor in retention of moisture, needing heavy fertilizers application and frequent irrigation. These soils are most suitable for the cultivation of crops in which underground part is of economic importance. However, the crops for which higher yield is more important should be grown in heavy soils, but in such soils, the crops usually take longer time to mature. Heavy soils often considerable plant nutrient reserves and also retain moisture for longer period, thus only small quantity of fertilizers is required, and irrigation is required relatively at long intervals but too heavy soils are not suitable for the cultivation of vegetable crops as they poor aeration and poor nutrient liberation capacity resulting in poor plant growth. Heavy soils produce rough and deformed roots with number of small fibers. Tubers produced in light soils generally have a more desirable shape and brighten skin colour than those are grown in heavy soils.

5.3.8.1 Soil Salinity

The crop growers have to grow crops in many areas around the globe by irrigating with water containing high salt content due to the scarcity of good quality water sources. The physiology, biochemistry as well as yield of leafy vegetable crops get affected by the soil salinity and salinity possess a major threat during production of these vegetables (Munns and Tester 2008). It's been estimated that soil salinity directly negatively affects 20% of irrigated agricultural land (Chinnusamy et al. 2005) and it negatively impacts around 1000 million ha of land, which is 7% of all land area (Szabolcs 1994).

Al-maskri et al. (2010) reported that soil salinity significantly influenced that number of leaves, plant fresh weight, shoot fresh and dry weight, shoot dry matter percentage, root fresh and dry weight, root dry weight percentage, leaf area and leaf area index of lettuce. Unlukara et al. (2008) found that the salinity led to continuous decrease in lettuce yield which went upto 60% yield loss. Omami (2007) reported *A. hypochondriacus* and *A. cruentus* to exhibit enhanced tolerance to salinity, as plants were able to survive in 200 mM NaCl treatment.

5.3.8.2 Heavy Metals

A number of studies reported the high concentration of heavy metals caused in inhibition of vegetative growth and crop productivity. Prolonged exposure of high concentration of heavy metals may lead to decrease in blood pressure and damage to numerous internal organs like liver, kidney, and lungs. It can also negatively influence a number of physical and neurological functions resembling sclerosis. The major heavy metals affecting the commercial cultivation of leafy vegetables are Cadmium, Copper, and lead. At higher concentration doses, cadmium (Cd) and copper (Cu) results in highly toxic impact and the plant growth attributing factors are significantly reduced in case of spinach and amaranthus (Chetan and Ami 2015). Accumulation of lead leads to the reduction in growth and nutrient uptake of the mineral ions such as Sodium, Potassium, Calcium, Magnesium, Iron, Copper and Zinc in spinach (Lamhamdi et al. 2013)).

5.4 Mitigation of Negative Impacts of Climate Change

Negative impact of the global warming can be mitigated by the adoption of sustainable strategies to enhance the production and productivity of the leafy vegetables. There exist numerous ways to continue growing crops even in the changing global climate scenario. Some of these measures are given below:

5.4.1 Strengthening of Crop Management Systems

The production of leafy vegetables cultivated in low land topography exposed to hot and humid climates can be enhanced with the help of numerous management methods. A number of biotic and abiotic stresses combating strategies have been developed by World Vegetable Centre, Taiwan in flood and drought prone as well as saline soil infested areas to produce high qualitative leafy vegetables.

5.4.2 Efficient Irrigation Management

Proper water management significantly positively affect the proper growth and development of leafy vegetable crops due to their high water content of their foliage. The critical stage of irrigation needs to be identified for individual leafy vegetables. The critical period varies according to the prevailing weather condition of the area, crop growth stage, water retention capacity of soil as well as soil texture.

5.4.3 Water Conserving Agronomical Practices

In order to protect the crop especially leafy vegetables susceptible to imbalanced water supply, a number of mitigation strategies are available. Those include the use of raised beds, utilization of rain shelter to protect the crop from heavy rains, and mulching which restricts water loss from soil. These strategies are reported to successfully save the crop from heavy rain, heat stress, excess moisture stress, etc. Growing of leafy vegetables inside protected structures in controlled condition to grow leafy vegetables all around the year.

5.4.4 Promotion of Climate-Resilient Leafy Vegetables

The existence of diverse germplasm is the preliminary requisite for the breeding programme aimed at mitigating the abiotic stresses caused by global warming. Use of abiotic stress resistant cultivars, use of efficient water management strategies and organic method of cultivation need to be more practiced by the crop growers to sustainably grow a crop with high yield and nutritional value.

5.4.5 High Temperatures Tolerance

In order to combine heat tolerance with high productivity, expansion of genome base by crossing the heat tolerant cultivars with high yielding accession followed by selection. A number of cultivars of a number of leafy vegetables have been developed to cultivate in high temperature areas. Various crops are altered and modified for ability to cope with heat tolerance. Minetto, X-01, Green Star, Magenta, Bronze Arrow Lose Leaf, Red Salad Bowl Oakleaf and Oakleaf Looseleaf Lettuce are heat tolerant lettuce cultivars, while Panama, Red Sun, Revolution, Bergamo cultivars are resistant to bolting; hence, recommended for high temperature regions. Similarly, the Japanese hybrids "Akarenso," "Alrite," "Alkame," and "Samba," spinach cultivars and 'Conquistador', 'Celery Paris' celery can be selected to grow in high temperature areas. Badi Chaulai, Pusa Lal Chauli *(A.tricolor)* of amaranthus also possess considerable tolerance to heat stress. Pusa Jyoti of Palak has wide adaptability to changing climate and recommended for all the year round cultivation.

5.4.6 Salinity Tolerance

A number of programs are underway to develop salt tolerant high yielding varieties. Although not much success has been achieved through conventional breeding programs, owing to physiologic and genetic complexity of the character (Flowers 2004). For the development of salt tolerant varieties, collection and evaluation of diverse germplasm, suitable method of screening, and successful transfer of salinity tolerance from the donor to the commercial elite genotypes are the main targets. Romanian lettuce cultivars and K_1, Quinoa amaranthus are suitable to cultivate in saline soil condition (Table 5.3).

Table 5.3 Classification of leafy vegetables and salad crops based on Salinity Tolerance

Less tolerant	Moderately tolerant	Tolerant
Celery	Amaranth	Palak
	Spinach	Lettuce

5.5 Conclusion

The key to successful and sustainable vegetable crop production is the process of developing of an understanding of environment and quantifying the agro-climatic requirements of different vegetable crops. Impact of anticipated climate variability on regional scale is the main challenge for a successful vegetable production. Leafy vegetable crops like lettuce, celery, parsley, spinach, etc., can be grown successfully in controlled environment, *viz.*, permanent protected structures or greenhouses where plant growth is not affected by outside environmental conditions. Components of climate like radiation, rainfall, temperature and wind affect microclimate modification within crop fields which are beyond human control. In order to bring forth the full potential of the highly scattered vegetable producing areas to sustainable vegetable producing regions, a great deal of understanding of agro-climatic resources of the region and microclimatic modifications and their interactions with vegetable crops is needed. Thus, by exploring the meteorological conditions through field microclimatic modifications by fulfilling the requirements of crops, the sustainable vegetable production may be achieved.

References

Abbass K, Qasim MZ, Song H, Murshed M, Mahmood H, Younis I (2022) A review of the global climate change impacts, adaptation, and sustainable mitigation measures. Environ Sci Pollut Res 29:42539–42559

Adenipekun CO, Oyetunji OJ (2010) Nutritional values of some tropical vegetables. J Appl Biosci 35:2294–2300

Ainsworth EA, Davey PA, Bernacchi CJ, Dermody OC, Heaton EA, Moore DJ et al (2002) A meta-analysis of elevated [CO2] effects on soybean (Glycine max) physiology, growth and yield. Glob Chang Biol 8(8):695–709

Al-Maskri AHMED, Al-Kharusi L, Al-Miqbali H, Khan MM (2010) Effects of salinity stress on growth of lettuce (*Lactuca sativa*) under closed-recycle nutrient film technique. Int J Agric Biol 12(3):377–380

Ashraf M, Foolad MR (2005) Pre-sowing seed treatment—a shotgun approach to improve germination, plant growth, and crop yield under saline and non-saline conditions. Adv Agron 88:223–271

Atherton JG, Farooque AM (1983) High temperature and germination in spinach. II Effects of osmotic priming. Sci Hortic 19(3–4):221–227

Aufhammer W, Czuczorova D, Kaul HP, Kruse M (1998) Germination of grain amaranth (*Amaranthus hypochondriacus* × *a.hybridus*): effects of seed quality, temperature, light, and pesticides. Eur J Agron 8(1–2):127–135

Berrie AM (1966) The effect of temperature and light on the germination of lettuce seeds. Physiol Plant 19(2):429–436

Cantwell MI, Kasmire RF (2002) Handling systems: flower, leafy, and stem vegetables. In: Postharvest technology of horticultural crops. Agriculture and Natural Resources Publication 3311. Oakland CA, pp 423–432

Chen Z, Shah Jahan M, Mao P, Wang M, Liu X, Guo S (2021) Functional growth, photosynthesis and nutritional property analyses of lettuce grown under different temperature and light intensity. J Hortic Sci Biotechnol 96(1):53–61

Chetan A, Ami P (2015) Effects of heavy metals (cu and cd) on growth of leafy vegetables-*Spinacia oleracea* and *Amaranthus caudatus*. Int Res J Environ Sci 4(6):63

Chinnusamy V, Jagendorf A, Zhu JK (2005) Understanding and improving salt tolerance in plants. Crop Sci 45(2):437–448

Clarkson JP, Fawcett L, Anthony SG, Young C (2014) A model for *Sclerotinia sclerotiorum* infection and disease development in lettuce, based on the effects of temperature, relative humidity and ascospore density. PLoS One 9(4):e94049

Cox EF, McKee JMT, Dearman AS (1976) The effect of growth rate on tip burn occurrence in lettuce. J Hort Sci 51(3):297–309

Downs RJ, Hellermers H (1975) Environment and the experimental control of plant growth. Academic Press Inc., London

Ejieji CJ, Adeniran KA (2010) Effects of water and fertilizer stress on the yield, fresh and dry matter production of grain Amaranth (*'Amaranthus cruentus'*). Australian J Agric Eng 1(1):18–23

Fasuyi AO (2006) Nutritional potentials of some tropical vegetable leaf meals: chemical characterization and functional properties. Afr J Biotechnol 5(1):49–53

Ferrocino I, Chitarra W, Pugliese M, Gilardi G, Gullino ML, Garibaldi A (2013) Effect of elevated atmospheric CO2 and temperature on disease severity of *fusarium oxysporum* f. sp. *lactucae* on lettuce plants. Appl Soil Ecol 72:1–6

Fiutak G, Michalczyk M (2020) Effect of artificial light source on pigments, thiocyanates and ascorbic acid content in kale sprouts (*Brassica oleracea* L. var. *sabellica* L.). Food Chem 330:127189

Flowers TJ (2004) Improving crop salt tolerance. J Exp Bot 55(396):307–319

Fu W, Li P, Wu Y (2012) Effects of different light intensities on chlorophyll fluorescence characteristics and yield in lettuce. Sci Hortic 135:45–51

Fukuda M, Matsuo S, Kikuchi K, Mitsuhashi W, Toyomasu T, Honda I (2009) The endogenous level of GA1 is upregulated by high temperature during stem elongation in lettuce through LsGA3ox1 expression. J Plant Physiol 166(18):2077–2084

Gacch RN, Kabaliye VN, Dhole NA, Jadhav AD (2010) Antioxidant potential of selected vegetables commonly used in diet in Asian subcontinent. Indian J Nat Prod Resour 1(3):306–313

Gazula A, Kleinhenz MD, Streeter JG, Miller AR (2005) Temperature and cultivar effects on anthocyanin and chlorophyll b concentrations in three related Lollo Rosso lettuce cultivars. HortScience 40(6):1731–1733

Giri A, Armstrong B, Rajashekar CB (2016) Elevated carbon dioxide level suppresses nutritional quality of lettuce and spinach. Am J Plant Sci 7(01):246–258

Grosch R, Kofoet A (2003) Influence of temperature, pH and inoculum density on bottom rot on lettuce caused by Rhizoctonia solani/Einfluss von Temperatur, pH und Inokulumdichte auf die Salatfäule verursacht durch Rhizoctonia solani. Zeitschrift für Pflanzenkrankheiten und Pflanzenschutz/J Plant Dis Protect 110(4):366–378

Gupta S, Lakshmi AJ, Manjunath MN, Prakash J (2005) Analysis of nutrient and antinutrient content of underutilized green leafy vegetables. LWT-Food Sci Technol 38(4):339–345

Hasegawa H, Fukasawa-Akada T, Okuno T, Niizeki M, Suzuki M (2001) Anthocyanin accumulation and related gene expression in Japanese parsley (Oenanthe stolonifera, DC.) induced by low temperature. J Plant Physio 158(1):71–78

Jenni S (2005) Rib discoloration: a physiological disorder induced by heat stress in crisp head lettuce. HortScience 40(7):2031–2035

Jie H, Kong LS (1998) Growth and photosynthetic characteristics of lettuce (*Lactuca sativa* L.) under fluctuating hot ambient temperatures with the manipulation of cool root-zone temperature. J Plant Physiol 152(4–5):387–391

Katzman LS, Taylor AG, Langhans RW (2001) Seed enhancements to improve spinach germination. HortScience 36(5):979–981

Kesari AN, Gupta RK, Watal G (2005) Hypoglycemic effects of *Murraya koenigii* on normal and alloxan-diabetic rabbits. J Ethnopharmacol 51:2603–2607

Khandaker L, Akond AM, Ali MB, Oba S (2010) Biomass yield and accumulations of bioactive compounds in red amaranth (Amaranthus tricolor L.) grown under different colored shade polyethylene in spring season. Sci Hortic 123(3):289–294

Khanna AK, Rizvi F, Chander R (2002) Lipid lowering activity of *Phyllanthus niruri* in hyperlipemic rats. J Ethnopharmacol 82(1):19–22

Kim SJ, Cho AR, Han J (2013) Antioxidant and antimicrobial activities of leafy green vegetable extracts and their applications to meat product preservation. Food Control 29(1):112–120

Kris-Etherton PM, Hecker KD, Bonanome A, Coval SM, Binkoski AE, Hilpert KF, Etherton TD (2002) Bioactive compounds in foods: their role in the prevention of cardiovascular disease and cancer. Am J Med 113(9):71–88

Kumar CP (2012) Climate change and its impact on groundwater resources. Int J Eng Sci 1(5):43–60

Kurt S, Tok FM (2005) Influence of inoculum concentration, leaf age, temperature, and duration of leaf wetness on Septoria blight of parsley. Crop Prot 25(6):556–561

Lafta A, Turini T, Sandoya GV, Mou B (2017) Field evaluation of green and red leaf lettuce genotypes in the Imperial, San Joaquin, and Salinas valleys of California for heat tolerance and extension of the growing seasons. HortScience 52(1):40–48

Lamhamdi M, El Galiou O, Bakrim A, Nóvoa-Muñoz JC, Arias-Estévez M, Aarab A, Lafont R (2013) Effect of lead stress on mineral content and growth of wheat (*Triticum aestivum*) and spinach (*Spinacia oleracea*) seedlings. Saudi J Biol Sci 20(1):29–36

Lefsrud MG, Kopsell DA, Kopsell DE, Curran-Celentano J (2005) Air temperature affects biomass and carotenoid pigment accumulation in kale and spinach grown in a controlled environment. HortScience 40(7):2026–2030

Leskovar DI, Esensee V, Belefant-Miller H (1999) Pericarp, leachate, and carbohydrate involvement in thermoinhibition of germinating spinach seeds. J Am Soc Hortic Sci 124(3):301–306

Li J, Hikosaka S, Goto E (2009, November) Effects of light quality and photosynthetic photon flux on growth and carotenoid pigments in spinach (*Spinacia oleracea* L.). In: VI International symposium on light in horticulture 907, pp 105–110

Loonat TA, Van den Heever E, Hammes P (2003) Effect of temperature on the germination of grain Amaranth. South Afr J Plant Soil 20(3):152–164

Mensah JK, Okoli RI, Ohaju-Obodo JO, Eifediyi K (2008) Phytochemical, nutritional and medical properties of some leafy vegetables consumed by Edo people of Nigeria. Afr J Biotechnol 7(14)

Minaxi RP, Acharya KO, Nawale S (2011) Impact of climate change on food security. Int J Agric Environ Biotechnol 4(2):125–127

Muhl QE, Du Toit ES, Robbertse PJ (2011) *Moringa oleifera* (horseradish tree) leaf adaptation to temperature regimes. Int J Agric Biol 13(6)

Munns R, Tester M (2008) Mechanisms of salinity tolerance. Annu Rev Plant Biol 59:651–681

Murshed M, Nurmakhanova M, Al-Tal R, Mahmood H, Elheddad M, Ahmed R (2022) Can intra-regional trade, renewable energy use, foreign direct investments, and economic growth reduce ecological footprints in South Asia? Energy Sources B: Econ Plan Policy:1–26. https://doi.org/10.1080/15567249.2022.2038730

Nguyen DT, Lu N, Kagawa N, Kitayama M, Takagaki M (2020) Short-term root-zone temperature treatment enhanced the accumulation of secondary metabolites of hydroponic coriander (*Coriandrum sativum* L.) grown in a plant factory. Agronomy 10(3):413

Nnamani CV, Oselebe HO, Agbatutu A (2009) Assessment of nutritional values of three underutilized indigenous leafy vegetables of Ebonyi state, Nigeria. Afr J Biotechnol 8(10):2321–2324

Noichinda S, Bodhipadma K, Mahamontri C, Narongruk T, Ketsa S (2007) Light during storage prevents loss of ascorbic acid, and increases glucose and fructose levels in Chinese kale (Brassica oleracea var. alboglabra). Postharvest Biol Technol 44(3):312–315

Omami EN (2007) Responses of Amaranth to salinity stress, Doctoral dissertation, University of Pretoria

Pogson BJ, Morris SC (1997) Consequences of cool storage of broccoli on physiological and biochemical changes and subsequent senescence at 20 C. J Am Soc Hortic Sci 122(4):553–558

Prior SA, Runion GB, Marble SC, Rogers HH, Gilliam CH, Torbert HA (2011) A review of elevated atmospheric CO2 effects on plant growth and water relations: implications for horticulture. HortScience 46(2):158–162
Priyadarshini P, Abhilash PC (2019) Climate action-based policy Administration in India: developments and challenges. Clim Change Environ Sustain 7(1):102–107
Proietti S, Moscatello S, Colla G, Battistelli Y (2004) The effect of growing spinach (Spinaciaoleracea L.) at two light intensities on the amounts of oxalate, ascorbate and nitrate in their leaves. J Hortic Sci Biotechnol 79(4):606–609
Proietti S, Moscatello S, Famiani F, Battistelli A (2009) Increase of ascorbic acid content and nutritional quality in spinach leaves during physiological acclimation to low temperature. Plant Physiol Biochem 47(8):717–723
Qian H, Liu T, Deng M, Miao H, Cai C, Shen W, Wang Q (2016) Effects of light quality on main health-promoting compounds and antioxidant capacity of Chinese kale sprouts. Food Chem 196:1232–1238
Rader HB, Karlsson MG (2006) Northern field production of leaf and romaine lettuce using a high tunnel. HortTechnology 16(4):649–654
Radoglou KM, Jarvis PG (1992) The effects of CO2 enrichment and nutrient supply on growth morphology and anatomy of *Phaseolus vulgaris* L. seedlings. Ann Bot 70(3):245–256
Rajeshkumar NV, Joy KL, Kuttan G, Ramsewak RS, Nair MG, Kuttan R (2002) Antitumour and anticarcinogenic activity of *Phyllanthus amarus* extract. J Ethnopharmacol 81(1):17–22
Rakshit A, Sarkar NC, Pathak H, Maiti RK, Makar AK, Singh PL (2009) Agriculture: a potential source of greenhouse gases and their mitigation strategies. IOP Conference Series: Earth and Environmental Science 6(24):20–33
Reynolds T, Thompson PA (1971) Characterisation of the high temperature inhibition of germination of lettuce (*Lactuca sativa*). Physiol Plant 24:544–547
Sakamoto M, Suzuki T (2015) Effect of root-zone temperature on growth and quality of hydroponically grown red leaf lettuce (*Lactuca sativa* L. cv. Red wave). Am J Plant Sci 6(14):2350
Sarker U, Lin YB, Oba S, Yoshioka Y, Hoshikawa K (2022) Prospects and potentials of underutilized leafy amaranths as vegetable use for health-promotion. Plant Physiol Biochem 182:104–123
Solankey SS, Kumari M, Akhtar S, Singh HK, Ray PK (2021) Challenges and opportunities in vegetable production in changing climate: mitigation and adaptation strategies. In: Solankey SS et al (eds) Advances in research on vegetable production under a changing climate, vol 1. Springer, Cham., ISBN 978-3-030-63497-1, pp 13–59
Sullivan MJ, Damicone JP, Payton ME (2002) The effects of temperature and wetness period on the development of spinach white rust. Plant Dis 86(7):753–758
Szabolcs I (1994) Prospects of soil salinity for the 21st century. Agrokém Talajt 43(1–2):5–24
Tamura A (2004) Effect of air temperature on the content of sugar and vitamin C of spinach and komatsuna. Hortic Res (Japan) 3(2):187–190
Tiryaki I (2009) Osmotic priming increases seed germination of *Amaranthus caudatus* L. at low temperature. Agrochimica 53(3):177–182
Unlukara A, Kurunç A, Kesmez GD, Yurtseven E (2008) Growth and evapotranspiration of okra (*Abelmoschus esculentus* L.) as influenced by salinity of irrigation water. J Irrig Drain Eng 134(2):160–166
Usman M, Balsalobre-Lorente D (2022) Environmental concern in the era of industrialization: can fnancial development, renewable energy and natural resources alleviate some load? Energy Policy 162:112780
Uysal A, Kurt S (2017) Influence of inoculum density, temperature, wetness duration, and leaf age on infection and development of spinach anthracnose caused by the fungal pathogen *Colletotrichumspinaciae*. Eur J Plant Pathol 149(4):1041–1052
Williamson G, Dupont MS, Su W, Heaney RK, Musk S, Roger G, Rhodes MJ (1997) Induction of glutathione S transferase activity in hepG2 cells by extracts of fruits and vegetables. Food Clzenz 2:157–160

Yamamura S, Ozawa K, Ohtani K, Kasai R, Yamasaki K (1998) Antihistaminic flavones and aliphatic glycosides from *Mentha spicata*. Phytochemistry 48:131–136

Zhang S, Li X, Sun Z, Shao S, Hu L, Ye M, Shi K (2015) Antagonism between phytohormone signalling underlies the variation in disease susceptibility of tomato plants under elevated CO2. J Exp Bot 66(7):1951–1963

Zhao X, Carey EE, Hall M (2003) Influence of protected environment and fertilizer source on growth and quality of lettuce and leafy greens during summer production in Kansas. HortScience 38:749. (abstr.)

Chapter 6
Impact of Climate Change on Perennial Vegetables Production and Mitigation Strategies

Rajesh Kumar, Lomash Sharma, Jitendra Kumar Kushwah, and Bahadur Singh Bamaniya

Abstract Scientific fraternity is concerned about the global climate change because the variations in important climatic factors may have a significant impact on vegetable crop production, jeopardising global and local food security. Climate change may also have a significant impact on global horticultural processes and productivity. The changes in the growth patterns, flowering and fruiting capabilities of many perennial vegetable crops are inevitable. These vegetable crops are likely to face significant challenges due to altered seasonal factors, impacting dormancy, acclimation, and subsequent reproduction behaviour. In the result of climate change, new pests and pathogens may become popular and harmful in the locations where their activity was previously restricted. Water and nutrients become scarce in some areas, reducing prospects for cultivating perennial vegetable crops. Climate change modifies genotype x environment interaction for physiological and economic features in the crop plants. However, the abundance of germplasm with the vast range of species and genetic diversity offers a chance to develop climate resilient varieties, which will reduce the impact of climate change and ensure economic benefits to farmers. Modification of current horticultural operations and increased practice of greenhouse technology are some of the options to mitigate the effect of climate change. This chapter discuss the mitigation strategies of perennial vegetables for sustainable horticulture through good agronomic practices, breeding for heat and drought resilience, stimulating new and innovative ideas like grafting technology and genetic engineering to increase perennial vegetable production and quality.

R. Kumar (✉) · L. Sharma · B. S. Bamaniya
Department of Horticulture, School of Life Sciences, Sikkim University, Gangtok, Sikkim, India
e-mail: rkumar@cus.ac.in

J. K. Kushwah
Department of Agriculture, Government of Uttar Pradesh, Auraiya, Uttar Pradesh, India

S. S. Solankey, M. Kumari (eds.), *Advances in Research on Vegetable Production Under a Changing Climate Vol. 2*, Advances in Olericulture,
https://doi.org/10.1007/978-3-031-20840-9_6

Keywords Climate change · Perennial vegetables · Drought · Breeding

6.1 Introduction

Climate change has become an undeniable fact of the world in the present scenario. The continuous change in the ecosystem is the results of global warming, affecting the human health and nature including climate change, glacier retreat and melting, rise of oceans levels, forest fires, ozone depletion, agricultural cropping system and water scarcity (Shabir 2010). Climate change is raising the concerns over variables such as temperature and rainfall, which in turn can severely affect crop growth and compromise agricultural production and in the end food security. The world's climate is dynamic, plenty of evidences point towards the fact that it has been in a state of constant flux ever since the planet was sufficiently solidified and the atmosphere was formed (Dixon 2009, 2012). Since the surrounding environment play an important role in regulating the biology of plants and animals, natural events serve as some of the best signifiers of climatic change. According to Inter-governmental Panel on Climate Change (IPCC 2000), prominent changes in climate has been observed since beginning of industrial revolution, which had its peak in the eighteenth and the nineteenth century; industrialization is a major contributor to global warming. This led to a surge in climate based research and the development on climate modelling technologies. Scientists agreed on that human activity have been a major factor resulting in these changes today. Greenhouse gases accumulate in the atmosphere through combustion of fossil energy and greenhouse gases (GHG) emissions, e.g., CO_2, N_2O, CH_4 etc. (IPCC 2013). It has been expected that twenty-first century increase the average temperature of the Earth by 2–4.5 °C (IPCC 2013).

Vegetables are being cultivated for herbaceous succulents, and prone to several abiotic stresses. Most of the vegetables are vulnerable to adverse climatic factors results in loss of crop production. Environmental conditions vary in every season and location play a major role in vegetable production (Singh and Singh 2013; Solankey et al. 2021). Research on adverse impacts of climate change on vegetable production hence becomes a priority. Continuous increasing in world population has concern regarding the stability of the global environment, which considerably increases the food demand. Factors such as water availability, weather, and soil fertility are affecting agriculture productivity (Noya et al. 2018; Solankey et al. 2019). These factors such as uncontrollable weather conditions and plant exposure to the changing environments reduce quality of products. The possible way to reduce the dependence on weather conditions to a certain level is the use of good and advanced agricultural practices like greenhouse production or controlled environment farming. To ensure high food security and to fulfil the consumer demands, adoption of strategies is required to combat the challenges by climate change. Growing of crops by limiting their life span will affect growth and development

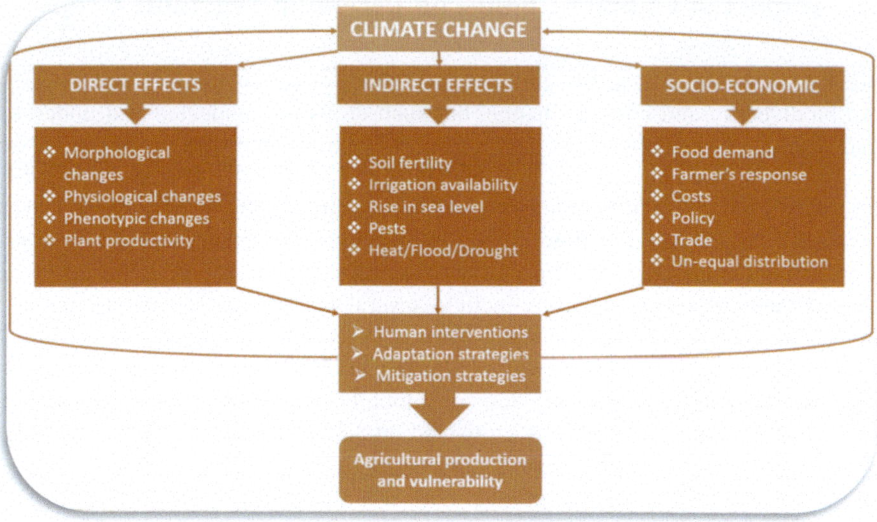

Fig. 6.1 Impact of climate change on agriculture. (Source: Raza et al. 2019)

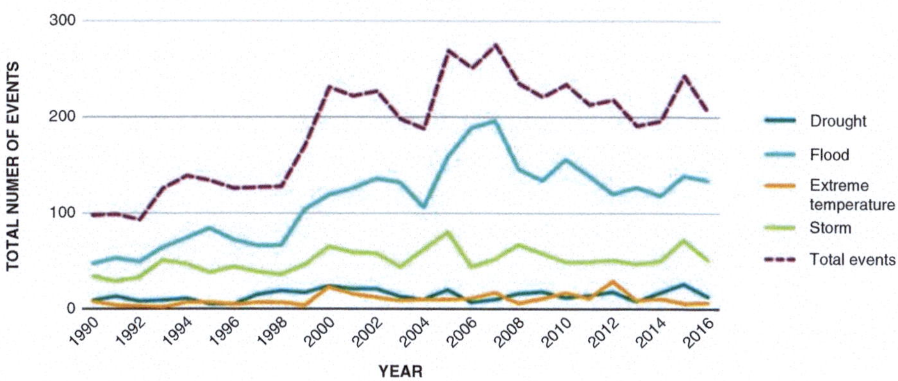

Fig. 6.2 Extreme events related to climate occurred during 1990–2016. (Source: FAO 2019 based upon data from Emergency Events Database)

because of decreased water availability and terminal heat stress, which cause a reduction in the horticultural production. Low and erratic rainfall in rainy season will primarily affect agriculture (Venkateswarlu and Shanker 2012). The effects of climate change on crop production are described in Fig. 6.1. Moreover, as per Food and Agriculture Organization, events of climate change increased dramatically as shown in Fig. 6.2.

6.2 Impact of Climate Change on Perennial Vegetable Crops

Perennial vegetables have the ability to deal with crises of nutrient deficiencies, crop biodiversity and climate change, but such crops are neglected and underutilized. Perennial vegetables have received little attention and only some individual crop species studied closely and in detail (Pimentel et al. 2012; Seemanthani 1964, Chadha and Patel 2007). Perennial vegetables are cultivated for their edible leaves (vegetative structures) and flower buds (reproductive structures). In the regions, where annual vegetables don't perform well, many perennial vegetables are best suited. Perennial vegetable crops include several halophytes, xerophyte, aquatic plants, and shade crops. Several shade tolerant Perennial vegetables are best suited to multi-strata agro forestry systems (Toensmeier 2007). Perennial vegetables as being region specific crops, very few are globally well known and traded crops. Perennial Vegetables are presently cultivated on 3.3 M ha, only 6% of the 52 M ha of total vegetable area globally. FAO reports 1.1 M ha in edible olives (*Olea europaea*) out of 10.6 M ha in total production across the globe, of which only 10% is utilized for table use (Lavee 1996). In India, only 1.5 M ha in asparagus (*Asparagus officinalis*), 0.6 M ha in avocado (*Persea americana*), and 0.1 M ha area is under globe artichoke (*Cynara scolymus*) production. India also has 38,000 ha area in the perennial drumstick (*Moringa oleifera*) (Ponnuswami 2012).

6.2.1 Elevated CO_2 Changes Yield and Quality of Produce

The advent of climate change is directly proportionate to the accumulation of greenhouse gases particularly CO_2, which emerge from the combustion of fossil fuels. Bisbis et al. 2018 revealed in a review that elevated CO_2 of atmosphere has the advantages by increase in CO_2 influx to the leaf, whereas reducing the stomata conductance concomitantly. This increases water use efficiency resulting in greater yield at a reduced water loss. Prolonged exposure to increased CO_2 may also influence quality of product, *e.g.*, enhancement in ascorbic acid, sugars, and greater antioxidant capacity, but probably it decreases some major and micronutrients availability such as magnesium, iron, and zinc in leafy vegetables (Dong et al. 2018).

6.2.2 High Temperatures Accelerate Plant Development

High temperatures fasten organ development and accelerate the plant development rate of vegetables, *e.g.*, flowering to fruiting. As a result, undesirable features may be encountered in the harvested products. These are the disadvantages of faster development of annual crops, which shorten the life cycle and reproductive system turns into lower yield potential. Productivity impacted negatively by temperature, by falling above or below specific thresholds at critical periods during crop development. Phenological development will be disrupted by photoperiod sensitivity of crops interacting with temperature. Some of the aspects affected by extreme temperatures during the reproductive stage which in turn accelerates the mechanism such as pollen viability and fruit formation (Hatfield et al. 2011).

6.2.3 Extreme Weather Events Reduce Yield and Quality of Produce

Extreme weather events are generally observed in terms of the season and average mean temperature of the month and include both extremely hot and low temperatures; as such extreme weather events include frost in winter as well as temperature above 30 °C in hot summer. It also includes hail or heavy rainfall with storms or drought; these events are directly connected to the temperature, rather form and intensify the precipitation or wind speed. An increasing frequency of extreme weather events is generally predicted by climate scenarios but these are with regional differences in their intensity and nature. Cultivating stress-resistant plants and to study their responses under different stresses is necessary for achieving food safety for an increasing global population. The plant response in relation to climatic stresses varies in the gene expression, physiological conditions, and metabolism. A very few reputed plant sensors have been recognized so far, though it was reported in many studies that plants are capable to sense any variation in surrounding environmental signal (Zhu 2016). The plant tissues are damaged and they respond consequently due to various environmental stresses, *e.g.*, transcriptional responses in specific tissues of roots against different stresses act differently (Dinneny et al. 2008). Factors such as drought, salinity, and chemicals effluence produce cellular signs like stress-responsive protein creation, increased associated solutes level, and more elevated antioxidant ratios. These primary stresses generate secondary stresses like osmotic and oxidative stress (Carvalho and Amancio 2019).

6.2.4 Effect of Temperature on Physiology of Perennial Vegetables

Different crop species at every stage of life cycle has varied responses to temperature and responses are based on phenological aspects primarily, *i.e.*, various stages of development of plant. The boundaries of observable growth are formed by a maximum and minimum temperature ranges for each species along with the increase in temperature to the optimum level, the growth (rate of node and leaf appearance) also increases. Compared to the reproductive development, the vegetative growth usually linked with relatively higher temperature requirement for most of the plant species.

Disruption of the pollination process limits the fruit bearing ability is a result of the exposure of the plants to extreme temperature ranges. A constant negative impact on plants is generally observed even though the degree of this impact varies among the species accordingly. The extreme effect on the atmospheric water vapour is one of the aspects of high temperature, is often overlooked. A decrease in the stomatal conductance results in high leaf temperature and a reduced photosynthesis is result of an increasing water vapour demand, which will lead to excess water transpiration through the leaf till the water supply will be restricted. Moisture stress can occur more rapidly if the plant is subjected to an extreme temperature because the plant becomes incapable to withdraw moisture from the deeper layer of soil to meet the increased atmospheric demand. The collective result of the hot air temperatures and the increasing atmospheric demand contribute towards negative effect of temperature extremes on the plant (Hatfield et al. 2011).

Climate change also manipulates the chilling requirements in some vegetable crops as a result of increasing temperatures throughout all the seasons. The influence of global warming inhibits winter dormancy may reduce the yield of many perennial vegetables. For example, asparagus goes into the dormancy and accumulates chilling of 0–7 °C over the winter period (Nie et al. 2016). Dormancy was consisting of two phases: (i) dormancy initiation and chilling temperatures accumulation, (ii) dormancy breakdown by high temperatures accumulation. Yields may be delayed and reduced as a result of a very high temperature during the first dormancy phase, which may extend the second dormancy phase length. In asparagus, with the increase of temperature by 3 °C, *i.e.,* from 2 to 5 °C, the duration of dormancy extends by 15 days (Nie et al. 2016). Moreover, the effects lead to reduced bud break and poor spear production.

It is quite clear from the finding of Kaufmann and Blanke 2017, that, the lack of chilling can be partially substituted by more forcing. The lack of cold hours in winter that induce flowering may be disallowed by higher temperatures in spring may be the result of climate change. Yamaguchi and Maeda 2015 successfully studied the breakdown of dormancy with the utilization of high temperature in Japan. A high temperature of 28 °C for 4 days or more is used to break dormancy during the forcing stage, resulted in rapid spear growth. It is revealed that high temperatures of 28 °C could break the dormancy of asparagus more quickly than chilling temperatures of 5 °C. However, negative effect on yield was also observed due to rapid sugar depletion when the high temperatures persisted for longer duration, *eg.,* 15–20 days.

6.2.5 *Influence of Drought*

Drought may be described as the absence or very less rainfall for a given period of time, sufficient enough for moisture depletion in the soil with a reduced water potential in the plant tissues. Climate change is expected to directly affect water availability and crop productivity, particularly in those vegetables affected by severe water stress conditions. Drought is the major cause of crop loss worldwide as it reduces average yield by 50% or more for most of the crop plants, it is a major crisis in the dry regions (Sivakumar et al. 2016). It has been suggested that a reduction in photosynthesis and photosynthetic assimilates amount allocated to floral parts is caused by moisture stress at flowering and this might cause an increase in the rate of abscission. Drought stress also causes metabolic impairment (Dias and Bruggemann 2010), apart from blocking the rate of photosynthesis by decreasing stomatal conductance (Yordanov et al. 2013). Solute concentration increases in the soil environment, which in turn leads to osmotic water out flow through plant cells, is caused by drought stress. This reduces productivity by inhibiting several physiological activities such as respiration and photosynthesis in most of the vegetables, caused by an increased moisture loss in plant cells (De la Pena and Hughes 2007). Moreover, water stress also reduces the biochemical capacity, which affects the availability and exploitation of sugar such as sucrose. This is indicated by a decrease in SPS (sucrose phosphate synthase) and inverse activity. A major role in the re-synthesis of sucrose is considered to be played by the SPS and it sustains the assimilatory carbon flux from source to sink (Isopp et al. 2008). The effect of drought is intensified and crop yield is further reduced by high temperature accompanied with drought that promotes evapo-transpiration and affects photosynthetic kinetics (Mir et al. 2012). Drought is considered as the major climatic stress for a lot of Indian states *viz.*, Rajasthan, Haryana, parts of Gujarat, and Andhra Pradesh (Mitra 2001). A 2/3rd of the Indian geographic area receives very less rainfall (lesser than 1000 mm), is categorized by erratic distributions. Among the net sown area of agricultural crops, about 140 million hectares (approx. 68%) is classified as susceptible to drought conditions and moreover, 50% of such area are reported as 'severe', where drought occur more regularly and frequently (National remote sensing centre, India).

In Tree tomato (*Solanum betaceum*) cool temperature induce flowering, whereas high temperatures coupled with drought negatively affects flowering as well as fruiting (Carrillo-Perdomo et al. 2015). Significant reduction in yield and starch content in the vegetables like cassava and sweet potato also takes place, which were earlier considered to be drought tolerant crops (Ravi and Mohankumar 2004; Ravi and Indira 1999). For tuber growth, mild water deficit is favourable, but growth of aerial parts and bulking of tuber both cease and become dormant under unfavourable water stress conditions. About 0.9 M ha area is affected and production is decreased by 40% in last five decades due to natural disasters like flood, drought and cyclones (Sivakumar et al. 2016). Although at extremely high temperatures of 33–40 °C under sufficient moisture, cassava may satisfactorily sustain its biomass, however at the temperature above 30 °C, sucrose synthesis translocates from the leaves and synthesis of starch in tubers is drastically affected. When the available soil moisture goes below 20%, sweet potato yields decreased as it drastically reduces the number of tubers during tuberization period. Lignifications of tubers are induced and tuber growth is hampered due to water stress during tuber initiation period.

6.2.6 Increase in Soil Salinity

Changes in weather patterns due to climate change resulted in the increased frequency of rainfall or drought above the normal range for more than a decade (Ayanlade et al. 2018). Variation in temperature and precipitation highly influences the soil salinity. It has been studied that the increased temperatures and decreased precipitation for long term (30 years) showed a direct relation with increase in salinity in arid soils due to decreased leaching of salt from the soil. Salinity is a serious problem in many crop growing areas as it reduces the growth as well as productivity of several vegetable crops. A reduction in the productivity of many perennial vegetables, which are mostly sensitive throughout the course of development, is caused by excessive soil salinity. It causes a reduction in length and mass of roots, at the same time they may become thinner or thicker. It also causes a delay or advancement at maturity depending on the species. The species of the vegetable crop and the variety within a species determines the degree of growth inhibition by salinity. The germination rate is slowed by salinity, and at higher levels it also reduces germination percentage. Environmental interactions like temperature, relative humidity, solar radiation and degree of air pollution also mediate the severity of salinity response (Shannon et al. 1994). Reduction in growth, leaf colour and changes in root/shoot ratio and rate of maturity are some of the osmotic effects of salinity. Ionic effects are manifested more easily in leaves and meristem damage produces the nutritional disorders symptoms. Thus, ion like Na^+ or Cl^- may accumulate in leaves or other plant parts in high concentrations and results in burning of leaves (Shannon and Grieve 1998). Ion-specific stress arising from altered ratios of K^+/Na^+, leads to an increase in concentrations of Na^+ and Cl^-, is injurious to the plants. It is caused by an initial water deficit due to high solute concentrations in the soil imposed by salinity. Some of the impact of salt stress on a plant is loss of turgidity, leaf abscission, reduced growth, decrease in photosynthesis, decrease in respiration, loss of cellular integrity, wilting, necrosis of tissues and ultimately plant death (Cheeseman 1988). The soil microbes' population in the rhizosphere and their interaction with roots are also affected by salinity. *Rhizobium spp.*, which is associated to legumes, can tolerate more salt than the legume itself, but plenty of evidence shown that nodulation process and fixation of N_2 by some of the legume crops are weakened by salinity (Lauchli 1984). It has been suggested that mycorrhiza symbioses increase the salt tolerance of the plants by improving phosphorus assimilation (Hirrel and Gerdemann 1980).

In an experiment, the number of buds per plant and bud growth was reduced in artichoke, as only fewer harvestable buds were obtained from plant when salinity level increased in the soil resulting in yield loss. Since bud circumference is reduced significantly with the increased levels of salinity, total yield reduction is accounted by the significant reduction in the weight of individual bud weight (Table 6.1) (Francois 1995). However, the harvesting traits are least affected in asparagus crop (Francois 1987).

Table 6.1 Bud yield traits of artichoke grown at six levels of salinity during two growing seasons. (Francois 1995)

Soil salinity (EC_e) (dSm^{-1})	Total bud yield ($t\ ha^{-1}$)	Avg. bud weight (g)	No. of buds/ plant	Avg. bud circumference (mm)
		1987		
4.6	16.7	211	6.1	289
6.6	16.8	203	6.6	287
7.4	16.1	206	1.3	287
8.7	10.0	190	5.9	282
10.6	9.0	179	6.3	279
11.6	5.3	178	5.4	274
Significance'	L^{**}	L^{**}	NS	L^{***}
		1988		
4.4	17.1	256	5.2	274
5.9	17.1	234	6.4	275
8.3	13.3	214	5.1	275
10.4	8.5	189	5.2	264
11.3	5.2	180	4.9	258
13.8	4.1	179	4.0	260
Significance'	L^{***}	$L^{***,Q*}$	L^{*}	L^{**}

^{z}L linear, Q quadratic
^{NS,*,**,***}: Significance at P=0.05, 0.01 and 0.005

With increasing soil salinity, the reduction in root pressure may arise; it would provide a mechanism for Ca movement to the inner bracts. As a result, Ca deficiency occurred with reduced root pressure and low transpiration. Necessity to maintain low levels of soil salinity is crucial for maximum yield even though artichoke tends to be more tolerant to salt than most of the other vegetable crops. Content of Mg and K in the leaf blades and midrib is also affected by salinity. The deficiency of Ca within the bud increases and the size of the buds decreases when salinity levels become too high (Francois et al. 1991).

6.2.7 *Flooding Interferes to Crop Growth and Production*

Being flood-susceptible plants, flooding causes serious problem to vegetable crops through hindrance in growth and development (Parent et al. 2008). Slow gas diffusion in the water and consumption by plant roots and microbes leads to oxygen deficiency, is normally associated with the occurrence of flooding condition. Many vegetables are highly susceptible to flooding and genetic variation to this trait is also limited, particularly in tomato, however the crops like asparagus and taro are least affected. Inhibition of aerobic processes which is caused due to oxygen deficit in the plant root zone is the most common damage by flooding in perennial vegetables. Endogenous ethylene is accumulated and causes harm to the plants in flooded condition (Drew 2009). A typical response of tomatoes to water logging is the rapid development of epinastic growth of the leaves and it also involves the activity of ethylene accumulation (Kawase 2011). With rising temperatures, the flooding symptoms become more severe; usually after a short flooding at high temperatures, rapid wilting is observed ultimately leads to death of the plants in tomato (Kuo et al. 2014).

Plant response to abiotic stresses depends upon the stage of development, length and intensity of stresses (Kumar 2017). Reduced stomatal conductance of a plant is one of the earliest plant physiological responses to soil flooding (Folzer et al. 2006). Flooding generally leads to leaf water potential increase, which decreases the stomatal conductance ultimately leads to reduction in carbon exchange rate and elevate the internal CO_2 concentration. Flooding affects the vegetative as well as reproductive growth of the plants negatively due to physiological malfunctioning (Gibbs and Greenway 2008). Chlorosis of leaves, reduced shoot and root growth, reduced accumulation of dry matter and yield reduction are some of the negative results of flooding on sensitive crop plants (Malik et al. 2002). The water-borne pathogens spread easier by flooding, as plants can be predisposed to infection by droughts and heat waves, alongside with dispersal of spores through wind (Pautasso et al. 2012).

6.2.8 Insect Responses to Climatic Change

Climate change also influenced the biology and ecology of insect pests (Jat and Tetarwal 2012). An increase in fecundity and earlier completion of life cycle in some insects with short life cycles such as diamond back moth and aphids are caused by increase in temperature. As a result, insect produced more generations per year than their actual rate (FAO 2009). In contrast, several years may be taken by some insects to complete their life cycle. Soil acts as an insulating medium that leads to changes in buffer temperature more than the air, some of the insect species that live in soil for some stages or whole of life cycle is likely to suffer more than above the surface of soil (Bale et al. 2010). Rising in the temperature leads to movement of insect species to higher latitudes, while in the tropical areas specific pest species might be adversely affected by higher temperature. Developmental and ovipositional rates, outbreaks of insects and introductions of invasive species are increased by high atmospheric temperature (Kumar et al. 2020), whereas the insect diversity in ecosystems, reliability of economic threshold (ET) levels, and efficacy of insect biological control by fungi is decreased (Das et al. 2011).

6.2.9 *Crop Adaptation to Extreme Overall Climate Stresses*

The climate undergoes severe alterations with the raise of the Earth's average temperature, as a result becoming abiotic stressful. Various threats are posed to naturally prevailing crop species, so environmental changes are of big concern (Espeland and Kettenring 2018). Heat and drought are the most predominant stresses under field conditions hence they have a considerable influence on crops (Pereira 2016). It is quite obvious that plants, for their normal growth, development and flowering require an optimum temperature. Temperature fluctuations highly influence plant physiology (Hatfield and Prueger 2015). The grain production and yield is severely affected by heat stress, whereas sterility is caused by chilling stress, and the morphophysiology of plants is negatively influenced by drought stress (Salehi-Lisar and Bakhshayeshan-Agdam 2016). Some of the impacts of these climatic stresses are severe agony on development and yield of plants, producing huge responses, including morphological, molecular, and physiological and biochemical modifications (Zandalinas et al. 2018). Overall, climate change and global warming have both negative and positive impacts on crop plants.

6.2.9.1 Positive Impact

1. Increased productivity from higher temperatures
2. Decline in moisture stress
3. Introduction of new crops
4. Acceleration in maturation rates
5. Prolonged growing season
6. Increased productivity from enhanced CO_2. (CO_2 fertilization only applies to some crops and will have small temporary benefits for higher latitude. (IPCC reports)

6.2.9.2 Negative Impact

1. Crop damage from extreme heat
2. Increased ground level ozone-toxicity to green plants
3. Stronger storms and floods
4. Warming stress
5. Waterlogged land
6. More soil erosion
7. Specialized mono-cropping, less adaptable
8. Decreased pesticides and herbicides efficiency
9. Planning problems due to less reliable forecast

10. Increased insect infestations
11. Torrential rain
12. Increased drought
13. Increased weed growth
14. Increased moisture stress
15. Increased crop diseases

6.3 Mitigation Strategies

6.3.1 Implications of Moringa Tree to Climate Change Mitigation

Since famine is connected to climate change, an important way to mitigate it by planting trees, which can sequester more carbon like *Moringa* (Amaglo 2013). Therefore, the need of implementation of climate smart policies can be built for more food systems development and combat climate change is of top most priority. *Moringa* tree has a great potential to not only store carbon, but also to improve the livelihoods of many farmers if it is grown on a much larger scale (Gedefaw 2015). According to Fuglie (2000), *Moringa oleifera* is called as "*Never Die Plant*", as it is highly adaptable to various soils and environmental extremes. Because of its numerous nutritional, medicinal and industrial values and greater adaptability, *Moringa oleifera* is proven as a suitable crop for mitigation of climate change (Ndubuaku et al. 2014). The tree acts as a good sink for CO_2 absorption and utilization because they produce heavy flushes even during the dry season. They contribute towards reduction of atmospheric CO_2 level, one of the major reasons of depletion of ozone layer and finally global warming. The tree is also considered to be an adaptable crop for climate change and to eliminate threats for food security (Ndubuaku et al. 2014).

When compared to the Japanese Cedar tree, Moringa tree may absorb fifty times higher CO_2 and 20 times higher than that of general vegetation according to the study of Villafuerte and Villafurte-Abonal (2009). A study on Moringa revealed that, a person emits 320 kg of CO_2 per year. A total of 23 Japanese Cedar trees absorbs this amount of CO_2 in 50 years of time; on the other hand, only 2 Moringa trees absorb this amount in 2 years. On an average one family car emits 2300 kg of CO_2/year; to absorb this CO_2 amount in 50 years, 160 Japanese Cedar trees are required; however only 10 Moringa trees are sufficient for this task in 2 years (Ndubuaku et al. 2014). Since Moringa tree sequesters more carbon through all plant parts, it may be a useful tool to combat global warming. Therefore, the impact of climate change can be mitigated by planting such an important tree on large scale.

6.3.2 Good Agronomic Practices

Several investigators have reported a lot of strategies for the farmers to deal with the climatic change for plant adaptation on the basis of series of experiments. The various useful approaches adopted by farmers to deal with climatic change are, the inclusion of practices such as modifying planting as well as harvesting time, selection of short duration crops, crop rotation, improved irrigation techniques and cropping schemes. These approaches are very beneficial for adaptability of the crop under abiotic stress conditions (Duku et al. 2018). To reduce the water logging of the field during rainy season, transparent plastic rain shelters may be used to prevent the direct impact on developing fruits for successful vegetable production as suggested by several researchers. The planting of vegetables on raised beds may increase the yield during rainy season because it improves drainage of water and reduces the anoxic stress to root zone (Welbaum 2015). Some of the other strategy to overcome the impact of climatic variation and to ensure good adaptability to the plants to assure food security is; sowing time modification, development and utilization of drought resistant varieties, and the introduction of new crops (Ali and Erenstein 2017). Implementation of techniques with the ability to boost development of crop plants under several climatic stresses is another plant adaptability approach. The soil moisture can be conserved by the practices like mulching with crop residue and with the use of plastic mulches. Excessive soil moisture in excessive rain also becomes a big problem in some instances; this could be overcome by the use of raised beds (La Pena and Hughes 2007). Some fundamental techniques to deal with climate stresses include the selection of sowing time, plant density, and better irrigation technology (Battisti et al. 2018). Use of fertilizers also plays a very vital role in the reduction of global warming effect and to support the plant for better adaptability. It is always advisable to maintain the soil fertility and increase productivity as it provides significant energy to the crop plants (Henderson et al. 2018).

6.3.3 Breeding Techniques

Plant breeding shows a lot of techniques in crop improvement under different abiotic stresses. By developing stress resistant cultivars, it provides a splendid option to ensure food safety and security and also helps plants to avoid various stresses throughout the crucial growth period of plant (Blum 2018). The use of genetic divergence analysis, inbreeding technology, assessment of germplasm, recombination and assortment to achieve plant perfection are the important parts for crop improvement. Morphological screening of germplasm not only identifies the lines with novel traits but also helps in the study of inheritance pattern of different vegetable crops (Kumar et al. 2018). For the breeding of new varieties based on morphological variability, genetic divergence study is considered to be very important method (Raza et al. 2018). Information on magnitude and nature of character association act as an important pre-breeding tool for efficient selection of novel resistant traits (Sharma et al. 2020).

Landraces are the important source for genetic studies, a wheat landrace maintained in data bank contains broader genetic base and is an important source for stress resistance as it is adjustable to varied climatic stress (Lopes et al. 2015). The screening procedure has identified 5 genotypes (CI-260, CE-534, CE-54, CI-848 and CI-308) and 9 land races (16, 7, 129, Narukku-3, TP White, CI-60, CI-4, CI-80 and CI-17) of cassava, are resistant to drought condition. Three sweet potato land races that have been identified as drought tolerant are VLS6, IGSP 14 and IGSP 10. A variety of sweet potato "Sree Bhadra" is found to be tolerant to drought is released by CTCRI, India (Annual Report 2013, 2014, 2015). Breeding for drought tolerance has also been tried in various seed spice crops (Malhotra 2009; Malhotra 2016). Molecular breeding alone or in integrated approach is the useful tool to develop cultivars resistant to abiotic stress using genomics technology like marker-assisted selection (MAS) by improving simple horticultural trait controlled by one or oligogenes for resistance (Kumar et al. 2021) and genome wide associated studies (GWAS) (Raza et al. 2019).

6.3.4 Genetic Engineering for Stress Resistance in Plants

Biotechnology is a powerful practice for manipulation of the genome of any organism for the benefits of mankind. The alteration of genetic structure through biotechnology is a potent and new strategy. The data from genetics are quite encouraging, may be exploited towards drought, salinity, heat and cold. A powerful finding to develop stress-resistant varieties is identification of stress-responsive TFs (transcriptions factors). These TFs can regulate the morphological characters of genes linked with different stresses in genetic engineered crops (Reynolds et al. 2015). There is a plenty of transgenic plants which are developed by genetic engineering to deal with the number of abiotic stresses. Compared to normal plants, genetically engineered plants exhibit considerable resistance against changing climate (Shah et al. 2016).

Various transcriptions factors are identified as plant-specific transcription factor which involves *AP2/ERFBP* group (Riechmann and Meyerowitz 1998). This group of *AP2/ERFBP* transcription factors is associated with different pathways related to plant growth and has role in responses towards abiotic stresses (Licausi et al. 2010). *AP2/EREBP* transcription factors are classified into 4 sub-families based upon their numbers and similarity. The sub-families consist of *DREB* (dehydration-responsive element-binding protein), *ERF*, *AP2* (Apetala 2), and *RAV* (related to *ABI3/VP1*). Two major subfamilies which are widely evaluated due to their function in biotic and abiotic responses of plant are *ERF* and *DREB* (Sharoni et al. 2011). In different water stress and chilling stress, the *DREB* has significant regulating ability (Stockinger et al. 1997).

6.3.5 Grafting Improves Stress Tolerance

Grafting the vegetables was first initiated during the twentieth century in East Asia, to minimize the effect of soil borne diseases like fusarium wilt, and to boost up the production of tomato, eggplant, and cucurbits, which are widely affected by the disease. Grafting in the present scenario became a regular practice in the vegetable cultivation in Korea, Japan and some other European countries (Martínez-Ballesta et al. 2010). It is rapid and alternative tool to the slow breeding technology aimed at increased environmental-stress resistance of the vegetable crops. Grafting serves as a promising tool for the modification of the root system for increasing its resistance to a range of abiotic stresses in the plants (Bhatt et al. 2013). Grafted plants are being utilized now to improve resistance towards drought, temperatures extremes, salinity and flooding, if suitable rootstocks are used (Martínez-Ballesta et al. 2010; Venema et al. 2008; He et al. 2009). However, degree of tolerance of rootstocks to various stresses also varies largely among different species, *viz.;* rootstocks from *Cucurbita spp.* are comparatively more resistant to salinity than *Lagenaria siceraria* (Matsubara 2012). It has been also revealed that melons when grafted using hybrid squash as rootstocks were found more tolerant to salt than the non-grafted plants (Yetisir et al. 2005). The cultivation of grafted plants in sweet pepper, eggplant and tomato and different cucurbits (watermelon, cucumber and pumpkin) has increased in past few years because of these beneficial effects of grafting (Hassell et al. 2008; Lee et al. 2010), alongside there is huge scope for such an efficient tool for various abiotic stress resistance in other perennial vegetables.

6.3.6 Developing Climate Resilient Vegetables

The most cost-effective option to encounter the changing climate is the use of improved and adapted germplasm in vegetables. However, most improved varieties represent a limited genetic variability including resistance to abiotic stresses. In order to achieve intensive and high input production systems, breeding the new varieties selected for the traits contributing to adaptation to low input and unfavourable environments would be of great importance. Genetic variation is an important factor for selection in improved and desirable parents from available germplasm for successful breeding programme (Debnath et al. 2020). The recognition of a new genetic variation for resistance to different climatic stresses could result in improved varieties well adapted to a broad range of climatic condition. The need is to identify and advance the genotypes with superior traits through greater combinations of the alleles at multiple loci. There is a great demand for the improved selection procedure to identify the superior genotypes with linked traits from wild species from the unfavourable environments to their cultivated varieties. Plants inhabitants to climates with varied seasonality are able to adapt more easily to changing environment (Pereira and Chaves 1995) and offer opportunity to identify the genes that provide such resistance.

6.4 Conclusion

At present time, with the challenge to meet food and nutritional security for growing population, the agricultural world is facing a difficult situation especially in the vegetable production. There is an urgent need to produce increasing amount of food from limited piece of land. With the result of increasing abiotic stresses and decline in favourable environment combined with the threat of global warming through greenhouse gases, the problem is aggravated at high extent. There is significant effect of raised temperature upon duration of crop, flowering and fruiting of perennial vegetables with very low economic yield and productivity. An immediate action is required to study climate change and its impact on crop growth, yield and quality, as the succulent and tender vegetables are very sensitive to heat, drought and flooding conditions. In order to determine the response of plants under abiotic stresses, there is an urgent necessity to explore the genetic basis for the mechanisms related to the response. To achieve better plant adaptation under abiotic stresses, physiological challenges present in the plants need to be studied and resolved on an urgent basis. Scientific community must focus upon adaptation technologies and the mitigation strategies for the crop plants. For inducing the resistance against abiotic stresses and to save agriculture in the future, there is a need to propagate improved cultural methods, cropping schemes, and different non-conventional strategies. Resilient crops under heat and drought can be developed through the help of breeding approaches. Certain significant strategies in the identification of the genes for changing climate are genome wide association studies (GWAS), genomic selection with high throughput phenotyping and genotyping.

Adaptation of improved vegetable production system in the changing climate will help in reducing malnutrition in developing nations through higher production and consumption of quality vegetables. There is an immediate requirement to develop sound adaptation strategies for mitigation of changing climate on productivity and nutritional quality of vegetables. Development of system for improved water use efficiency under hot and dry condition should be emphasised. Soil moisture can be conserved through mulching with crop residues and plastic mulches. In order to overcome excessive soil moisture due to flooding, crops can be grown on raised beds. It is necessary to identify germplasm with tolerance to high temperatures, drought and other climatic hazards, with the ability to produce higher yield. Genetic population developed to identify and introgress genes providing resistance to climatic stresses and to generate the tools for genetic engineering. It is necessary to continuously enhance conservation agriculture and to protect vegetable crops from extreme environmental conditions in the developing countries. Technical, socioeconomic as well as political considerations must be included in an effective extension tactic that needs to be in place. Finally, the key components of a sustainable adaptation strategy to climate change are capacity building and education.

References

Ali A, Erenstein O (2017) Assessing farmer use of climate change adaptation practices and impacts on food security and poverty in Pakistan. Clim Risk Manag 16:183–194. https://doi.org/10.1016/j.crm.2016.12.001

Amaglo N (2013) Moringa as a climate change mitigation strategy. Hunan Agriculture University, Faculty of Food Science and Technology, Changsha, p 410128

Anonymous (2013) Annual report. Central Tuber Crop Research Institute, Thiruananthpuram

Anonymous (2014) Annual report. Central Tuber Crop Research Institute, Thiruananthpuram

Anonymous (2015) Annual report. Central Tuber Crop Research Institute, Thiruananthpuram

Ayanlade A, Radeny M, Morton JF, Muchaba T (2018) Rainfall variability and drought characteristics in two agro-climatic zones: an assessment of climate change challenges in Africa. Sci Total Environ 630:728–737. https://doi.org/10.1016/j.scitotenv.2018.02.196

Bale JS, Masters GJ, Hodkinson ID, Awmack C, Bezemer TM, Brown VK, Butterfield J, Buse A, Coulson JC, Farrar J, Good JE (2010) Herbivory in global climate change research: direct effects of rising temperature on insect herbivores. Glob Change Biol 8(1):1–16. https://doi.org/10.1046/j.1365-2486.2002.00451

Battisti R, Sentelhas PC, Parker PS, Nendel C, Gil MD, Farias JR, Basso CJ (2018) Assessment of crop-management strategies to improve soybean resilience to climate change in southern Brazil. Crop Pasture Sci 60(2):154–162. https://doi.org/10.1071/CP17293

Bhatt RM, Rao NKS, Harish DM (2013) Significance of grafting in improving tolerance to abiotic stresses in vegetable crops under climate change scenario. In: Climate-resilient horticultural adaptation and mitigation strategy, pp 159–175

Bisbis MB, Gruda N, Blanke M (2018) Potential impacts of climate change on vegetable production and product quality-a review. J Clean Prod 170:1602–1620. https://doi.org/10.1016/j.jclepro.2017.09.224

Blum A (2018) Plant breeding for stress environments. CRC Press, Boca Raton

Carrillo-Perdomo E, Aller A, Cruz-Quintana SM, Giampieri F, Alvarez-Suarez JM (2015) Andean berries from Ecuador: a review on botany, agronomy, chemistry and health potential. J Berry Res 5(2):49–69. https://doi.org/10.3233/JBR-140093

Carvalho LC, Amancio S (2019) Cutting the Gordian knot of abiotic stress in grapevine: from the test tube to climate change adaptation. Physiol Plant 165(2):330–342. https://doi.org/10.1111/ppl.12857

Chadha KL, Patel VB (2007) Prospect of indigenous perennial plants as source of vegetable. In: International Conference on Indigenous Vegetables and Legumes. Acta Hortic 752:49–54. https://doi.org/10.17660/ActaHortic.2007.752.4

Cheeseman JM (1988) Mechanisms of salinity tolerance in plants. Plant Physiol 87(3):547–550. https://doi.org/10.1104/pp.87.3.547

Das DK, Singh J, Vennila S (2011) Emerging crop pest scenario under the impact of climate change-a brief review. J Agric Phys 11:13–20

De la Pena R, Hughes J (2007) Improving vegetable productivity in a variable and changing climate. J SAT Agric Res 4:1–22

Debnath A, Kumar R, Prasad SK, Sharma N, Kushwah JK (2020) Multivariate analysis for genetic diversity estimation among tomato (*Solanum lycopersicum* L.) genotypes. Econ Environ Cons 26(3):1208–1211

Dias MC, Bruggemann W (2010) Limitations of photosynthesis in *Phaseolus vulgaris* under drought stress: gas exchange, chlorophyll fluorescence and Calvin cycle enzymes. Photosynthetica 48(1):96–102. https://doi.org/10.1007/s11099-010-0013-8

Dinneny JR, Long TA, Wang JY, Jung JW, Mace D, Pointer S, Barron C, Brady SM, Schiefelbein J, Benfey PN (2008) Cell identity mediates the response of *Arabidopsis* roots to abiotic stress. Science 320(5878):942–945. https://doi.org/10.1126/science.1153795

Dixon GR (2009) The impact of climate and global change on crop production. In: Letcher TM (ed) Climate Change, pp 307–324. https://doi.org/10.1016/b978-0-444-53301-2.00017-8
Dixon GR (2012) Climate change- impact on crop growth and food production, and plant pathogens. Can J Plant Path 34(3):362–379
Dong J, Gruda N, Lam SK, Li X, Duan Z (2018) Effect of elevated CO_2 on nutritional quality of vegetables: a review. Front Plant Sci 9:924. https://doi.org/10.3389/fpls.2018.00924
Drew MC (2009) Plant responses to anaerobic conditions in soil and solution culture. Curr Adv Plant Sci 36:1–14
Duku C, Zwart SJ, Hein L (2018) Impacts of climate change on cropping patterns in a tropical, subhumid watershed. PLoS One 13(3):0192642. https://doi.org/10.1371/journal.pone.0192642
Espeland EK, Kettenring KM (2018) Strategic plant choices can alleviate climate change impacts: a review. J Environ Manag 222:316–324
FAO (2009) Global agriculture towards: issues brief, Rome
FAO (2019) FAO Stat online database. http://www.fao.org/faostat/en#home
Folzer H, Dat JF, Capelli N, Rieffel D, Badot PM (2006) Response of sessile oak seedlings (*Quercus petraea*) to flooding: an integrated study. Tree Physiol 26(6):759–766. https://doi.org/10.1093/treephys/26.6.759
Francois LE (1987) Salinity effects on asparagus yield and vegetative growth. J Am Soc Hortic Sci 112:432–436
Francois LE (1995) Salinity effects on bud yield and vegetative growth of artichoke (*Cynara scolymus* L.). Hort Sci 30(1):69–71. https://doi.org/10.21273/HORTSCI.30.1.69
Francois LE, Donovan TJ, Maas EV (1991) Calcium deficiency of artichoke buds in relation to salinity. Hort Sci 26(5):549–553. https://doi.org/10.21273/HORTSCI.26.5.549
Fuglie LJ (2000) New uses of *Moringa* Studied in Nicaragua ECHO Development Notes 68
Gedefaw M (2015) Environmental and medicinal value analysis of *Moringa* (*Moringa oleifera*) tree species in Sanja, North Gondar, Ethiopia. Am J Contemp Sci Res 4802:20–35
Gibbs J, Greenway H (2008) Mechanisms of anoxia tolerance in plants, growth, survival and anaerobic catabolism. Funct Plant Bio 30:1–47
Hassell RL, Memmott F, Liere DG (2008) Grafting methods for watermelon production. Hort Sci 43(6):1677–1679. https://doi.org/10.21273/HORTSCI.43.6.1677
Hatfield JL, Prueger JH (2015) Agro ecology: implications for plant response to climate change. Crop Adapt Clim Change:27–43
Hatfield JL, Boote KJ, Kimball BA et al (2011) Climate impacts on agriculture: implications for crop production. Agron J 103(2):351–370. https://doi.org/10.2134/agronj2010.0303
He Y, Zhu Z, Yang J, Ni X, Zhu B (2009) Grafting increases the salt tolerance of tomato by improvement of photosynthesis and enhancement of antioxidant enzymes activity. Environ Exp Bot 66(2):270–278. https://doi.org/10.1016/j.envexpbot.2009.02.007
Henderson B, Cacho O, Thornton P, van Wijk M, Herrero M (2018) The economic potential of residue management and fertilizer use to address climate change impacts on mixed smallholder farmers in Burkina Faso. Agric Syst 167:195–205
Hirrel MC, Gerdemann JW (1980) Improved growth of onion and bell pepper in saline soils by two vesicular-arbuscular mycorrhizal fungi. Soil Sci Soc Am J 44(3):654–655. https://doi.org/10.2136/sssaj1980.03615995004400030046x
IPCC (2000) Summary for policymakers. In: Nebojsa D et al (eds) Emissions scenarios: a special report of IPCC Working Group III, Nakic enovic. Cambridge University Press, Cambridge
IPCC (2013) The physical science basis. Summary for policymakers. In: Stocker TF et al (eds) Contribution of Working Group I to the Fifth Assessment Report of the Intergovernmental Panel on Climate Change. Cambridge University Press, Cambridge
Isopp H, Frehner M, Long SP, Nosberger J (2008) Sucrose-phosphate synthase responds differently to source-sink relations and to photosynthetic rates: *Lolium perenne* L. growing at elevated pCO_2 in the field. Plant Cell Environ 23:597–607
Jat MK, Tetarwal AS (2012) Effect of changing climate on the insect pest population National Seminar on sustainable agriculture and food security: challenges in changing climate

Kaufmann H, Blanke M (2017) Performance of three numerical models to assess winter chill for fruit trees- a case study using cherry as model crop in Germany. Reg Environ Chang 17(3):715–723. https://doi.org/10.1007/s10113-016-1064-6

Kawase M (2011) Anatomical and morphological adaptation of plants to water-logging. Hort Sci 16:30–34

Kumar SN (2017) Climate change and its impacts on food and nutritional security in India. Agric Clim Change Threats Strat Policy 1:48

Kumar R, Kumar R, Prasad BD, Solankey SS, Kumar J, Bamaniya BS (2018) Genetic variation study using morphological and DNA marker based genotyping in bottle gourd (*Lagenaria siceraria* (Mol.) Standl.). Curr J Appl Sci Tech 31(2):1–10. https://doi.org/10.9734/CJAST/2018/45876

Kumar R, Singh MP, Bag N, Sharma L, Bamaniya BS (2020) Invasive pest *Tuta absoluta* (Meyrick) on Tomato- Alert in Sikkim. Ind J Entomol 82(3):593–594

Kumar R, Chhetri S, Bamaniya BS, Kushwah JK, Verma RK (2021) Molecular tools for bacterial wilt resistance in tomato. In: Solankey SS, Shamim MD (eds) Biotic stress management in tomato, in the series- innovation in horticultural science. Apple Academic Press, CRC press, pp 25–42

Kuo DG, Tsay JS, Chen BW, Lin PY (2014) Screening for flooding tolerance in the genus *Lycopersicon*. Hort Sci 17:76–78

La Pena RD, Hughes J (2007) Improving vegetable productivity in a variable and changing climate. International Crops Research Institute for the Semi-Arid Tropics 4:1–22

Lauchli A (1984) Salt exclusion: an adaptation of legumes for crops and pastures under saline conditions. In: Staples RC, Toenniessen GH (eds) Salinity tolerance in plants. Strategies for crop improvement. Wiley, New York

Lavee S (1996) World olive encyclopaedia. International Olive Oil Council 1:61–106

Lee JM, Kubota C, Tsao SJ, Bie Z, Echevarria PH, Morra L, Oda M (2010) Current status of vegetable grafting: diffusion, grafting techniques, automation. Sci Hort 127(2):93–105. https://doi.org/10.1016/j.scienta.2010.08.003

Licausi F, Giorgi FM, Zenoni S, Osti F, Pezzotti M, Perata P (2010) Genomic and transcriptomic analysis of the AP2/ERF superfamily in *Vitis vinifera*. BMC Genomics 11:719. https://doi.org/10.1186/1471-2164-11-719

Lopes MS, El-Basyoni I, Baenziger PS, Singh S, Royo C, Ozbek K (2015) Exploiting genetic diversity from landraces in wheat breeding for adaptation to climate change. J Exp Bot 66(12):3477–3486. https://doi.org/10.1093/jxb/erv122

Malhotra SK (2009) Standardization of screening procedure for drought tolerance in coriander. In: 5th world congress of cellular- molecular biology Indore, India vols 2-6(v). School of Biotech, Indore, India, p 15

Malhotra SK (2016) Water soluble fertilizers in horticultural crops- an appraisal. Indian J Agric Sci 86(10):245–256

Malik AI, Colmer TD, Lambers H, Setter TL, Schortemeyer M (2002) Short-term water-logging has long-term effects on the growth and physiology of wheat. New Phyto 1153(2):225–236. https://doi.org/10.1046/j.0028-646X.2001.00318.x

Martínez-Ballesta MC, Alcaraz-Lopez C, Muries B, Mota-Cadenas C, Carvajal M (2010) Physiological aspects of rootstock–scion interactions. Sci Hort 127(2):112–118

Matsubara S (2012) Studies on salt tolerance of vegetables, 3: salt tolerance of rootstocks. Scientific Reports of the Faculty of Agriculture, Okayama University

Mir RR, Zaman-Allah M, Sreenivasulu N, Trethowan R, Varshney RK (2012) Integrated genomics, physiology and breeding approaches for improving drought tolerance in crops. Theor Appl Genet 125(4):625–645. https://doi.org/10.1007/s00122-012-1904-9

Mitra J (2001) Genetics and genetic improvement of drought resistance in crop plants. Curr Sci 80(6):758–763

Ndubuaku UM, Ndubuaku TC, Ndubuaku NE (2014) Yield characteristics of *Moringa oleifera* across different ecologies in Nigeria as an index of its adaptation to climate change. Sustain Agric Res 3(1):95–100

Nie LC, Chen YH, Liu M (2016) Effects of low temperature and chilling duration on bud break and changes of endogenous hormones of asparagus. Euro J Hort Sci 81(1):22–26. https://doi.org/10.17660/eJHS.2016/81.1.3

Noya I, Gonzalez-Garcia S, Bacenetti J, Fiala M, Moreira MT (2018) Environmental impacts of the cultivation-phase associated with agricultural crops for feed production. J Clean Prod 172:3721–3733. https://doi.org/10.1016/j.jclepro.2017.07.132

Parent C, Capelli N, Berger A, Crèvecoeur M, Dat JF (2008) An overview of plant responses to soil water-logging. Plant Stress 2:20–27

Pautasso M, Doring TF, Garbelotto M, Pellis L, Jeger MJ (2012) Impacts of climate change on plant diseases-opinions and trends. Euro J Plant Patho 133(1):295–313. https://doi.org/10.1007/s10658-012-9936-1

Pereira A (2016) Plant abiotic stress challenges from the changing environment. Front Plant Sci 7:1123. https://doi.org/10.3389/fpls.2016.01123

Pereira JS, Chaves MM (1995) Plant responses to drought under climate change in Mediterranean-type ecosystems. Ecol Stud, In, pp 140–160. https://doi.org/10.1007/978-1-4612-4186-7-7

Pimentel D, Cerasale D, Stanley RC, Perlman R, Newman EM, Brent LC, Mullan A, Chang DT (2012) Annual vs. perennial grain production. Agric Ecosys Environ 161:1–9. https://doi.org/10.1016/j.agee.2012.05.025

Ponnuswami V (2012) Advances in production of Moringa, vol 625. All India Co-ordinated Research Project-Vegetable Crops, Horticultural College and Research Institute, Tamil Nadu Agricultural University, Periyakulam, p 604

Ravi V, Indira P (1999) Crop physiology of sweet potato. Hortic Rev 23:277–338

Ravi V, MohanKumar CR (2004) Cassava based multiple cropping systems. Hortic Rev 30:355–500

Raza A, Mehmood SS, Ashraf F, Khan RSA (2018) Genetic diversity analysis of *brassica* species using PCR-based SSR markers. Gesunde Pflanzen 71(1):1–7. https://doi.org/10.1007/s10343-018-0435-y

Raza A, Razzaq A, Mehmood SS, Zou X, Zhang X, Lv Y, Xu J (2019) Impact of climate change on crops adaptation and strategies to tackle its outcome: a review. Plan Theory 8:34. https://doi.org/10.3390/plants8020034

Reynolds M, Tattaris M, Cossani CM, Ellis M, Yamaguchi-Shinozaki K, Saint-pierre C (2015) Advances in wheat genetics: from genome to field. Exploring genetic resources to increase adaptation of wheat to climate change. Springer, p 355-368

Riechmann JL, Meyerowitz EM (1998) The AP2/EREBP family of plant transcription factors. Biol Chem 379(6):633–646. https://doi.org/10.1515/bchm.1998.379.6.633

Salehi-Lisar SY, Bakhshayeshan-Agdam H (2016) Drought stress in plants: causes, consequences, and tolerance. In: Drought stress tolerance in plants. Springer, pp 1–16

Seemanthani B (1964) A study of practices and problems in the cultivation of some perennial vegetables in Madras State. South Indian Hort 12:1–7

Shabir O (2010) A summary case report on the health impacts and response to the Pakistan floods of 2010. PLoS Curr:5

Shah SH, Ali S, Hussain Z, Jan SA, Ali GM (2016) Genetic improvement of tomato (*Solanum lycopersicum*) with AtDREB1A gene for cold stress tolerance using optimized agrobacterium-mediated transformation system. Int J Agric Biol 18(3):471–482. https://doi.org/10.17957/IJAB/15.0107

Shannon MC, Grieve CM (1998) Tolerance of vegetable crops to salinity. Sci Hortic 78(1–4):5–38. https://doi.org/10.1016/S0304-4238(98)00189-7

Shannon MC, Grieve CM, Francois LE (1994) Whole-plant response to salinity. In: Wilkinson RE (ed) Plant-environment interactions. Marcel Dekker, New York, pp 199–244

Sharma N, Bhutia KD, Kumar R, Prasad SK, Debnath A, Sharma MM (2020) Polygenic variation for morphological and biochemical traits of brinjal genotypes (*Solanum melongena* L.) and its wild relatives. Bioscan 15(3):303–309

Sharoni AM, Nuruzzaman M, Satoh K, Shimizu T, Kondoh H, Sasaya T, Choi IR, Omura T, Kikuchi S (2011) Gene structures, classification and expression models of the AP2/EREBP transcription factor family in rice. Plant Cell Physiol 52(2):344–360. https://doi.org/10.1093/pcp/pcq196

Singh S, Singh DR (2013) Vegetable crops in Andaman and Nicobar Islands. Advances in Vegetable Crops. IIVR, Varanasi

Sivakumar R, Nandhitha GK, Boominathan P (2016) Impact of drought on growth characters and yield of contrasting tomato genotypes. Madras Agric J 103:78–82

Solankey SS, Prakash S, Pawan S, Randhir K (2019) Vulnerability of vegetable crops to Climate Change. In: Singh P et al (eds) National Seminar on "Recent Advances in Agriculture for Sustainable Rural Development (RAASRD-2019)", pp 64–70. March 17–19, 2019 at VKSCoA, Dumraon (BAU, Sabour), Bihar; ISBN: 978-93-85675-11-9

Solankey S.S., Kumari M., Akhtar S., Singh H.K. and Ray P.K. (2021). Challenges and opportunities in vegetable production in changing climate: mitigation and adaptation strategies *in:* advances in research on vegetable production under a changing climate Vol. 1. (Eds. S.S. Solankey *et al.*), springer nature., Switzerland., ISBN 978-3-030-63497-1. Pp. 13-59

Stockinger EJ, Gilmour SJ, Thomashow MF (1997) Arabidopsis thaliana CBF1 encodes an AP2 domain-containing transcriptional activator that binds to the C-repeat/DRE, a cis-acting DNA regulatory element that stimulates transcription in response to low temperature and water deficit. Proc Natl Acad Sci 94(3):1035–1040. https://doi.org/10.1073/pnas.94.3.1035

Toensmeier E (2007) Perennial vegetables: from artichoke to "Zuiki" Taro, a Gardener's guide to over 100 delicious, easy-to-grow edibles. Chelsea Green, White River Junction

Venema JH, Dijk BE, Bax JM, Van Hasselt PR, Elzenga JTM (2008) Grafting tomato (*Solanum lycopersicum*) onto the rootstock of a high-altitude accession of *Solanum habrochaites* improves suboptimal-temperature tolerance. Environ Exp Bot 63(1–3):359–367. https://doi.org/10.1016/j.envexpbot.2007.12.015

Venkateswarlu B, Shanker AK (2012) Dry land agriculture: bringing resilience to crop production under changing climate. In: Crop stress and its management: perspectives and strategies. Springer, pp 19–44

Villafuerte LR, Villafurte-Abonal L (2009) Data taken from the forestry Agency of Japan in Moringa. Apples of Gold Publishing, Singapore

Welbaum GE (2015) Vegetable production and practices. Centre Agril Biosci Int, p 476

Yamaguchi T, Maeda T (2015) Effect of high-temperature treatments on the breaking of dormancy in one-year-old Asparagus (*Asparagus officinalis* L.). Environ Cont Boi 53(1):23–26. https://doi.org/10.2525/ecb.53.23

Yetisir H, Çaliskan ME, Soylu S, Sakar M (2005) Some physiological and growth responses of watermelon [*Citrullus lanatus* (Thunb.) Matsum. and Nakai] grafted on to *Lagenaria siceraria* to flooding. Environ Exp Bot 58(1–3):1–8

Yordanov I, Velikova V, Tsonev T (2013) Plant responses to drought, acclimation, and stress tolerance. Photosynthetica 38(2):171–186

Zandalinas SI, Mittler R, Balfagon D, Arbona V, Gomez-Cadenas A (2018) Plant adaptations to the combination of drought and high temperatures. Physiol Plant 162(1):2–12. https://doi.org/10.1111/ppl.12540

Zhu JK (2016) Abiotic stress signalling and responses in plants. Cell 167(2):313–324. https://doi.org/10.1016/j.cell.2016.08.029

Chapter 7
Impact of Climate Change on Underexploited Vegetable Crops Production and Mitigation Strategies

Hemant Kumar Singh, Shashank Shekhar Solankey, Pankaj Kumar Ray, Prakash Singh, Md. Shamim, Raj Narain Singh, and Anjani Kumar

Abstract Several underexploited species, because they are hardy and naturally able to grow in harsh conditions, can be used as alternatives to staple crops in places where temperature (cold/heat), moisture (drought/flooding), poor soil fertility, soil salinity, and other factors limit crop production. Better crop resilience in the face of biotic and abiotic challenges under the changing climatic situation may be achieved by greater usage of these vegetables, which will contribute to diversity in agricultural production systems. Because they are in high demand locally and can be grown commercially in either rural or urban settings, these underutilised vegetables provide prime opportunities for increased utilisation of crop biodiversity in horticulture. The underexploited species are also adapted to many tropical conditions,

H. K. Singh (✉)
Krishi Vigyan Kendra (Bihar Agricultural University, Sabour, Bhagalpur), Kishanganj, Bihar, India

S. S. Solankey
Department of Vegetables & Floriculture, Agricultural Research Institute, Patna, Bihar, India

P. K. Ray
Krishi Vigyan Kendra (Bihar Agricultural University, Sabour, Bhagalpur), Saharsa, Bihar, India

P. Singh
Department of Plant Breeding & Genetics, Veer Kunwar Singh College of Agriculture (VKSCOA) (Bihar Agricultural University, Sabour, Bhagalpur), Dumraon, Buxar, Bihar, India

Md. Shamim
Department of Molecular Biology and Genetic Engineering, Dr. Kalam Agricultural College (Bihar Agricultural University, Sabour, Bhagalpur), Kishanganj, Bihar, India

R. N. Singh
Directorate of Extension Education, Bihar Agricultural University, Bhagalpur, Bihar, India

A. Kumar
ICAR-Agricultural Technology Application Research Institute (ATARI), Patna, Bihar, India

S. S. Solankey, M. Kumari (eds.), *Advances in Research on Vegetable Production Under a Changing Climate Vol. 2*, Advances in Olericulture,
https://doi.org/10.1007/978-3-031-20840-9_7

diseases and pests. Therefore, they can be a prominent genetic resistance source for cultivated species against various pests and diseases. These species have vast potential for sustainable yield and better income generation. A number of species, including the yam (*Dioscorea esculenta*), caahua (*Chenopodium pallidicaule*), bambara groundnut (*Vigna subterranea*) and tepary bean (*Phaseoulus acutifolius*) are being cultivated as substitute crops for various regions around the world because of adapted to semi-arid environments. These neglected and underutilized species NSU are more durable and more suited to grow in marginal environments than conventional staple crops, offering cost-effective and practical food production alternatives. This issue features indigenous and minor veggies that reduce hunger, malnutrition, and improve health. It may also help academics, researchers, and farmers.

Keywords Underexploited vegetables · Temperature · Moisture · Salinity · Stress · Biodiversity

7.1 Introduction

Vegetable crops that are neither extensively traded nor produced on a commercial basis may be considered underused. Vegetables are usually herbaceous and succulent in nature so far, easily affected by abiotic as well as biotic stresses with minor changes in the climate (Solankey et al. 2021). Underutilized vegetables are those that have been used historically but are not well recognised or appreciated for their fruit, fibre, fodder, oil, or medicinal characteristics. Despite India's various agro-climatic conditions, which are conducive to cultivating a total of sixty (60) common and thirty (30) less common vegetable crops are grown, little focus has been paid to underexploited vegetables (Kumar et al. 2018).

The Indian subcontinent has one of the richest emporia in the world, with more than 2500 plant species that are used in traditional medicine and are also sources of food. One of the most abundantly endowed with a wide variety of genetic resources is the Indian subcontinent. Only roughly 200 of the over 2.5 million species of flowering plants found throughout the globe are known to have been domesticated, despite the fact that about 3000 of these species are considered to be sources of food. It is estimated that there are over 400 different species of vegetable crops across the world. It is believed that approximately 80 different species of major and minor vegetables first appeared in India.

However, these species have the potential to provide nutritional security, health, income and environmental benefits that are now being underutilised. The consumption of such foods is frowned upon in certain societies. As rich stores of vitamins, minerals, protein, and other plant nutrients, indigenous vegetables and underused legume crops might play an important role in attaining nutritional security. For sustainable food production, underutilised traditional vegetables like African night shade (*Solanum scabrum*), Amarnath (*Amaranthus Spp.*), Asian brinjal (*Solanum *melogena*), African brinjal (*Solanum aethiopicum*), Malabar spinach (*Basella alba*), Drumstick (*Moringa oleifera*), Water spinach (*Ipomoea aquatica*), Winged bean (*Psophocarpus tetragonolobus*) as well as many gourd species lessen the environmental influence of food production. These resilient crops have evolved to restricted soil and environment and need little inputs (Hughes and Ebert 2013).

In India, children under the age limit of 3–4 years are most likely to suffer from malnutrition deficiency disease. People in the area have been eating underutilised vegetable crops for a long time, and they are knowledgeable of the nutritional and therapeutic benefits these crops provide. These are inexpensive and may be obtained easily. Evidently, research and conservation efforts pertaining to these crops of healthy vegetables that are extensively eaten have paid comparatively little attention to them. One example of the latter is the cultivation of underutilized indigenous vegetables, which are better able to adapt to changes in the local climate. In addition to protecting impoverished rural children in India, especially those under the age of three to four, from malnutrition deficient illness, these crops are resilient to a wide range of weather and soil conditions. UNFCC defined climate change as "a change of climate which is attributed directly or indirectly to human activity or due to natural variability". People are aware of the nutritional and therapeutic advantages of underutilised vegetable crops, which have an ancient historical background of local usage.

Uncultivated but edible plant species have provided a means of subsistence for the poor for generations, but they face an uncertain future as a consequence of climate change threats such as shifts in rainfall patterns, temperature, relative humidity, radiation, the weeds-pests-diseases complex, and overall changes in the trends of climatic elements. If you want to understand through the Fig. 7.1 given below, then it is clear that the problem caused by climate change should be diagnosed in time and take action. The interconnection of climate, ecosystems, biodiversity, and human societies (Fig. 7.1) incorporates natural, ecological, social, and economic

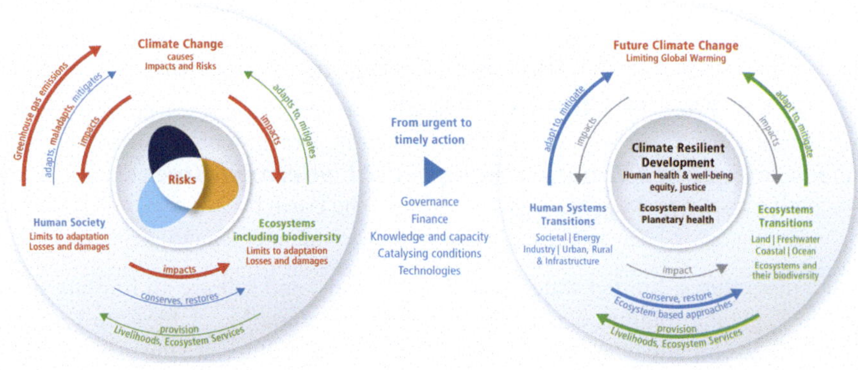

Fig. 7.1 From climate risk to climate resilient development: climate, ecosystem (including biodiversity) and human society as coupled system (IPCC 2022)

sciences more than prior IPCC 2022 assessments. Climate change consequences, hazards, and adaptation are assessed against non-climatic world trends.

Main interaction and trends **b**. Options to reduce climate risk and establish resilience.

Temperature, rainfall, relative humidity, atmospheric gas composition, etc., over a extensive time period and bigger geographical region. Risk, which combines effect size and likelihood, incorporates uncertainty in climate change, exposure, impacts, and adaptation. Rising temperatures, shifting precipitation patterns, increased UV radiation, and more severe weather events like droughts and floods are serious challenges to tropical vegetable production (Tirado et al. 2010). Vegetable crops are vulnerable to climate vagaries, and rapid temperature rises and erratic precipitation may disrupt normal growth, blooming, pollination, fruit development, and crop output (Afroza et al. 2010). Climate change and its fluctuation affect agriculture, notably vegetable crops. Short growing seasons may reduce fruit and vegetable yield due to heat stress and low water availability. Climate change and fluctuation have increased uncertainties and dangers, limiting vegetable output. Climate change may raise vegetable prices. Climate change promotes the spread of infections and the emergence of novel insect pests, fungal, bacterial, and viral illnesses. To address expanding needs in a world of diminishing land, water, and climate change, location-specific and knowledge-intensive climate-smart vegetable treatments are required (Malhotra and Srivastva 2014). This paper reviews climate change's effects on vegetable production and management. India's climate and soil are ideal for growing a variety of underexploited vegetables. Consequently, the Indian government has taken certain initiatives to raise awareness about underexploited vegetables. Some GOs and NGOs have been working to enhance the production of underused vegetables through various development programmes. We can therefore conclude that the production of underutilised and phytochemistry vegetables will meet the shortage of per capita consumption availability and thus solve the nutritional problems while also creating employment and raising the income generation of rural society and finally contributing to the national economy.

7.2 Vegetable Production and Management Practices in the Face of Climate Change

7.2.1 Impact of Climatic Changes on Underexploited Vegetable Production

High temperatures, limited water availability, floods, and salt are all limiting problems in vegetable output, according to several workers. Vegetables are typically vulnerable to climatic extremes; as a result, high temperatures are the primary cause of poor yields, and the effects of climate change will only be exacerbated in the future. Climate change, particularly increasing carbon dioxide (CO_2) concentrations and rising temperatures, may have a significant impact on the status of a pest species. Temperature affects insect pests together with CO_2. Climate change may also harm natural enemies. As the world warms, understanding the population dynamics of plant insect-pests and their natural enemies in traditional vegetable production systems will be a critical and basic problem. Indigenous minor vegetable crops (locally used underutilised vegetable crops), especially green vegetable crops, and trait-specific germplasm require more study.

Prolonged drought and increasing pest infestations are two of the detrimental effects of climate change on farms and crops, which in turn has serious economic consequences (Bifulco and Ranieri 2017). While most agricultural products suffer as a consequence of climate change, vegetables are more vulnerable due to their short production cycles and herbaceous nature (Abewoy 2018). (Prasad and Chakravorty 2015). While indigenous veggies play a crucial part in creating a climate-resilient route, there is a lack of data on both the quantitative and qualitative levels to back up this claim (Padulosi et al. 2011; Pachauri et al. 2014; Chivenge et al. 2015). Underutilized indigenous vegetables (UIVs) may play a vital role in climate-resilient vegetables, despite their limited percentage of agro-food systems (Scheelbeek et al. 2018). The expected detrimental effect of climate change on indigenous vegetables cannot be disregarded, and more recent study demonstrates that the influence of climate change on indigenous vegetables is both beneficial and bad in a qualitative sense (Chepkoech et al. 2018).

7.2.1.1 Temperature

According to the IPCC (2014), significant climate change has already happened since the 1950s and the second half of this century is projected to see a rise in global temperature of 0.4–2.6 °C (depending on future greenhouse gas emissions). This high temperature is the cause of erratic rainfall pattern, melting of arctic glacier, higher sea level, appearance of severe droughts, storms, salt-water and floods inundation, varied wind patterns, heat waves, changes in ocean currents, acidification, forest fires and induce hole in ozone (Minaxi et al. 2011; Kumar 2012) have been a major threat for vegetable production in the tropical and arid zones (Tirado et al. 2010; Abdelmageed et al. 2014).

A greater contribution of greenhouse gas (GHG) emission is from agriculture and allied sectors. These emissions might rise even more in the future, if intensive agricultural practices will be performed as well as rising demand for animal products. Between 2005 and 2050, the demand for livestock-related goods is predicted to increase by 70%. Extreme heat and dryness during crop blossoming may limit output gains in certain places. Gradual temperature and CO_2 rises may improve agricultural yields.

Heat waves may lead to heat stress in both plants and animals, which has a detrimental effect on how much food is produced. When the plants are in the blooming stage, which is a single, crucial stage, extreme heat waves are very bad for crop output since they may result in no seeds at all. Heat stress in animals may reduce their fertility and production. Lettuce plants form premature flowering and seeds without forming heads when get exposure to higher temperature (21°–27 °C). In general, leafy vegetables, which are typically found growing wild or as weeds, are simple to cultivate, have minimal production costs, and provide a high yield. Because they are highly resistant to the effects of climate change and one of the least expensive vegetables now on the market, they may be accurately defined as a crop that is both climate smart and nutrition smart, and they should be promoted and exploited on a worldwide scale. India is the home of a large number of leafy vegetables, such as *Moringa oleifera, Basella alba, Murraya koenigii*, etc., as well as a number of other green vegetables that are less well-known and are only available during specific times of the year. However, most of vegetables require optimum temperatures of 25 °C. Optimum temperature lies between maximum and minimum temperature range presented in Table 7.1. Extreme temperature above 40 °C causes desiccation whereas temperature below 0 °C witness chilling and freezing injury in many vegetables.

Higher temperatures other than optimum witness affect the development of transition phase like leafy head, bulb, tuber, fleshy stem, fleshy root, head and curd and malfunction in reproductive growth that leads to poor flowering and seed setting cruciferous vegetables (Wheeler et al. 2000). Higher temperatures faster the development rates resulting improper organ development in vegetables e.g., transition phase have undesirable traits (Table 7.2). Warm temperatures throughout all seasons can interfere the chilling requirements in temperate vegetables. With global warming, winter dormancy may be inhibited which may results in reduced yield and quality in many vegetables.

Table 7.1 Maximum and minimum temperature for production of underexploited vegetables

Temperature (°C)	Vegetables
Maximum	
32	Zucchini, Squashes and Sweet Corn
30	Leeks and Shallots
27	Beans
24	Swede, Parsnip, Lettuce, Celery, Cabbage, Brussels Sprouts and Artichokes
Minimum	
5	5 Spinach, Beetroot, Brussels sprout, Parsnips, Turnips and Swedes
7	Artichokes, Celery, Leeks, Shallots and Lettuce
10	Zucchini, Butter squash and Beans
12	Sweet corn

Adapted from Maynard and Hochmuth 1997

Table 7.2 Impact of high temperature for minor vegetables

Degree of sensitivity	Vegetables	Effect due to prolong period
Very sensitive	Brussels Sprout	Hollow stem, leafiness, absent of curd, poor quality heads, blindness, brown head, uneven cluster development, loose clusters
Sensitive	Beans	Poor pollen viability, improper pollination, abscission of flower buds fibrous pod
	Persnip	Inferior root quality
	Lettuce	Thermo-dormancy of seeds Bolting, loose heads, bitterness, tip burn, small & light heads,
	Celery	Less quality stems Garlic Cool season crop effected
	Leek	Cool season crop effected
Tolerate	Squashes	Improper pollination
	Shallot	Tolerate
	Parsley	Tolerate
	Asparagus	Fibrousness, feathering, delayed spears growth and over branching
	Sweet corn	Pollen blast, poor pollination and fruits set on cob

Modified from Pramanik et al. (2020)

Table 7.3 Effect of low temperature on minor vegetables

Degree of sensitivity	Vegetables	Effect due to prolong period
Very sensitive	Beans	Pitting and russeting, poor root system, reduced photosynthesis and yield.
	Zucchini & Button squash	Decay, especially Alternaria rot
	Sweet corn	Poor germination, and low percentage emergence
Sensitive	Lettuce	Russeting, Anthracnose, external frost damage, external cracking
Moderate tolerance	Chard	Poor germination, and low percentage emergence
	Snow peas	6 weeks after emergence and flowering
	Celery	8 weeks after emergence and blooming
	Parsley	Effect of all parts of plants
	Leek	10 weeks after emergence
	Parsnip	Emergence stage
	Asparagus	Dull, gray-green, limp tips

Modified from Pramanik et al. (2020)

7.2.1.2 Impact of Low Temperature on Vegetables

Plants suffer low temperature which includes chilling and freezing temperature and the extent of damage varies considerably according to different of crops. Vegetables suffer chilling injury with temperature ranges from 0 °C–10 °C and freezing injury below 0 °C which cause significantly damage to crops (as presented in Table 7.3).

Tropical and subtropical vegetables are most sensitive to chilling and freezing injury, even with small exposure to freezing conditions, whereas temperate vegetables often tolerate to mere freezing condition unless it is too severe. Lettuce (*Lactuca sativa*) is a cool season crop that requires a cool and damp environment (Table 7.3). Temperatures below 24 °C in sunny circumstances are ideal for growth and development and the production of high-quality lettuce. A minimum average temperature of 18 °C is required for effective lettuce cultivation.

A study found that night temperature is more important than day temperature in floral induction and bolting. By experiencing night temperatures of 21 °C as opposed to 16 °C, lettuce var. Iceberg flowers 21 days early. Some varieties with such genotypic composition are resistant to bolting, while experiencing greater night temperatures of more than 15 °C. Though lettuce evolved in temperate regions of the globe, cold temperatures around 0 °C are particularly damaging to growth and development, resulting in damage or delayed maturity as a consequence of chilling injury, resulting in a drop in yield or crop failure. Wurr et al. (1996) discovered that increasing the temperature from 16.3 to 21.1 °C causes early faster growth rates and shorter crop duration, reducing lettuce output by 17 percent. Premature seed head development begins with a quick increase in temperature over the ideal temperature, resulting in lower quality heads and low yields. Temperature has a good effect on lettuce seed germination. Lovatt et al. (1997) discovered that soil temperatures over 24 °C limit seed germination in lettuce, while soil temperatures above 30 °C severely inhibit germination. Kretschmer (1978) discovered that temperatures ranging from 15 to 20 °C resulted in the greatest percentage germination (90%) in lettuce. High soil temperature and relative humidity are closely connected to Ca availability, resulting in deficiency and tip burn in lettuce.

Thompson (1986) discovered that with crucial weather variance, sweet corn fails to provide optimal output, but normal or below-average temperatures in late summer produce the highest yield. Sweet corn plants that are exposed to ideal temperatures, especially during the silking and tasselling stages, produce more (Commuri and Jones 2001). Shaw (1983) discovered that higher temperatures over 32 °C reduce sweet corn yields by more than 4% each day for a period of up to 3–4 weeks after pollination. Ledencan et al. (2008) found that extreme drought and heat reduced kernel water and sugar content.

7.2.1.3 Impact of Flood

Vegetables, particularly those with shallow root systems, are very vulnerable to floods. More than a third of the world's irrigated land is flooded. Heavy rain, improper irrigation, unlevelled land, poor drainage, or heavy soil may cause it. The majority of vegetables, particularly those with shallow root systems, are very delicate and easily damaged by water. In low-lying, rain-fed regions, waterlogging is a significant issue that has a negative impact on the growth and production of crops. Oxygen deprivation is the primary factor that contributes to damage caused by water logging. Because oxygen deprivation inhibits nutrient and water absorption, plants begin to wilt even when they are surrounded by an excessive amount of water. When circumstances are waterlogged, plant roots must work harder to get oxygen since the air in the soil has been replaced by water. Therefore, drowning substantially inhibits root respiration, making it difficult for roots to continue their normal activities of nutrition and water intake. This results in necrosis of the root, which opens the door for infection with pathogens carried by the soil (Decoteau 1988). The severity of the symptoms associated with flooding rises as temperatures rise; quick withering and mortality of tomato plants are often seen after a brief high temperature when period of flooding is present (Kuo et al. 1982). Diseases become more severe as a result of flooding. Discoloration, root rot, and plant mortality before their time are signs of sick roots. The injury limits the root systems' capacity to acquire minerals or carry out other crucial tasks for thc shoot. Thc two widespread illnesses Phytophthora and Pythium are the ones that harm roots in poorly drained soil the most. During the second, third, and fourth days after flooding, pea plant roots were found to have higher ABA concentrations, which caused the stomata in the leaves to partly shut (Zhang and Davies 1987). To get into the mesophyll cells where it may be utilised for photosynthesis, CO_2 must first diffuse into those spaces within the leaf via the stomata. The light-independent processes are stopped when the stomata shut because the CO_2 levels in the leaf decrease quickly. After then, photosynthesis is stopped. According to research by Igwilo and Udeh (1987) on the effects of water stagnation on yam (*Dioscorea spp.*) vines over 24, 48, and 72 hours, the leaf degeneration progressed from newly developing lesions on the lower leaves to necrotic patches to full necrosis. When asparagus (*Asparagus officinalis* L.) plants were continually flooded for 48 hours after being transplanted at 2 and 3 weeks, compared to two 8-hour floods separated by a 16-hour drainage phase, the growth and survival of the transplanted seedlings were delayed (Falloon et al. 1991). When anaerobic conditions continue for many days, the chemical characteristics of the soil also alter. After that, when the water level recedes, the lower leaves lose their chlorophyll, crop development is halted, and surface root growth multiplies. The leaf is also particularly susceptible to waterlogging; during a period of waterlogging, alterations in the leaf's respiration, chlorophyll concentration, and photosynthetic assimilation have been observed. Under waterlogging circumstances, chlorophyll fluorescence properties in particular are often affected. After 5 days of flooding, the bitter melon's respiratory capacity was just 28% of what it was before the floods (Liao and Lin 1995). Under anaerobic circumstances, plant transpiration is impacted, and prolonged water logging causes root death from a lack of oxygen.

A key issue for vegetable farmers during the rainy season in tropical regions where high temperatures during the daytime cause quick withering and death of plants is to mitigate the effects of floods. Grafting onto the rootstock of Solanum melongena improved tomato disease and flooding tolerance (Palada and Wu 2007). Reversing soil erosion, rebuilding degraded land, improving drought and floods resilience, increasing water use efficiency (WUE), water conservation techniques and agro-genetic biodiversity all contribute to organic farming's ability to adapt to climate change (IAASTD 2008; IFOAM 2009).

7.2.1.4 Impact of Drought

The principal causes of agricultural drought include a lack of precipitation, high temperatures, and/or winds that are above normal, all of which induce moisture loss from soils and plants. Drought's impact on agricultural productivity is influenced by a wide range of factors, including social and ecological vulnerabilities, the availability of irrigation, the types of crops grown, and other factors. Because of the sensitivity of vegetable production and their inherent weakness, smallholder farmers are particularly vulnerable. In vegetable cultivation, drought and water stress have been highlighted as key constraints. Drought in agriculture occurs when soil water availability to plants in a given location drops below a level that is inadequate for crop development and growth, and consequently negatively impacts the final crop production. Succulent fruits and vegetables, by definition, contain more than 90% water. Succulent leafy crops like amaranthus, palak, and spinach are hard to grow in a dry climate, thus farmers are forced to compromise on quality and quantity (AVRDC 1990). Consequently, the quality and output of vegetables are highly influenced by water; drought circumstances have a significant impact on vegetable production. Drought length, severity, and timing all have a role in how much of an impact drought has. When water stress occurs, electrolyte leakage, leaf relative water content, chlorophyll content, vegetative development, and fruit output and quality are all significantly reduced. While enlarging storage organs (root or bulb), drought stress may negatively affect both the production and quality. Most susceptible to drought stress are cucumbers, melons, pumpkin and squashes (including lima and snap), lima bean snap beans peas peppers, sweet corn, and tomatoes. As a result, water shortages, particularly during the vegetative development phase, have the potential to significantly impact yields and quality of vegetables (Table 7.4).

Dryness stress during flowering and fruit setting affects fruit quality and quantity as well as yield by decreasing the number of fruits and seeds, with the latter resulting in a reduction in overall output. Water deficit conditions have an impact on the physiological processes that are primarily responsible for plant growth and development. Plants display a variety of drought stress defence mechanisms at the molecular, cellular, and whole plant levels. In order for a plant to thrive, it must get the right amount of water. Rain and irrigation are the primary methods of supplying water to most soils. When water is scarce, crops are only watered when absolutely necessary. The following crops must be watered at certain times throughout their growth cycle.

For instance, Luoh et al. (2014) discovered that in situations of water stress, the fresh edible yield of three AIVs is decreased. Additionally, indigenous foods like amaranth have been shown to be drought tolerant (Van den Heever and Slabbert 2007). Given that the major mechanism of drought adaptation for stress recovery was likely a change in stem biomass, breeding for characteristics associated with drought tolerance should focus on dark-adapted quantum yield (Fv/Fm) with a rich source of genetic variety (Jamalluddin et al. 2021). According to a study by Sambo (2014), unused crops might provide scientists a plentiful source of genetic material for modifying. It is possible to develop strong, drought-tolerant crops using this genetic material. Recent research suggests that these underutilised native crops

Table 7.4 Critical stages of drought stress and its impact on vegetable crops

Impact of water stress	Critical period for watering	Vegetables
Toughness of leaves, accumulation of nitrates and poor foliage growth.	Throughout growth and development of plant	Leafy vegetables
Reduce spear size and quality as well as increased fiber content, results tougher, lower quality spears.	Spear production and fern (foliage) development	Asparagus
Toughness of leaves, stunted plant growth, burning of tip.	Consistently throughout development	Lettuce
Leaf become slight greyish cast, blossom drop, poor seed viability.	Pollination and pod development	Lima bean
Inadequate moisture levels results blossoms drop as well as sett pods fail to fill and poor seed viability.	Flowering and pod enlargement	Snap bean
Tassel and shed pollen before silks on ears are ready for pollination, lack of pollination and fertilization results missing rows of kernels, poor yields, or even eliminate ear production as well as poor seed viability.	Silking, tasselling and ear development	Sweet corn

Adapted from Bahadur et al. (2010) and Kumar et al. (2012)

might survive in places with limited water resources, and that their untapped potential could hold the key to the future security of food and nutrition (Mabhaudhi et al. 2016).

Especially in the major areas of each crop's diversity, there are a number of neglected and underutilised species (NSU) that have long been used for food by traditional populations. Although globalisation of food systems has led to a homogenization of crops cultivated throughout the globe at the expense of practice, it has also directed to introductions of crop in nations outside of the traditional areas of variety (Khoury et al. 2016). Therefore, nations use "foreign crops" that were brought from locations with a different diversity profile than their own, showing that introduction of crop is a process that has occurred throughout the history of cultivated crops. Therefore, it is important to support and promote research on neglected and underutilised species to learn more about their usefulness and to advocate their usage as other crops to develop more resilient cropping systems. Cowpea (*Vigna unguiculata*), quinoa (*Chenopodium quinoa*), millet (*Pennisetum glaucum*), buckwheat (*Fagopyrum sp.*) and sweet potato (*Ipomoea batatas*), research showed that certain plants can adapt to water scarcity, and they've subsequently been introduced as commercial crops. (Table 7.5). Exploring and introducing these NUS into agricultural systems in different regions around the world is a testament to the global efforts surrounding them, proving their importance as a strategy for future needs relating to nutritional enhancement, crop diversification and adaptation to changing climates.

A highly varied agricultural production system is efficient for supporting food and nutritional security in water-scarce regions is made possible by fully realising

Table 7.5 Neglected and underutilized species introduced as drought-tolerant crops in several countries

Origin of crop	Vegetables	Registered variety by countries
Andean region (Risi and Galwey 1984)	Quinoa (*Chenopodium quinoa* Willd.)	France & Germany (1), Australia & Ukraine (2), USA (3), Denmark & Canada (4) Netherlands (5), (Jacobsen 2017 and UPOV 2020)
American highland tropical and subtropical regions (Di Fabio and Parraga 2017)	Amaranth(*Amaranthus cruentus* L., *Amaranthus hypochondriacus* and *Amaranthus caudatus* L.)	Germany (1 *Ah*, 1 *Acr*), Russian (7 *Ah*, 8 *Ac*, 6 *Acr*), Slovakia (1 *Ac*, 2 *Acr*), Romania (1 *Ac*, 1 *Acr*), Ukraine (1 *Ac*), Czech Republic (1 *Acr*), The Netherlands (1 *Ac*), Poland (2 *Acr*), New Zealand (1 *Acr*) (UPOV 2020)
Southern Africa (Padulosi and Ng 1997)	Cowpea (*Vigna unguiculata*)	Poland (1), Portugal (1), Bulgaria (2), Romania (3), Korea (4), Moldova (6), Turkey (7), China (7), Australia (8), Brazil (13) (UPOV 2020)
North of South America and central America (Munoz-Rodríguez et al. 2018)	Sweet potato (*Ipomoea batatas*)	Romania (2), Switzerland (3), Ukraine (4), Slovenia (8), Israel (11), South Africa (29), China (42), USA (89), (UPOV 2020)
Andean region (Atchison et al. 2016)	Andean Lupin (*Lupinus mutabilis*)	Czech Republic (1), The Netherlands (1), Germany (1) (UPOV 2020)

Adapted from Rosero et al. (2020)

the promise of NUS (Singh et al. 2020). A number of species, including the teff (*Eragrostis tef*), yam (*Dioscorea esculenta*), drumstick (*Moringa oleifera*), taro (*Colocasia esculenta*), tepary bean (*Phaseoulus acutifolius*) and canahua (*Chenopodium pallidicaule*). Due to their natural adaptability to arid semi-arid locations (centre of origin or diversity), among others, might serve as alternative crops for different places throughout the world (Khoury et al. 2016) to contribute to better agricultural diversification and water-efficient farming practises.

The precise water management of individual plants is a key aspect of our system's complicated phenotyping technique. This was accomplished via the use of computer-controlled irrigation systems to keep the soil water capacity at a predetermined level (20 and 60 percent for water limited and well-watered conditions, respectively). Using this method, it was feasible to keep track of how much water each plant used over time. Our plant phenotyping platform's computer-controlled water delivery system was used to provide regulated watering conditions (Cseri et al. 2013; Feher-Juhasz et al. 2014). Mulching, proper watering, appropriate nutrient delivery, the use of rhizobacteria that promote plant development, and the choice of climate-resistant cultivars are some of the treatments and management techniques being utilised to alleviate drought stress.

7.2.1.5 Impact of Salinity

Most crops, including vegetables, cannot be produced in many parts of the globe due to excessive soil salt content, especially in arid and semi-arid environments (FAO 2002). Globally, 20 percent of cultivated land and 33% of irrigated land are damaged and degraded by salinity, which is a serious issue for agricultural production. About 950 million hectares of land in arid and semi-arid areas are salt-affected due to natural and anthropogenic causes (Aronson 1985). Many agricultural crops, including most vegetables, which are especially sensitive during the plant's ontogeny, suffer from excessive soil salt, which lowers their yield. Allium species—with the exception of chive—are grown for their bulbs and sometimes the base of the flattened leaf blades. For garnish and flavour, chive is only utilised in the leaf blades. Onion, garlic, leek, and chive are often thought to be salt sensitive based on production loss, although only onion and garlic have strong evidence (Fig. 7.2) Shannon and Grieve (1999). Most vegetable crops have poor salt tolerance (ECt; 1 to 2.5 dS m-1 in saturated soil extracts), and when salty water is used for irrigation, this salt tolerance reduces.

Plants respond to salt in a variety of ways, including changes in morphology, physiology, and metabolism. Plants'rates of photosynthesis and respiration were lowered. Protein, fatty acid, and total carbohydrate concentrations were all reduced by salinity, while the concentration of amino acids, notably proline, was significantly elevated. At the cellular, tissue, and whole-plant levels, halophytes are responsive to salinity (Aslam et al. 2011). When it comes to salt stress, halophytes have the finest germplasm and can flourish in very salty circumstances (Mishra and Tanna 2017). Osmotic and oxidative stress may harm plant development in salty soils across the globe, affecting agricultural yields in the process (Joseph et al. 2011). Due to the presence of harmful ions, salt causes water and ionic imbalance in plants. Salt stress stunts plant development and darkens the leaf colour (Rani et al. 2019). The vast majority of vegetable crops have a limited tolerance for saline water that is regularly administered to them. Crops may be sensitive, moderately

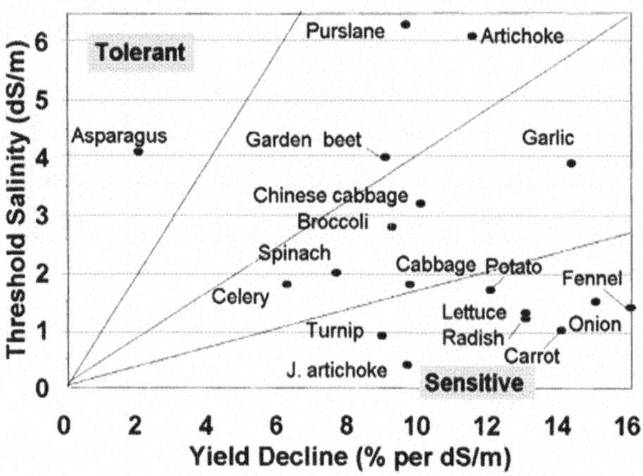

Fig. 7.2 Salt tolerance of several vegetable species as rated by the salinity threshold and percent yield decline (Shannon and Grieve 1999)

Table 7.6 Vegetable crops tolerant to salinity (ECe dS/m)

Sensitive crops (Up to 4)	Semi-tolerant (4–6)	Tolerant crops (6–8)
Bean, fennel, radish, **celery**, peas, brinjal,	Carrot, onion, cauliflower, muskmelon, chilli, water melon cucumber, pumpkin, okra, potato, bottle gourd, **lettuce**, bell pepper, **artichoke**	Fennel, palak, spinach, turnip, carrot, bitter gourd, onion (seed crops), cabbage, beet root, **asparagus**

Modified from Maas et al. (1990) and Hanson et al. (2006)

sensitive, moderately tolerant, or tolerant to salt. Last are unsuitable crops. Most vegetable crops are environmentally sensitive (Maas et al. 1990 and Hanson et al. 2006) Table 7.6. It has long been believed that asparagus is the vegetable crop that can tolerate salt the best.

7.3 Conclusion

Strategies for mitigating technologies that are climate-resilient Several attempts have been launched to reduce the potential effects of climate change on vegetable production as well as on the national economy. Vegetable crop genetic improvement is a suitable adaptation method to face the challenges of climate change. Wild relatives are a significant source of genes and gene complexes that must be assessed and exploited with the right methods and instruments in the context of climate change. More focus should be put on trait discovery than on trait specific assessment in order to enhance the use of germplasm. In this study, we looked at two techniques to enhancing crops' resistance to water scarcity: I bringing underutilised and neglected species into cropping systems, and (ii) exploiting natural drought stress tolerance genes discovered in crop wild relatives and landraces. To combat harsh climatic pressures, they include effective management techniques, steps to increase water and nutrient usage efficiency and biological nitrogen fixation, genotype selection for greater adaptability, and genetic engineering. methods in management to increase vegetable output. Another method of using plant biodiversity to combat climate change is by grafting a sensitive scion cultivar onto a resistant rootstock. Agronomic techniques including mulching, organic farming, cropping systems that sequester carbon, and agroforestry provide a variety of potential options for mitigating the effects of climate change on vegetable production. Protected farming and post-harvest technologies may be important strategies for overcoming climate change issues. For optimal production and the development of technology for underexploited vegetables, awareness and information distribution about the impacts of soil and environmental elements at the producers and researchers are essential.

References

Abdelmageed AH, Gruda N, Geyer B (2014) Effects of temperature and grafting on the growth and development of tomato plants under controlled conditions. Rural Poverty Reduction through Research for Development and Transformation

Abewoy D (2018) Review on impacts of climate change on vegetable production and its management practices. Adv Crop Sci Tech 6(330):7

Afroza B, Wani KP, Khan SH, Jabeen N, Hussain K (2010) Various technological interventions to meet vegetable production challenges in view of climate change. Asian J Hort 5:523–529

Aronson J (1985) Economic halophytes—a global review. In: Plants for arid lands. Springer, Dordrecht, pp 177–188

Aslam R, Nazish B, Nabgha-e-Amen MM, Waseem S (2011) A critical review on halophytes: salt tolerant plants. J Med Plant Res 5(33):7108–7118

Atchison GW, Nevado B, Eastwood RJ, Contreras-Ortiz N, Reynel C, Madrinan S, Filatov DA, Hughes CE (2016) Lost crops of the Incas: origins of domestication of the Andean pulse crop tarwi, Lupinus mutabilis. Am J Bot 103:1592–1606

AVRDC (1990) Vegetable production training manual. Asian Vegetable Research and Training Center. Shanhua, Tainan, pp 447

Bahadur A, Kumar R, Mishra U, Rai A, Singh M (2010) Physiological approaches for screening of tomato genotypes for moisture stress tolerance. In: National Conference of plant physiology (NCPP-2010). BHU, Varanasi, p 142

Bifulco M, Ranieri R (2017) Impact of drought on human health. Eur J Intern Med 46:e9–e10

Chepkoech W, Mungai NW, Stober S, Bett HK, Lotze-Campen H (2018) Farmers' perspectives impact of climate change on African indigenous vegetable production in Kenya. Int J Clim Change Strateg Manag 10(4):551–579

Chivenge P, Mabhaudhi T, Modi AT, Mafongoya P (2015) The potential role of neglected and underutilised crop species as future crops under water scarce conditions in Sub-Saharan Africa. Int J Environ Res Public Health 12(6):5685–5711

Commuri PD, Jones RJ (2001) High temperatures during endosperm cell division in maize: a genotypic comparison under in vitro and field conditions. Crop Sci 41:1122–1130

Cseri A, Sass L, Torjek O, Pauk J, Vass I, Dudits D (2013) Monitoring drought responses of barley genotypes with semi-robotic phenotyping platform and association analysis between recorded traits and allelic variants of some stress genes. Aust J Crop Sci 7:1560–1570

Decoteau D (1988) Plant physiology: environmental factors and photosynthesis. Greenhouse Glazing & Solar Radiation Transmission Workshop on Center for Controlled Environment Agriculture. Rutgers University, Cook College, pp 1–6

Di Fabio A, Parraga G (2017) Origin, production and utilization of Pseudocereals. Wiley, Hoboken, pp 1–27

Falloon PG, Greathead AS, Mullen RJ, Benson BL, Grogan RG (1991) Individual and combined effect of flooding phytopthora rot and metalaxyl on asparagus establishment. Plant Dis 75:514–518

FAO (2002) Working with local institutions to support sustainable livelihoods. Food and Agriculture Organization. Rome, Italy

Feher-Juhasz E, Majer P, Sass L, Lantos C, Csiszar J, Turoczy Z (2014) Phenotyping shows improved physiological traits and seed yield of transgenic wheat plants expressing the alfalfa aldose reductase under permanent drought stress. Acta Physiol Plant 36:663–673

Hanson B, Grattan A, Fulton A (2006) Agricultural salinity and drainage; Davis, California Irrigation Program WMS (Water Management Series) 3375; University of California: Oakland, CA, USA, pp 1–159

Hughes JDA, Ebert AW (2013) Research and development of underutilized plant species: the roleof vegetables in assuring food and nutritional security. In: Massawe F, Mayes S, Alderson P (eds) Proceedings of the 2nd international symposium on underutilized plant species: crops forthe future—beyond food security. International Society for Horticultural Sciences (ISHS), Korbeek-Lo, pp 79–91

IAASTD (2008) International assessment of agricultural knowledge, science and technology for development: agriculture at a crossroads. Island Press, Washington, DC
IFOAM (2009) Organic agriculture–a guide to climate change and food security. International Federation of Organic Agriculture Movements EU group
Igwilo N, Udeh ACD (1987) Effect of water logging yam vines on vegetative growth and tuber yield. Ann Appl Biol 110:391–398
International Panel on Climate Change (IPCC) (2014) Fifth Assessment Report (AR5) [(accessed on 17 October 2014)]. Available online: http://www.ipcc.ch/report/ar5/
IPCC (2022) Summary for policymakers [H.-O. Portner DC, Roberts ES, Poloczanska K, Mintenbeck M, Tignor A, Alegria M, Craig S, Langsdorf S, Loschke V, Moller A Okem (eds.)]. In: Climate Change 2022: Impacts, Adaptation, and Vulnerability. Contribution of Working Group II to the Sixth Assessment Report of the Intergovernmental Panel on Climate Change [H.-O. Portner DC, Roberts M, Tignor ES, Poloczanska K, Mintenbeck A, Alegria M, Craig S, Langsdorf S, Loschke, V, Moller A, Okem B, Rama (eds.)]. Cambridge University Press. In Press, ISBN 978-92-9169-159-3. pp. 1–35
Jacobsen SE (2017) The scope for adaptation of quinoa in northern latitudes of Europe. J Agron Crop Sci 203:603–613
Jamalluddin N, Massawe FJ, Mayes S, Ho WK, Singh A, Symonds RC (2021) Physiological screening for drought tolerance traits in vegetable Amaranth (Amaranthus tricolor) germplasm. Agriculture 11:994
Joseph B, Jini D, Sujatha S (2011) Development of salt stress-tolerant plants by gene manipulation of antioxidant enzymes. Asian J Agric Res 5(1):17–27
Khoury CK, Achicanoy HA, Bjorkman A, Navarro-Racines C, Guarino L, Flores-Palacios X, Engels JMM, Wiersema JH, Dempewolf H, Sotelo S (2016) Origins of food crops connect countries worldwide. Proc R Soc B Biol Sci 283:20160792
Kretschmer M (1978) Temperature and lettuce seed germination. Acta Hortic 83:167–174
Kumar AJ, Deuri R, Sharma P, Singh SP (2018) Underutilized vegetable crops and their importance. J Pharmacogn Phytochem 7(5):402–440
Kumar R, Shekhar SS, Major S (2012) Breeding for drought tolerance in vegetables. Veg Sci 39(1):1–15
Kumar SV (2012) Climate change and its impact on agriculture: a review. Int J Agric Environ Biotechnol 4(2):297–302
Kuo DG, Tsay JS, Chen BW, Lin PY (1982) Screening for flooding tolerance in the genus Lycopersicon. Hort Sci 17(1):6–78
Ledencan T, Sudar R, Simic D, Zdunic Z, Brkic A (2008) Effects of the agroecological factors on sweet corn quality. Cereal Res Commun 36:1411–1414
Liao CT, Lin CH (1995) Effect of flood stress on morphology and anaerobic metabolism of *Momordia charantia*. Environ Exp Bot 35(1):105–113
Lovatt J, Heisswolf S, Carey D, Henderson C, O'Brien R, Deuter P (1997) Lettuce information kit. Department of Primary Industries, Queensland
Luoh JW, Begg CB, Symonds RC, Ledesma D, Yang RY (2014) Nutritional yield of African indigenous vegetables in water-deficient and water-sufficient conditions. Food Nutr Sci 5(9):812–822
Maas EV (1990) Crop salt tolerance. In: Tanji KK (ed) Agricultural salinity assessment and management, Ch. 13, ASCE, Manuals & Reports on Engineering, 71:262–304
Mabhaudhi T, O'Reilly P, Walker S, Mwale S (2016) Opportunities for underutilised crops in southern Africa's post–2015 development agenda. Sustainability 8(4):302
Malhotra SK, Srivastva AK (2014) Climate smart horticulture for addressing food, nutritional security and climate challenges. In: Srivastava AK (ed) Shodh Chintan scientific articles. ASM Foundation, New Delhi, pp 83–97
Maynard DM, Hochmuth GJ (1997) Knott's handbook for vegetable growers. Wiley
Minaxi RP, Acharya KO, Nawale S (2011) Impact of climate change on food security. Int J Agric Environ Biotechnol 4(2):125–127

Mishra A, Tanna B (2017) Halophytes: potential resources for salt stress tolerance genes and promoters. Front Plant Sci 8:829

Munoz-Rodríguez P, Carruthers T, Wood JR, Williams BR, Weitemier K, Kronmiller B, Ellis D, Anglin NL, Longway L, Harris SA (2018) Reconciling conflicting phylogenies in the origin of sweet potato and dispersal to polynesia. Curr Biol 28:1246–1256.e12

Pachauri RK, Allen MR, Barros VR, Broome J, Cramer W, Christ R (2014) Climate change 2014: synthesis report contribution of working groups I, II and III to the fifth assessment report of the intergovernmental panel on climate change, p 151

Padulosi S, Ng NQ (1997) Origin, taxonomy, and morphology of *Vigna unguiculata* (L.) Walp. In: Singh BB, Mohan Raj DR, Dashiell KE, Jackai LE (eds) Advances in cowpea research. Sayce Publishing, Devon, pp 1–12

Padulosi S, Heywood V, Hunter D, Jarvis A (2011) Underutilized species and climate change: current status and outlook crop adaption climate change. pp 507–521

Palada MC, Wu DL (2007) Increasing off-season tomato production using grafting technology for peri-urban agriculture in Southeast Asia. Acta Hortic 742:125–131

Pramanik K, Pradhan J, Mohopatra PP, Jena C, Acharya KL (2020) Impact of temperature: an important climate changing factor on vegetable crops. In: Agriculture and forestry: current trends, perspectives. Immortal publications, Vijayawada, pp 105–134

Prasad BVG, Chakravorty S (2015) Effects of climate change on vegetable cultivation-a review. Nat Environ Pollut Technol 14:923–929

Rani S, Sharma MK, Kumar N (2019) Impact of salinity and zinc application on growth, physiological and yield traits in wheat. Curr Sci. (00113891) 116(8)

Risi C, Galwey NW (1984) The Chenopodium grains of the Andes: Inca crops for modern agriculture. Adv Appl Biol 10:145–216

Rosero A, Granda L, Berdugo-Cely JA, Samajova O, Samaj J, Cerkal R (2020) A dual strategy of breeding for drought tolerance and introducing drought-tolerant, underutilized crops into production systems to enhance their resilience to water deficiency. Plan Theory 9(10):1263

Sambo BE (2014) Endangered, neglected, indigenous resilient crops: a potential against climate change impact for sustainable crop productivity and food security. J Agric Vet Sci 7(2):34–41

Scheelbeek PF, Bird FA, Tuomisto HL, Green R, Harris FB, Joy EJ (2018) Effect of environmental changes on vegetable and legume yields and nutritional quality proc. Natl Acad Sci 115(26):6804–6809

Shannon MC, Grieve CM (1999) Tolerance of vegetable crops to salinity. U.S. salinity laboratory, department of agriculture, agricultural research service, 450 W. big springs road, Riverside, CA 92507, USA. Sci Hortic 78:5–38

Shaw RH (1983) Estimates of yield reductions in corn caused by water and temperature stresses. Crop reactions to water and temperature stresses in humid, temperate climates. Westview, Boulder, pp 49–65

Singh M, Malhotra N, Sharma K (2020) Buckwheat (Fagopyrum sp.) genetic resources: what can they contribute towards nutritional security of changing world? Genet Resour Crop Evol 67:1–20

Solankey SS, Kumari M, Kumar M, Silvana N (2021) The role of research for vegetable production under a changing climate future trends and goals. In: Solankey SS et al (eds) Advances in research on vegetable production under a changing climate Vol. 1. Springer Nature, Cham., ISBN 978-3-030-63497-1, pp 1–12

Thompson LM (1986) Climatic change, weather variability, and corn production. Agron J 78:649–653

Tirado MC, Clarke R, Jaykus LA, Mc Quatters Gollop A, Frank JM (2010) Climate change and food safety: a review. Food Res Int 43:1745–1765

UPOV (2020) International Union for the Protection of New Varieties of Plants. PLUTO: plant variety database. Available online: https://www.upov.int/pluto/en/

Van den Heever E, Slabbert MM (2007) Proceedings of the 1st international conference on indigenous vegetables and legumes prospectus for fighting poverty, hunger and malnutrition. Acta Hortic:752–281

Wheeler TR, Craufurd PQ, Ellis RH, Porter JR, Vara Prasad PV (2000) Temperature variability and the yield of annual crops. Agri Ecosyst Environ 82:159–167

Wurr DCE, Fellows JR, Phelps K (1996) Investigating trends in vegetable crop response to increasing temperature associated with climate change. Sci Hortic 66:255–263

Zhang J, Davies WJ (1987) ABA in roots and leaves of flooded pea plants. J Exp Bot 38(4):649–659

Chapter 8
Impact of Climate Change on Tuber Crops Production and Mitigation Strategies

K. Madhusudhan Reddy, Randhir Kumar, and S. Bhargav Kiran

Abstract Indian agriculture has made quick progress toward reaching food self-sufficiency. Horticulture has become the finest alternative for diversification in order to address the needs for food, nutrition, and healthcare in addition to offering better returns on farmland and improved employment opportunities. With little investment in horticulture, India has become the world's second-largest producer of fruits and vegetables, increasing production, productivity, and export. Interventions made by technology and development activities are responsible for this shifting environment. The problems we face today, nevertheless, are considerably more serious than they were in the past, and they must be resolved strategically while harnessing advances in science and technology. Our effectiveness in addressing difficulties that might require investment and coordinated efforts made in an integrated manner is attested to by our past accomplishments. By the year 2050, it is predicted that the world's food demand will have doubled, and research to boost agricultural productivity has been hailed as the solution to feeding an expanding global population. Although cereal crops have historically been the primary and preferred food source in many nations, they are currently under a tremendous deal of strain to meet the growing demand due to climate change, deficient soil conditions, and an increase in disease and insect incidences. The greatest cereal substitute for now is a tuberous root crop like cassava since it can withstand extended periods of drought, disease, and pest pressure while still providing the poor agricultural communities with profitable yields. Because of these qualities, cassava is a crucial crop for ensuring food security in Africa. A progressive change in the existing state

K. M. Reddy
Department of Horticulture (Vegetable and Floriculture), Bihar Agricultural College (Bihar Agricultural University, Sabour), Bhagalpur, Bihar, India

R. Kumar (✉)
Agricultural Research Institute (Bihar Agricultural University, Sabour, Bhagalpur), Patna, Bihar, India

S. B. Kiran
Division of Vegetable Science, Indian Agricultural Research Institute, Pusa, New Delhi, India

S. S. Solankey, M. Kumari (eds.), *Advances in Research on Vegetable Production Under a Changing Climate Vol. 2*, Advances in Olericulture,
https://doi.org/10.1007/978-3-031-20840-9_8

of the climate over a few decades is referred to as climate change; for instance, the Sahara used to have a rainy climate and now has a dry one. Human activity alters the global composition of the atmosphere as well as the natural climate fluctuation found over comparable time periods responsible for climate change. Unpredictable rainfall that starts early or late, poor rainfall distribution, and little rainfall are some of the changes. Extremes of too hot or too cold are present as well as moderate and too hot or too cold temperatures. Climate change, in general, refers to long-term modifications in global weather patterns including temperature increases over time, variations in rainfall, and storm activity that may result from greenhouse effects and ongoing deforestation. The term "climate change" is now only used to describe alterations to the current climate, specifically the increase in the global average superficial temperature. Obviously, the climate is continually changing, and the signal that these changes are taking place can be assessed on a variety of temporal and spatial dimensions. Climate can be thought of as the combination of intricate weather patterns that have been averaged over a sizable portion of the planet and described in terms of both the mean of weather and attributes.

Keywords Climate change · Temperature · Mitigating tactics · Tuber crops

8.1 Introduction

Among tuber crops, potatoes are grown in India during a thermally favourable window, and the availability of this window is predicted to be significantly impacted by climate change. Due to climatic restrictions, thermal sensitive potato production is primarily restricted to the North Indian plains (approximately 80% of the region) when the soil is irrigated. Shortening of the actual growing season would result from higher temperatures at the beginning of the season. According to the early findings of the simulation studies, productivity in Punjab, Haryana, and Western U.P. is anticipated to increase by 7% and 4%, respectively, in the years 2020 and 2050. (Singh et al. 2005). Due to the frequent occurrence of frost in these places, the growing season is now short. However, as winters become warmer, frost situations may become less frequent, resulting in longer growing seasons and increased output. States like Bihar, West Bengal, and Gujarat are anticipated to see output decline by 2–19% and 9–55% in 2020 and 2050, respectively. This is because the winters in these regions are already balmy, and sharp rise in winter temperatures would have a negative impact on production per unit area. However, cassava and sweet potatoes are thought to be drought-tolerant, considerable tuber production and starch content reductions do occur, and responses to water deficiency stress and high temperature stress conditions vary among cultivars. For the growth of tubers, mild water-deficit stress (WDS) is beneficial. Drought in the first 3 months of cassava inhibits tuber start and bulking. According to studies, during the last 4 months of the crop growth cycle, drought conditions were observed to diminish tuber output by roughly 28–42%

(Ramanujam 1985). When soil moisture availability falls below 20%, sweet potato yields decrease, and the tuber initiation period is the most vulnerable because of its impact on tuber number. Water stress during the tuber initiation period causes tubers to lignify, which inhibits tuber growth. The Trivandrum-based Central Tuber Crops Research Institute has released the drought-tolerant sweet potato cultivar "Sree Bhadra". While cassava can sustain vegetative growth and biomass at high temperatures (33–40 °C) with adequate soil moisture, temperatures above 30 °C will have an impact on the synthesis of starch in tubers and the export of sucrose from leaves. In simulated tests run at ICAR-CTCRI, Thiruvananthapuram for yield parameters of sweet potatoes resulted that, yields were increased by 1.26 t/ha when the mean temperature rise by 1 °C between 19.24° and 20.24 °C, whereas it declined by 0.12 t/ha when the mean temperature rise by 1 °C between 28.24° and 29.24 °C.

8.2 Effect of Temperature on Tuber Crops

The increase in size and development of potatoes are impacted by temperature. The stated cardinal temperatures for net photosynthesis is nearest (0–7 °C), optimal (16–25 °C), and superlative (40 °C) (Kooman and Haverkort 1995). The establishment phase is also impacted by temperature, particularly soil temperature. Low temperature slows the growth of sprouts (Firman and Allen 1992). Febricity or calefaction more than ideal also inhibit the growth of sprouts (Midmore and Ingram 1984). According to reports, subapical necrosis, which stops sprout growth, is the cause of the decreased emergence rate at temperatures greater than those considered optimal (McGee et al. 1986). High temperatures will change the crop's morphological features, causing etiolated growth and smaller size compound leaves and branched leaflets, which would lower the LAI (Ewing 1997). Kirk and Marshall (1992) found that the rate of leaf emergence is directly related to temperature between the ranges of 9 and 25 °C, with no further growth happening above 25 °C. They also found that 25 °C is the optimal temperature for leaf expansion (Benoit et al. 1986). Because leaves do not fully expand at very high temperatures as less light is reflected. Despite the fact that there was less leaf area in warm temperatures, Prange et al. (1990) found that the leaf dry weight was unaffected, characterising that there was less expansion of leaf. As a result, hot climate-adapted genotypes have reduced specific leaf area (Midmore and Prange 1991. Therefore, it has been anticipated in potato models that leaf expansion will increase linearly up to 24 °C before decreasing linearly to 0 at 35 °C (Kooman and Haverkort 1995). High temperatures also have an impact on stem elongation. High day time and low night time temperatures induce stem elongation, which is practically linear up to 35 °C (Manrique 1990), while high night time temperatures encourage branching (Moreno 1985). Since yield is a function of intercepted radiation and radiation/ light utilisation efficiency, shorter ground-cover duration, which has been shown to positively correlate with yield, at high temperatures decreases production (Vander Zaag and Demagante 1987). Early establishment of a full canopy cover and maintenance of it for an extended period of time guarantee increased radiation absorption and higher yield. Potato variety Kufri Surya can tolerate up to 25 °C night temperature, moreover Kufri Sheetman, Kufri Dewa is frost tolerant varieties developed by CPRI, Shimla. While Cassava varieties H-97 & Sree Sahya and sweet potato variety Sree Nandini are drought tolerant (Solankey et al. 2021).

8.3 Effect of Elevated CO_2 on Tuber Crops

Increased CO_2 content has been shown to have a favourable impact on growth and yield in controlled tests using growth chambers and open top chambers. Only a few negative effects have been observed. Increased CO_2 had no effect on the number of tubers, but the mean tuber weight increased mostly because of a rise in the number of cells without changing the volume of cells in the tubers (Collins 1976; Donnelly et al. 2001; Chen and Setter 2003). But more staff have seen an increase in the number of tubers (Miglietta et al. 1997; Craigon et al. 2002). Increased CO_2 concentration has been demonstrated to quicken the onset and flowering of tubers but hasten the senescence of leaves, according to and Vaccari et al. (2001). It was determined that the relationship between leaf senescence and ambient CO_2 levels was linear up to 660 ppm. It has been shown that further high CO_2 concentrations lower the amount of chlorophyll in leaves, especially later in the growing season following tuber commencement (Lawson et al. 2001; Bindi et al. 2002). Stomatal conductance was found to be reduced by 59% in potatoes when CO_2 concentration was doubled, but the rise in net photosynthetic rate, which rose by about 53%, was unaffected by this decrease. This resulted in a 16% decrease in transpiration rate and an 80% increase in immediate transpiration efficiency (Olivo et al. 2002). It has been reported that reducing evapotranspiration (ET) by up to 12–14% can save water (Olivo et al. 2002; Magliulo et al. 2003).

8.4 Increased CO_2 and Temperature's Impact on Tuber Crops Productivity

Without any modifications, the INFOCROPPOTATO model (Singh et al. 2005) was employed to know the impact of climate change on potato production. It was observed that at high CO_2 of 550 ppm and 1 °C rise in temperature, potato production will increase by 11.12%. However, according to the future climate predictions for India, the temperature will likely rise by 3 °C in 2050 and there will be a 550 ppm atmospheric CO_2 concentration (IPCC 2007). According to this scenario, potato production will decrease by 13.72% in 2050. Moreover, without adaptation, the expected 400 ppm of CO_2 and 1 °C increase in temperature in 2020 will result in a 3.16% decrease in potato yield.

8.5 Impact Quality of Tuber Crops

Though the quality of potatoes is expected to be significantly damaged in terms of marketable grade of tubers and internal illnesses, CO_2 enrichment does not seem to counteract the negative impacts of increased temperature on tuber yield. By raising starch and dry matter levels while lowering glycoalkaloid and nitrate concentrations, increased CO_2 enhanced the quality of tubers (Schapendonk et al. 2000; Donnelly et al. 2001; Vorne et al. 2001). High CO_2 levels cause tubers to lose nearly all of their nutritional value, which raises the risk of cooking-related discoloration, according to Cao and Tibbitts (1992) and Fangmeier et al. (2002). (Vorne et al. 2001).

Manihot esculenta, sometimes referred to as cassava, mandioca, manioc, and yuca, is a perennial woody plant in the spurge family (Euphorbiaceae). Despite being a South American native, cassava is grown all across the tropics and subtropics. Cassava has been termed "the drought, war, and famine crop" because it can thrive in challenging environments (like arid soils with low fertility) and can be harvested whenever needed (rather than only at the end of the growing season), providing a reserve of food in times of famine and conflict. In addition, it has been suggested that cassava may be more climate change resistant than other major crops.

8.6 Cassava Production in the World

In 1961, 9.6 million hectares of cassava were used to produce 78.5 million tonnes of cassava, with around 44% of the production coming from Africa (FAOSTAT 2019). 322 million tonnes of cassava were produced globally in 2017 on 26 million hectares of land (FAOSTAT 2019). Despite the fact that Africa produces more than 58% of the world's cassava and takes up more than 75% of the land area dedicated to its cultivation, the average fresh root yield of cassava is 8.9 tonnes per hectare, which is much lower than the global average of 11.9 tonnes per hectare and the yield seen in Asia (13.3 tonnes per hectare) (FAOSTAT 2019). Cassava is grown in more than 105 nations; Nigeria, the Democratic Republic of the Congo, Brazil, and Indonesia are the largest producers. Smallholder farmers should expect yields of 1–10 tonnes of fresh roots per hectare of cassava. However, fresh root yields may be possible to reach 75–80 t ha^1 by adopting good agricultural practices (Anikwe and Ikenganyia 2017).

Cassava production has risen considerably during the past 60 years. Africa and Asia saw the most pronounced growth between 1996 and 2017. (Fig. 8.1a). Throughout contrast, cassava production increased more gradually in Latin America. The significant increase in the cassava production can be partially explained by an increase in the amount of land harvested in Africa as farmers come to appreciate the crop's economic value (Fig. 8.1b), and partially by significant yield gains in Asia as a result of new improved cultivars and improved agronomic practises (Fig. 8.1c) (Aye and Howeler 2017). Cassava output and demand are expected to rise as a result

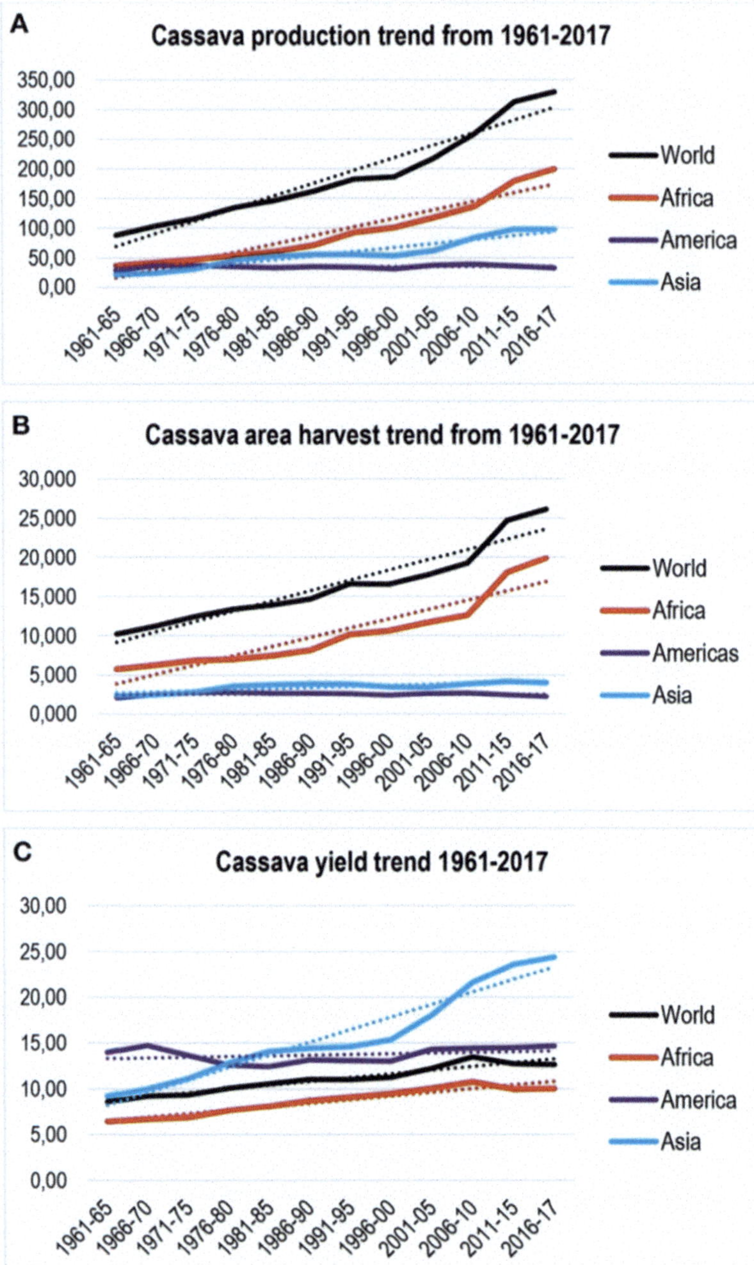

Fig. 8.1 Cassava production (**a**) area harvest (**b**) and yield (**c**) trend from 1961 to 2017. (Source FAOSTAT 2019)

of the crop's capacity to withstand drought and generate reasonable yields on marginal and low-fertility soils.Between 30°N and 30°S, ranging from tropical to subtropical regions of Africa, Asia, and Latin America have considerable annual cultivation of cassava (*M. esculenta*) (Duangpatra 1988). One of the major problems facing the globe now is climate change. It is defined as significant deviations from the mean values of meteorological variables, such as temperature and precipitation, for which averages have been computed over a lengthy period of time. Global climate has been reportedly changed significantly in the recent decades as a result of increased human activity that altered the makeup of the earth's atmosphere.

Agriculture is the industry which is mostly affected by climate change and has the largest economic impact due to its vast scale and severe sensitivity to climatic conditions. Variations in environmental factors like temperature and rainfall have a significant impact on crop productivity. Depending on the crop, the region, and the intensity of the parameter change, the effects of altered precipitation patterns, rising temperatures, and variation in CO_2 concentration. When the temperature rises, the yield is observed to drop, although the impact of the increasing precipitation is likely mitigated or lessened.

Agricultural type, climate scenario, and the CO_2 concentration affects all influence the crop productivity as seen in Iran as a result of climatic conditions. A decrease in precipitation or an increase in temperature has been found to considerably lower farmers' net income in Cameroon. This issue, along with poor policy-making, has resulted in low demand for Cameroon's agricultural exports, which has led to volatility in the nation's income. The temperature in Veracruz, Mexico, has an effect on coffee yield, according to statistical research. It is observed that coffee production will not remain economically feasible for the growers in the upcoming years due to indications of a 34% decline in current production.

Depending on the region and irrigation method, climate change has different effects on crop productivity. The environment may suffer as a result of the expansion of irrigated areas, which can increase crop yield. The rise in temperature is projected to reduce crop productivity because many crops now have shorter growing seasons. If both the temperate and tropical regions experience a warming of 2 °C, it is projected that the combined production of wheat, rice, and maize will fall. Climate change typically has a higher impact on tropical regions due to the proximity of tropical crops to their high-temperature regimes and the resulting high-temperature stress during high temperatures.

8.6.1 Biotic and Abiotic Stresses

Furthermore, humid and warm climates are more likely to have insect pests and diseases. In addition to temperature and rainfall, other factors that affect agricultural yields include humidity and wind speed. Without these factors, it is possible to overestimate the climate change cost. Furthermore, it has been discovered that until

2100, change in the climate is anticipated to decrease China's yields of cereals like wheat, maize, and rice by 18.2612.13, 45.1011.55, and 36.2510.75%, respectively.

Since the turn of the century, extreme weather events have increased in frequency in the Netherlands, having a considerable impact on wheat yield. The severity of the wheat production drop was dictated by the week in which an extreme weather change has been occurred.

Due to climate change, it has been predicted that there would be more droughts in the near future in the majority of the world's areas, with a rise in the area impacted by drought from 15.4 to 44.0% by the year 2100. The region considered to be most vulnerable is Africa. Major crops' yields in drought-affected areas are predicted to drop by more than 50% by 2050 and by over 90% by 2100.

8.6.2 Physiological Adaptation of Plant

Dramatic changes in these variables are more likely to have an impact on physiological alterations than changes in mean climate because plant-water relations are more responsive to temperature and precipitation variations. Depending on the plant's species and stage of growth, different plants react differently to climatic change. Different plants have thresholds that are specific to each species, and each species' responses—including root lengthening, changes to the angle at which roots develop, and decreased yield.

There was a 0.42–0.02 K raise in air temperature as a result of decreased transpiration in plants and higher CO_2 level in the atmosphere. The warming of terrestrial surfaces can be increased by 3.33 0.03 K as a result of the direct radiative effect and the indirect physiological effect of elevated CO_2.

As atmospheric CO_2 levels grow, more crops should be able to be harvested, and different crops will have different effects on how plants develop. In the absence of harsh conditions, C_3 and C_4 crops are forecast to consume less water even if C_3 crops are projected to produce more. However, it's likely that rising temperatures and shifting precipitation will offset these beneficial benefits of increased CO_2.

8.6.3 Cassava Farmers' Knowledge About Climate Change

According to Table 8.1's findings, cassava farmers were distributed according to their knowledge of climate change, with the majority (98.2%) reporting delayed rainy season onset and early rainy season cessation, high rainfall (91.3%), high temperatures (89.1%), intense sunshine (84.0%), and longer dry seasons (52.2%). While few (35.2%) saw stormy weather, there were few instances of low rainfall (27%), a longer rainy season (13%) or severe winds (2.6%). The outcome suggested that a high proportion of farmers in the region were aware of global warming. This supports the finding that farmers are aware of climate change made by Umunnakwe

Table 8.1 Awareness and impact of climate change on cassava

Observed changes	Frequency	Percentage
Dry seasons are longer	120	52.2
Longer rainy season	30	13.0
There is high rainfall	210	91.3
There is high temperature	205	89.1
There is low rainfall	62	27.0
There is low temperature	25	10.9
Delayed onset of rains/early cessation.	226	98.2
Intensive sunshine	193	84.0
Stormy weather	81	35.3
Strong wind	06	2.6

Table 8.2 Distribution of farmers based on information about climate change adaption measures from various sources

Sources of information	Frequency	Percentage
Extension agent	112	48.7
Research institution	24	10.4
Fellow farmers	211	91.7
Friends/relations	219	95.2
Farmers' co-operative	10	4.3
Extension bulletins	34	14.8
Radio	228	99.1
Television programme	201	87.4
Internet	31	13.
Social organizations	16	7.0

et al. (2014) and Arimi (2014). Implicitly, the farmers' level of knowledge is expected to make them more willing to use adaptation techniques.

According to Table 8.2 results, radio accounted for the majority (99.1%) of the sources of information used by cassava growers when researching adaption options. Television (87.4%), fellow farmers (91.7%), friends/relatives (95.2%), and extension agents (48.7%) were all closely behind this. Extension bulletins (14.8%), the internet (13.5%), research organisations (10.4%), weather stations (7.0%), and farmers cooperatives (4.3%) are some additional sources. The findings showed that farmers accessed many information sources, which may have influenced their understanding of climate change and usage of adaption tactics. This supports the finding by Farauta et al. (2011) reported that mass media has a tendency to reach huge audiences more quickly.

8.6.4 Adaptation Strategies to Climate Change

The distribution of cassava farmers is the result of their use of climate change adaptation strategies, including the use of pest and disease resistant varieties (M = 2.8), crop rotation (M = 2.5), adjustments to planting and harvesting dates (M = 2.7), mulching/cover cropping (M = 2.9), increased use of family labour (M = 2.8), intensive use of organic manure (M = 2.6), mixed cropping (M = 2.7), more frequent weeding (M = 2.7), and changes to timing. Construction of drainage systems (M = 1.6), liming/acidification (M = 1.9), land rotation (M = 2.3), terracing (M = 1.7), tree planting (M = 2.1), water storage in ponds (M = 1.7), use of weather forecasts (M = 1.9), joining of cooperative society (M = 1.9), use of available credit facilities (M = 1.9), contouring (M = 1.4), use of different tillage systems (M = 1.4 The results showed that an average cassava farmer employed several adaptation techniques to lessen the consequences of climate change in the region, yielding a grand mean of M = 2.1. The findings of Obayelu et al. (2014), Adebisi Adelan and Oyesola (2013), and Enujeke and Ofuoku (2009) that farmers have not completely utilised the majority of adaptation techniques to climate change due to the high cost of these technologies were corroborated by the results. Due to the lack of information, such as weather forecasts in the event that planting dates were changed, and the accessibility of extension agents to train and introduce these adaptation tactics to them, there was a low degree of use.

8.6.4.1 Relativity in the Use of Climate Change

Strategies for adaptability results indicated the analysis of variance test performed to see if farmers in the study area had different climate change adaption tactics. In contrast to the tabulated results (F-tab = 1.871), the test resulted in an F-cal of 1.46 at V1 = 21 and V2 = 108 degrees of freedom, a significant value of 0.32, and a 0.05 significant threshold. Therefore, the claim that there are no appreciable differences between cassava farmers' adaptation tactics in the three zones of Abia State is accepted. This conclusion suggests that farmers in the three zones experienced the same effects of climate change and, as a result, they utilised similar practises, concepts, technology, and adaptation strategies.

8.6.4.2 Mitigation

Farmers' assessments of the severity and hazard posed by climate change are the main drivers behind voluntary mitigation. The adaption is however impacted by the accessibility of relevant information. The number of people who are affected by water stress will also decline thanks to mitigating measures, but those who are still at risk will need adaptation measures.

In order to implement climate-resilient technology, farmers can employ both conventional and agro-ecological management strategies, such as crop biodiversification, soil and water management practices.

These management techniques guarantee enhanced soil health, better soil quality, and a reduction in soil erosion, which results in resilient soils and cropping systems that ultimately provide food security in the face of climate change.

The educational initiatives that focus on regional, practical, and local aspects and can be monitored by individual behaviour are the most successful in raising awareness of climate change for ecological development.

The fact, that most of the farmers favoured adaptations but only a small minority favoured GHG reduction shows the importance of focusing on programmes that have adaptation and mitigation components.

The main adaptation strategies for mitigation can be divided into three categories: resource conservation technologies, agricultural system technology, and socio-economic or policy initiatives. Small and marginal farmers are less able to adjust to climate change due to a lack of information, which makes them more susceptible to losses.

The lack of management measures and the financial repercussions of climate change make African farmers particularly vulnerable.

Changes in sowing dates are just one agronomic tactic that can be utilised to lessen the consequences of climate change.

Crop yield loss is at its lowest when farmers in sub-Saharan Africa use sequential cropping methods and alter the sowing dates to suit the climate.

The agroforestry sector can lower atmospheric GHG levels and support Kenyan small farmers as they adapt to climate change. There are various simple techniques for lowering GHG emissions, such as alternative drying of rice, mid-season drainage, better diets for cattle, increasing the efficacy of N utilisation, and soil carbon.

The spread of technology will have a major impact on how farmers adapt to climate change. Climate change can be mitigated by using easy adaptation strategies like changing of planting time and variety. The primary priorities are capacity building, public research assistance, and market integration.

The study came to the conclusion that adaptation techniques are essential components in dealing with climate change because they assist farmers in achieving their goals for food production, revenue generation, and livelihood in the face of climatic change and harsh weather. As a result, the study came to the conclusion that information on adaption tactics, weather forecasts, and early warning systems are essential for those efforts. Cassava farmers should have access to these climate change data so, they can adapt to the changing environment and address the issues with food security that are crucial to the agendas for economic growth, sustainable development, and poverty reduction. Consequently, the following suggestions were made:

1. Cassava producers should establish a strong association of cassava farmers so that they may come together to discuss the effects of climate change, adaptation tactics, obstacles to effective adaptation, and potential solutions to the problem. This is due to the fact that everyone involved must work together to combat climate change.
2. The various levels of government must put in place deliberate policies to provide trainings for cassava farmers on early weather prediction/ forecasting system, and farmers' access to meteorological data.
3. The public and commercial sectors, non-governmental organisations (NGOs), and the media should be more actively involved in promoting climate change adaptation solutions through various sources, particularly the radio. This will encourage cassava growers to use them.
4. For a long-term solution, the government should incorporate climate change concerns and adaptation strategies into the national development plan because the risks posed by climate change not only pose a threat to the development of agriculture (the food supply), but also to the overall growth and well-being of the nation.
5. It's crucial to improve the nation's extension services so they can better advise farmers about farm-level adaption options and other pertinent climate information. To raise cassava farmers' knowledge of climate change and adaptation options and to encourage cassava farmers' use of adaptation measures, extension agents should step up their educational campaigns, seminars, and follow-up visitations.
6. It is important to clearly identify the roles of the public and private sectors, as doing so will make it easier to commercialise crop technology developed through public sector research. The ARC launched a project to coordinate cassava R&D in South Africa in response to the lack of coordination among the many parties and the crop's growing importance. In an effort to map the future of South African cassava R&D and commercialization of the crop, a number of stakeholder engagements have been organised.

8.6.4.3 Awareness and Promotion for Tuber Crops Production

Recurrent and severe droughts have an impact on the local production of staple foods including maize, wheat, and potatoes. Investigating alternate, climate-resilient solutions has elevated in importance. It is essential to increase public awareness of the current environmental issues and potential solutions. Symposia, conferences, and seminars where professionals from national and international institutions may discuss their research findings on cassava should have a significant influence in terms of raising awareness. Key decision-makers, farmers, growers, and processors can use these conferences as vital platforms to acquire direct information from authorities in countries where cassava is a prominent crop. It is advisable to get funding from both domestic and international institutions.

Additionally, in order for academics, decision-makers, growers, and processors to comprehend the crop's economic importance, it is crucial to use a promotional centre to introduce them to various cassava products.

A commodity-based association represents the entire value chain, including producers, consumers, processors, traders, importers, exporters, input distributors, and transporters. These organisations may play a significant role in advocating for legislative and administrative changes as well as furthering the particular interests of their members. The South Africa Cassava Industry Association (CIASA) was established under the umbrella of Department of Trade, Industry, and Competition (DTIC) for cassava value chain address and organisation at every steps (IDC 2017). Despite its aggressive efforts, CIASA is currently unable to organise cassava research or the development of a cassava value chain in South Africa. It could be argued that the group was established too soon, when it lacked governmental backing and a critical mass.

The CIASA needs to be strengthened and revitalised in order to adequately represent the whole spectrum of stakeholders and the SA cassava value chain. The CIASA constitution should be reviewed, and its mission should be clarified, according to recommendations made during a meeting with R&D stakeholders. This will benefit the whole SA cassava industry, including researchers who were overlooked by the DTIC. The 2016–2017 Industrial Policy action plan (IPAP), which was informed by consultations that resulted in the development of CAISA, contains cassava starch to boost trade activities in the industrialization of cassava.

8.6.5 Market Expansion and Product Variety

Despite the fact that there is now little demand for cassava as a food in South Africa, there is potential to develop cassava products that are both economical and desirable to customers there. Brazil has developed a number of culinary items using cassava and has a healthy home market (Demiate and Kotovicz 2011). High-quality cassava flour can be used as a replacement for wheat flour and as a source of food starch for creating bread, pastries, cookies, and biscuits. Given the immense potential of cassava in the world's starch market, both the production of premium culinary starch and lower-priced industrial items ought to be given top priority. Climate change and global warming's indirect effects:

8.7 Drought

Potatoes need an adequate water supply because of their shallow root system. The roots of the potato plant are normally 40–50 cm deep (Beukema and Van der Zaag 1990). Potato is quite susceptible to drought, especially at the start of the tuber, which causes a notable reduction in tuber output. Drought stress affects the root: shoot ratio and how much dry matter goes to the root, shoot, leaf, and stem depending on the stage of development of the plant (DS). The ratio of roots to shoots increases during a drought while the production of dry matter declines, indicating a shift in the growth equation in favour of roots. Additionally, roots of plants grown in drought-prone settings are typically weaker. Both responses enable stressed plants to benefit from the available resources.

8.8 India's Regional Vulnerability to Climate Change

The entire region of the Indian plains, which is mostly used for potato irrigation, is in danger, the second-largest producer of potatoes in India and the state with the highest production, appears to be quite vulnerable. West Bengal has moderate winters and a narrow "window" of appropriate growing season; any increase in temperature will have a negative influence on productivity and cause issues with storage and post-harvest handling of food under warmer temperatures. Bihar and Uttar Pradesh, which make up the majority of the nation's potato crop, are other vulnerable states. The states of Punjab and Haryana as well as nearby regions in northern part of western Uttar Pradesh and Rajasthan, where winters are particularly harsh and frost occurs occasionally, may see some benefits from global warming.

8.9 Extreme Weather Event Observations

1. Winter time rains experienced at planting time impact emergence, delay planting, and reduce tuber output.
2. Heavy rains throughout the growing season that cause flooding have an impact on tuber yield.
3. Heavy rains during harvest season induce rotting in the field and in make shift potato mounds.
4. Early in the crop season, a evercast sky and showers always promote the attack of the disease known as late blight, which results in a significant decrease in yield.
5. In West Bengal, Uttar Pradesh, and Bihar, the tuber yield was decreased in 2008 due to winters that were relatively warmer.

8.10 Measures of Adaptation to Climate Change and Global Warming

1. Using crop residue as mulch for few days after planting; substituting micro irrigation system for furrow and basin systems.
2. Change the cultural management of farming systems based on potatoes.
3. Crop residue management on farms and conservation tillage.
4. The development and expansion of cold storage facilities as well as ac transit between production and consuming centres.
5. Supporting higher pest and water control costs.
6. Weather insurance for the expensive to cultivate cash crop of potatoes. To produce ware and seed potatoes in warm climates, intensify research, teaching, and development in these fields.
7. Modifying the date of planting and implementing the measures for integrated pest management (IPM).

8.10.1 Future Research Approaches

1. Development of early warning for disease forecasting. Quantification of regional susceptibility and impact evaluation.
2. Breeding heat-tolerant, short-lived cultivars that take advantage of biodiversity is a top aim.
3. Breeding cultivars that are disease, salinity, and drought resistant.
4. Early planning for potential relocation and the discovery of additional potato-growing regions.
5. More effective agronomic management of water and fertiliser use.
6. The creation of agro-techniques for cropping systems based on potatoes in warm climates.
7. The creation of late blight and virus resistance cultivars.
8. Adjusting the timing of chemical sprays based on the population of newly emerging pathogens.

8.11 Conclusion

In the past century, potatoes, which are native to temperate regions and are often cultivated under long-day circumstances in pleasant and chilly summers in Europe and America, were imported to India and adapted to tropical hot conditions. In India's rather chilly winters, the crop is primarily restricted to the Indo-Gangetic plains. 84% of India's total potato production is contributed by the crop planted at winter season in the north Indian plains. Here, the crop is entirely irrigated. In the tropics, high temperatures have an impact on growth and development. High temperatures decrease photosynthesis in potatoes, although tuber formation and the partitioning of assimilates or product of photosynthesis to tubers are more temperature-sensitive processes. Most diseases will likely be unaffected directly by the rising CO_2, as many fungi that live in soil can withstand CO_2 levels that are more than 10- or 20-fold higher. However, it's likely that rising CO_2 concentrations will affect plant diseases by altering the physiology and architecture of the hosts. In the Indo-Gangetic plains, Wart, powdery scab, black scurf, and common scab diseases may become less widespread as a result of global warming, but viral, late blight, charcoal rot, and bacterial wilt are likely to become more common. On early blight, it has little effect. Climate change is being highlighted by apical leaf curl virus infestation in the Indo-Gangetic lowlands, unexpected leaf hopper burns outbreaks in Gujarat in 2006–2007, and late blighting severe epiphytotic forms in Karnataka and Maharashtra in 2006–2007 and 2008. With an increase in temperature, it is anticipated that the biology of insect pests and vectors will change significantly. Climate change and global warming will have a huge impact on India's potato growth story. These factors will have an impact on all aspects of production and profitability as well as seed production, storage, marketing, and processing of this perishable vegetatively propagated commodity. If adequate adaption methods are not developed for prompt application and execution, potato growth estimates in India may be slowed or even reversed under the impact of future climate change scenarios. Under future climate change and global warming scenarios, temperature increase and atmospheric CO_2 increase simultaneously. By 2020 and 2050, respectively, it is predicted that global warming will cause India's potato production to fall by 3.16% and 13.72% from current levels. Climate change will have a direct impact on potato production, but it will also have a number of indirect implications on the crop's supply, storage, use, and acreage in the future. It is crucial to regularly monitor weather and climate changes as well as disease outbreaks and insect pest infestations.

References

Adebisi Adelan O, Oyesola OB (2013) Socio-economic factors affecting adaptation strategies of selected horticultural farmers to climate change in Osun state Nigeria. J Agric Biol Sci 8(4):344–350

Anikwe MAN, Ikenganyia EE (2017) Ecophysiology and Production Principles of Cassava (*Manihot* sp) in Southeastern Nigeria

Arimi K (2014) Determinants of climate change adaptation strategies used by rice farmers in southwestern, Nigeria. J Agric Rural Dev Trop Subtrop 115(2):91–99

Aye TM, Howeler R (2017) Integrated crop management for cassava cultivation in Asia. In: Clair H (ed) Achieving sustainable cultivation of cassava. Volume 1: cultivation techniques. Burleigh Dodds series in agricultural science. Burleigh Dodds Science Publishing, Cambridge, p 424

Benoit GR, Grant WJ, Devine OJ (1986) Potato top growth as influenced by day-night temperature differences 1. Agron J 78:264–269

Beukema HP, Van der Zaag DE (1990) Introduction to potato production. Wageningen, The Netherlands, p 224
Bindi M, Hacour A, Vandermeiren K, Craigon J, Ojanperä K, Selldén G, Högy P, Finnan J, Fibbi L (2002) Chlorophyll concentration of potatoes grown under elevated carbon dioxide and/or ozone concentrations. Eur J Agron 17(4):319–335
Cao W, Tibbitts TW (1992) Temperature cycling periods affect growth and tuberization in potatoes under continuous light. HortScience 27:344–345
Chen C-T, Setter TL (2003) Response of potato tuber cell division and growth to shade and elevated CO2. Ann Bot 91(3):373–381. https://doi.org/10.1093/aob/mcg031
Collins WB (1976) Effect of carbon dioxide enrichment on growth of the potato plant. Hortic Sci 11:467–469
Craigon J, Fangmeier A, Jones M, Donnelly A, Bindi M, de Temmerman L, Persson K, Ojanperä K (2002) Growth and marketable-yield responses of potato to increased CO_2 and ozone. Eur J Agron 17:273–290
Demiate IM, Kotovicz V (2011) Cassava starch in the Brazilian food industry. Cienc Tecnol Aliment 31:388–397
Donnelly A, Craigon J, Black CR, Colls JJ, Landon G (2001) Does elevated CO_2 ameliorate the impact of O3 on chlorophyll content and photosynthesis in potato (Solanum tuberosum)? Physiol Plant 111:501–511
Duangpatra P (1988) Soil and climatic characterization of major cassava growing areas in Thailand. In: Howeler R, Kawano K (eds) Cassava breeding and agronomy research in Asia. Proc. workshop., 26-28 October 1987. CIAT, Bangkok, Bangkok, pp 157–184
Enujeke EC, Ofuoku AU (2009) Determinants of adaptation to climate change among arable crop farmers in Edo-state, Nigeria and its implications for extension service. Int J Adv Biotechnol Res 2(2):220–227
Ewing EE (1997) Heat stress and tuberization stimulus. Am J Potato 58:31–49
Fangmeier A, De Temmerman L, Black C, Persson K, Vorne V (2002) Effects of elevated CO_2 and/or ozone on nutrient concentrations and nutrient uptake of potatoes. Eur J Agron 17:353–368
FAOSTAT (2019) Food and agriculture organization of the United Nations statistics (FAOSTAT). FAO
Farauta BK, Egbule CL, Idrisa YL, Agu VC (2011) Farmers' perceptions of climate change and adaptation strategies in northern Nigeria: an empirical assessment. African Technology Policy Studies Network Nairobi Kenya., pp 16–19
Firman D, Allen PJ (1992) An evaluation of small seed for ware-potato production. J Agric Sci 118(2):185 193
IDC (2017) A study on market potential for increased industrial starch production in South Africa. Industrial Development Cooperation, Pretoria
IPCC (2007) Impacts, adaptation and vulnerability: contribution of working group II to the fourth assessment report of the intergovernmental panel on climate change. Cambridge University Press, Cambridge/New York
Kirk WW, Marshall B (1992) The effect of temperature on the initiation of leaf primordia in developing potato sprouts. J Exp Bot 36:1634–1643
Kooman P, Haverkort AJ (1995) Modelling development and growth of the potato crop influenced by temperature and day length: LINTUL-POTATO. Curr Issues Prod Ecol 3:41–59
Lawson T, Craigon J, Black CR, Colls JJ, Tulloch A-M, Landon G (2001) Effects of elevated carbon dioxide and ozone on the growth and yield of (Solanum tuberosum L.) grown in open-top chambers. Environ Pollut 111:479–491
Magliulo V, Bindi M, Rana G (2003) Water use of irrigated potato (Solanum tuberosum L.) grown under free air carbon dioxide enrichment in Central Italy. Agric Ecosyst Environ 97:65–80
Manrique LA (1990) Growth and yield of potato grown in the greenhouse during summer and winter in Hawaii. Commun Soil Sci Plant Anal 21:237–249
McGee E, Jarvis MC, Duncan HJ (1986) The relationship between temperature and sprout growth in stored seed potatoes. Potato Res 29:521–524

Midmore DJ, Ingram KT (1984) Simulation of potato crop growth and development. Crop Sci 24:21–27

Midmore DJ, Prange RK (1991) Ann Bot 69(1):13–20

Miglietta F, Magliulo E, Bindi M, Lanini M (1997) The effect of free air carbon dioxide enrichment (FACE) on potato (*Solanum tuberosum* L.) design and performance of the CO_2 fumigation system. Glob Chang Biol 4:163–172

Moreno U (1985) Environmental effects on growth and development of potato plants. In: Potato physiology. Academic Press Inc, London, pp 481–501

Obayelu OA, Adepopu AO, Idowu T (2014) Factors influencing farmers choices of adaptation to climate change in Ekiti state, Nigeria. J Agric Environ Int Dev 108(1):3–16

Olivo N, Martinez CA, Oliva MA (2002) The photosynthetic response to elevated CO_2 in high altitude potato species *(Solanum curtilobum)*. Photosynthetica 40(2):309–313

Prange RK, McRae KB, Midmore DJ, Deng R (1990) Reduction in potato growth at high temperature: role of photosynthesis and dark respiration. Am Potato J 67:357–369

Ramanujam T (1985) Leaf density profile and efficiency in partitioning dry matter among high and low yielding cultivars of cassava (Manihot esculenta Crantz). Field Crop Res 10:291–303

Schapendonk AH, van Oijen M, Dijkstra P, Pot CS, Jordi WJ, Stoopen GM (2000) Effects of elevated CO_2 concentration on photosynthetic acclimation and productivity of two potato cultivars grown in open-top chambers. Funct Plant Biol 27:1119–1130

Singh JP, Govindakrishnan PM, Lal SS, Aggarwal PK (2005) Increasing the efficiency of agronomy experiments in pottao using INFOCROP-POTATO model. European Potato J 48(3):131–152

Solankey SS, Kumari M, Kumar M, Silvana N (2021) The role of research for vegetable production under a changing climate future trends and goals. In: Solankey SS et al (eds) Advances in research on vegetable production under a changing climate Vol. 1. Springer Nature, Cham., ISBN 978-3-030-63497-1, pp 1–12

Umunnakwe VC, Pyasi VK, Pande AK (2014) Factors influencing involvement in agricultural livelihood activities among rural youth in Jabalpur district of Madhya Pradesh, India. Int J Agric Policy Res 2(8):288–295

Vaccari F, Miglietta F, Magliulo E, Giuntoli A, Cerio L, Bindi M (2001) Free air CO_2 enrichment of potato (Solanum tuberosum L.): photosynthetic capacity of leaves. Ital J Agron 5:3–10

Vander Zaag P, Demagante AL (1987) Potato (Solanum spp.) in an isohyperthermic environment. I. Agronomic management. Field Crop Res 17:199–217

Vorne V, Ojanperä K, De Temmerman L, Bindi M, Högy P, Jones MH, Lawson T, Persson K (2001) Effects of elevated carbon dioxide and ozone on potato tuber quality in the European multiple-site experiment 'CHIP-project'. Eur J Agron 17:369/381

Chapter 9
Impact of Climate Change on Vegetable Seed Production and Mitigation Strategies

Durga Prasad Moharana, Pragnya Paramita Mishra, Sarvesh Pratap Kashyap, Menka Pathak, D. R. Bhardwaj, Keshav Kant Gautam, and Anand Kumar Singh

Abstract The climate extremes and variability pose a severe threat to agriculture and hence to food and nutritional security. Climate change obstructs the path towards achieving the Sustainable Development Goal of a hunger-free and nutritionally-secured world. Seed security plays a pivotal role in safeguarding global food sustainability. Vegetable seed production, both in terms of quantity and quality, is primarily affected by various climatic vagaries like extreme temperatures, elevated CO_2, water stress, etc. There is an urgent need to address this insecurity pertaining to vegetable seed production to ensure a self-reliant global seed system. In this chapter, a thorough elaboration of vegetable seed production vis-à-vis climate change and various mitigation strategies has been presented.

Keywords Climate change · Seed production · Vegetable · Resilient crop · Adaption · Mitigation

D. P. Moharana (✉) · A. K. Singh
Department of Horticulture, Institute of Agricultural Sciences, Banaras Hindu University, Varanasi, Uttar Pradesh, India

P. P. Mishra
Department of Floriculture and Landscaping, College of Agriculture, Odisha University of Agriculture and Technology, Bhubaneswar, Odisha, India

S. P. Kashyap
Bio Zenith Scientific Research Pvt. Ltd., Mirzapur, Uttar Pradesh, India

M. Pathak
Department of Vegetable Science, College of Agriculture, Odisha University of Agriculture and Technology, Bhubaneswar, Odisha, India

D. R. Bhardwaj · K. K. Gautam
Division of Crop Improvement, ICAR-Indian Institute of Vegetable Research, Varanasi, Uttar Pradesh, India

S. S. Solankey, M. Kumari (eds.), *Advances in Research on Vegetable Production Under a Changing Climate Vol. 2*, Advances in Olericulture, https://doi.org/10.1007/978-3-031-20840-9_9

9.1 Introduction

The earth's climate system has been challenged by severe threats of undesirable and unavoidable changes. All the major realms of the climate system, *viz.*, atmosphere, cryosphere, biosphere, and ocean, are already facing the vagaries of these changes. The frequencies of extreme climate events with higher intensity and longer duration are rising. The high concentrations of greenhouse gases (GHGs) (especially carbon dioxide), warmed-up and acidified ocean with a declined level of oxygen, high atmospheric temperature, sea level rise, etc. are the resultant of the climate extremities (Fig. 9.1) (IPCC 2013).

Plants, just like any other organisms on the earth, are adversely affected by climate change, and being sessile in nature, immediate escape from the effect is not entirely feasible (He et al. 2018). The growth, development, and reproduction of the crop plants are highly influenced by the climate change. Nowadays, the climate change is becoming one of the significant research challenges to be addressed by plant researchers.

Moreover, with the ever-increasing population of the globe, there is a high chance of food insecurities due to a decline in food production and a rise in demand. All four aspects of the food security, viz., availability, accessibility, utilization, and stability are affected by the climate change (Al et al. 2008; Schmidhuber and Tubiello 2007). In 2020, the prevalence of undernourishment scaled to about 9.9% and around 768 million people around the globe have faced hunger (WHO 2021). Vegetables, along with fruits, play a crucial role in ensuring global food and nutritional security. During the COVID-19 pandemic, there was a decline in the consumption of these protective foods due to lesser availability, aberrant rise in price,

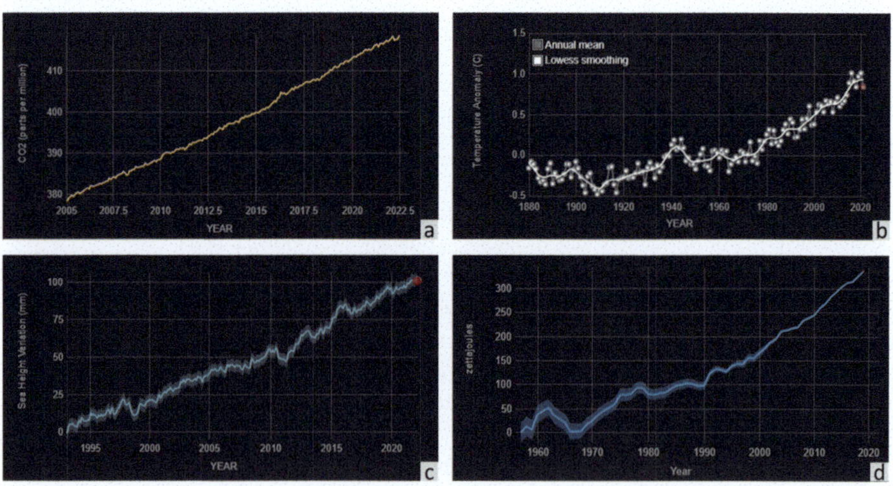

Fig. 9.1 The effects of climate change (**a**) rise in the atmospheric carbon dioxide (CO_2) (**b**) change in the global surface temperature (**c**) change in the sea level, and (**d**) change in the ocean heat content. (Source: https://climate.nasa.gov/)

poor quality, less market visits concerning the contamination, etc. (Litton and Beavers 2021). Seed industry is one of the keystones of the global food security and climate change is having a detrimental effect on the seed security (Hampton et al. 2016; Bruins 2009). In this chapter, we will elaborately discuss about the consequences of climate change with special reference to vegetable seed production.

9.2 Climate Change, Plants, and Food Production

Plants are highly dependent on various climatic factors for the completion of their life cycles (Fig. 9.2). The abnormal changes in the climate have resulted in knock-on effects on the plants affecting their growth, development, reproduction, and resilience (Eisenach 2019). The ecophysiology, diversity, distribution, and interactions (with other species) of the plants are affected by the elevated CO_2 (eCO_2), rise in temperature, and irregularities in the precipitation (Parmesan and Hanley 2015; IPCC 2014). Any anomaly in the climatic parameters hinders the survival and sustainability of the plants.

Climate change can accelerate the occurrences of genetic erosion, narrowing down the plant diversity, and the ability of the wild species and wild relatives of a plant to survive in their current ecosystem is also affected. An elevation in the atmospheric concentration of CO_2 and ozone, along with intensified rainfall, may have a negative impact on the plant structure and physiology. This will result in increased vulnerability of the plants towards the incidence of various diseases and pests. Similarly, there is also a high chance of breakdown of resistance in plants due to the rise in the temperature (Velásquez et al. 2018). Rise in the temperature also hastens the plant metabolism and alters the rate of respiration, organ development,

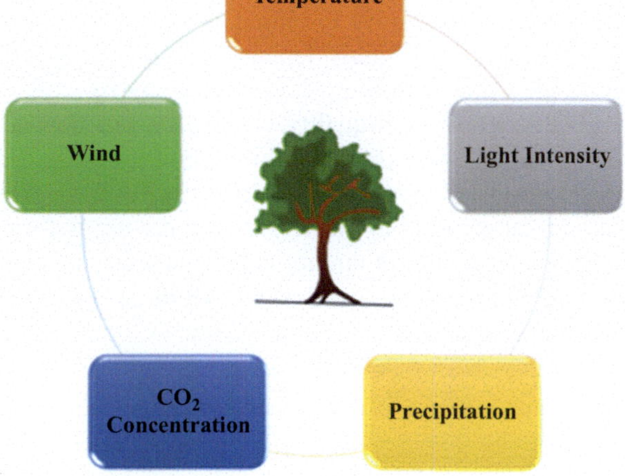

Fig. 9.2 Climatic factors affecting plant growth

senescence, and source-sink relationships (Bita and Gerats 2013; Farrar and Williams 1991). In response to high temperature, crop suitability and production decrease, especially in sub-tropical and tropical areas. The yield and product quality are compromised under the heat stress as it affects the fruit set and quickens the life cycle of the annual vegetables restricting the time for photoassimilation (Mbow et al. 2019). The plants may get weakened by heavy rainfall and storms, and the chance of infection by pathogens is aggravated. The already-stressed plants, due to the various climatic vagaries, are more prone to the attack of different pests and diseases (Skendžić et al. 2021).

Agriculture is highly vulnerable to climate change (Fig. 9.3). To match food production with the demand of the growing global population, the farmers have to tussle with the climatic vagaries. Also, the growers have to deal with various biotic and abiotic stresses that are triggered by the weather extremities like increased levels of GHGs, extreme temperatures, unpredictable and uneven rainfalls, etc.

Crop productivity is highly affected by changes in the temperature (Zhao et al. 2017; Hatfield and Prueger 2015; Ray et al. 2012) and water stress conditions (Wu et al. 2022; Vema et al. 2022; Sadras et al. 2016). Temperature rise leads to increased crop water demand (Salman et al. 2021). There is a drastic reduction in the production of important crops with the rise in temperature (Tito et al. 2018). The high frequencies of heavy rainfalls and droughts, temperature extremes, salinity, etc. are expected to severely hamper global crop productivity, which may cause undernourishment and even starvation (Dhankher and Foyer 2018).

9.3 Vegetable Seed Production: A Climatic Perspective

The seed industry is the foundation for food security in the world. Without high-quality seeds, other inputs and developed technologies are insignificant as the seed is the basic and vital unit of agriculture. In the last five decades, the seed yield of the majority of crops has increased by about 1–3% per year (Bruins 2009). Seed production is very sensitive to climate and weather conditions. Various parameters of seed production are highly dependent on the climatic variables prevailing during the entire life cycle of the crop (Maity et al. 2016).

Production of high-quality vegetable seeds depends on various climatic factors such as temperature (high and low), light (photoperiod and intensity), precipitation (distribution and total), wind (velocity and direction), etc. Various vegetables (tropical and temperate) have different climatic needs for efficient seed production. For the production of high-quality vegetable seeds, all the steps starting from seed sowing to seed processing, are crucial. All these stages are quite vulnerable to climatic anomalies.

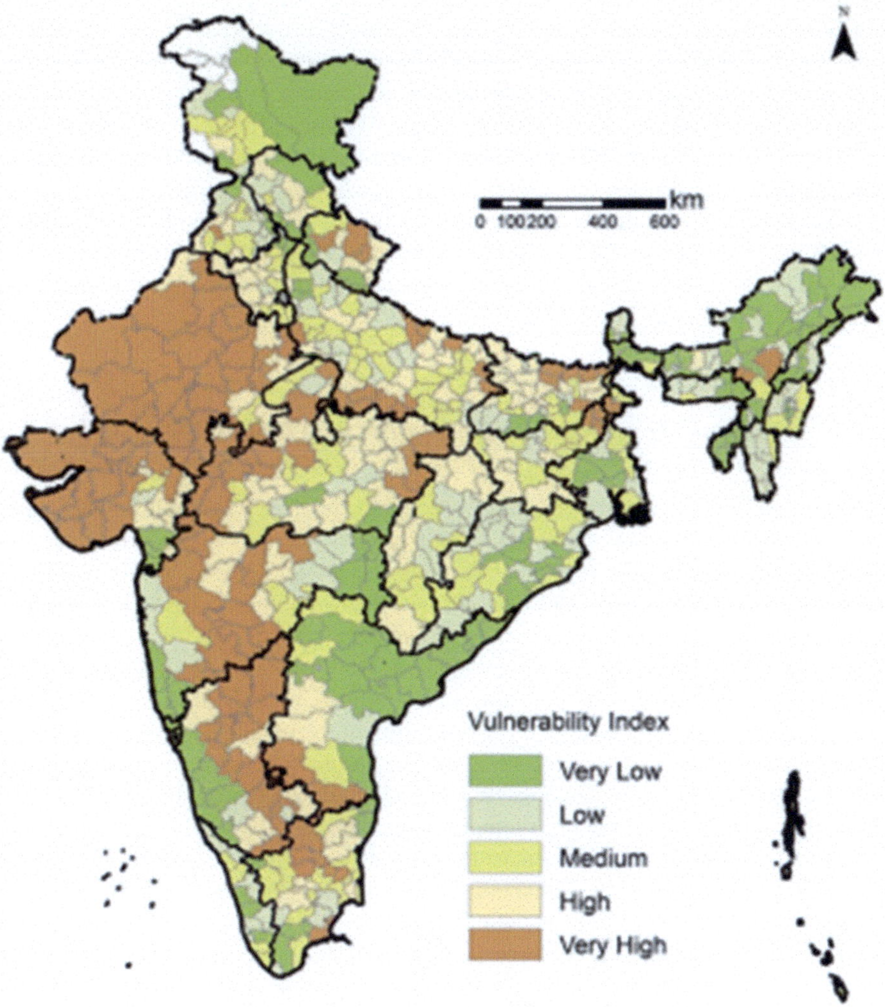

Fig. 9.3 Vulnerability of Indian agriculture to climate change (2021–2050). (Source: Rama Rao et al. 2013)

9.4 Climate Change Versus Vegetable Seed Production

The changing climate, mainly the erratic rise in temperature and irregular rainfall patterns, are the major limiting factors for vegetable productivity (Abewoy 2018). The successful production of good quality vegetable seeds is largely hampered due to the changing climate (Hampton et al. 2016; Singh et al. 2013). The reproductive phenology of the vegetable crops has been affected by the changing climatic conditions. Plants show high sensitivity toward the temperature, photoperiod, and moisture conditions pertaining to the phenology and reproductive behaviour (Meena et al. 2015a, b; Rathcke and Lacey 1985). In case of entomophilous vegetable crops, seed setting essentially relies on unhindered pollination by insect pollinators like honeybees, bumble bees, flies, native bees, etc. With the change in climate factors, the foraging activity and behaviour of these pollinators might get disturbed (Maity et al. 2016).

In response to the high-temperature level, a forced maturity and low seed quality can be observed during vegetable seed production. The climate warming may lead to the shifting of the season of seed germination from spring to autumn in alpine species (Mondoni et al. 2012).

9.4.1 Effect of Temperature Fluctuations AND eCO_2

Rising temperatures are likely to reduce the life cycle of the vegetables by affecting the flowering and fruiting phenology (Ramírez and Kallarackal 2015; Hatfield and Prueger 2015). During the reproductive stage, most vegetables can only tolerate a narrow temperature change, and beyond that, the yield and seed setting get hampered. Rise in temperature accelerates the anthesis process and lessens the crop period, whereas low temperature causes pollen sterility leading to poor pollen germination, and ultimately hampers the seed setting (Chakrabarti et al. 2011). The seed germination is reduced due to the restriction of the plant's capability to provide required assimilates that are essential for the production of storage compounds responsible for seed germination. Physiological deterioration of seeds is also induced or aggravated in the high-temperature conditions (Hampton et al. 2013).

The rise in temperature influences the chilling requirement of different vegetables, mainly temperate ones. In asparagus, a rise in temperature by 3 °C, increases the dormancy period by 15 days leading to delayed yield and reduced spear production (Nie et al. 2016). If the high temperature persists for a more extended period, i.e., 2–3 weeks, there will be an escalation in the sugar depletion, causing yield penalties (Yamaguchi and Maeda 2015).

In tomato, a rise in the temperature causes bud drop, poor pollen production, abnormal flower development, dehiscence, ovule abortion, poor viability, reduced carbohydrate availability, and other reproductive abnormalities leading to reduced fruit set, smaller and low-quality fruits (Hazra et al. 2007; Singh et al. 2015; Cruz-Ortega et al. 2002). Above the temperature of 25 °C, the pollination and fruit set in tomato are affected (Thamburaj and Singh 2001). A slight rise in temperature may reduce the availability of carbohydrate at the crucial phases of plant development in tomato, leading to yield penalties (Sato et al. 2006). Seed germination and pollen tube development are affected due to low temperatures during the growth period leading to reduced fruit set (Pardossi et al. 1992).

Similarly, in chilli, flower drop, ovule abortion, poor fruit set, and fruit drop is caused by high temperature (Arora et al. 2010). Fertilization in pepper is highly sensitive to high temperatures as suggested by the inhibited fruit set, due to the rise in post-pollination temperatures (Erickson and Markhart 2002).

In the case of seed production of late cauliflower in Kullu valley, the fluctuation in temperature during the month of February drastically affected the curd formation, and seed yield was reduced by 49.17–100% (Gill and Singh 1973).

In North Western Himalayas area, the seed yield of cabbage is affected up to a larger extent attributed to the changes in various climatic parameters (Kumar et al. 2009).

High temperature significantly affects germination in cucurbits. At 42 °C, seeds of most of the cucurbits do not germinate (Kurtar 2010). Variations in temperature lead to delayed ripening of fruits and reduced sweetness in melons (Abewoy 2018). Vegetative growth increases in Cucurbitaceous vegetable crops like ash gourd, bottle gourd, pumpkin, etc. during the warm, humid climate resulting in poor female flower production (Ayyogari et al. 2014).

In case of potato, elevated temperature causes the suppression of tuber formation, minimization of fitness of seed tubers, reduction in carbon assimilation towards tuber starch, and acceleration of leaf senescence (Sonnewald et al. 2015; Hancock et al. 2014; Wolf et al. 1991; Fahem and Haverkort 1988; Menzel 1985; Ewing 1981). In potato cv. Superior, at increased temperature level, both canopy net photosynthetic rate and stomatal conductance reduced drastically leading to decrease in both tuber yield and biomass production (Lee et al. 2020).

During the growth period of celery and lettuce, the unprecedented rise in temperature results in premature production of seed heads (bolting) leading to poor production of quality heads and yield reduction (Prasad and Chakravorty 2015).

In Chinese cabbage cultivars, the proline content and cell membrane permeability slightly increase under high-temperature conditions (Datta 2013). To avoid the premature bolting in case of heat-sensitive Chinese cabbage cultivars, daily mean temperature should be above 18 °C during the raising phase (Guttormsen and Moe 1985).

In red kidney bean cv. Montcalm, high temperature negatively affects the seed setting and causes a reduction in seed number (Prasad et al. 2002). High temperature during flowering results in pollen sterility and reduced seed setting in cowpea (Ahmed et al. 1992), and soybean (Djanaguiraman et al. 2011; Salem et al. 2007).

Similarly, vegetable crops, primarily from tropical and subtropical regions, are facing the vagaries of low-temperature stress (chilling and freezing). The crop growth and productivity are affected, eventually leading to crop failure. The geographical distribution and survivability of plants are also disturbed due to low-temperature stress (Xin and Browse 2000).

In potato, tuber yield is penalized by 10–50% by the frost injury considering the intensity and phase of occurrence (Atayee and Noori 2020). In case of tomatoes for processing industries and early vegetables, frost damage during spring is hazardous (Yadav 2010). Tomatoes and soybeans express the signs of damage caused due to exposure to low temperatures of 10–15 °C (Lynch 1990). In bell pepper and tomato, low temperature also induces malformation of flowers and fruits (Adams et al. 2001; Aloni et al. 1999).

The concentration of atmospheric CO_2 is increasing at an alarming rate (Fig. 9.1a). At an increased CO_2 level, the seed mass has been reported to experience both a rise and decline in C_3 plants, but in case of C_4 plants, there is no change. In legume crops, the increase is higher than in non-legumes (Hampton et al. 2013).

Environmental vagaries like increased CO_2 after a certain extent along with rising temperature, have significantly affected the Solanaceous vegetables penalizing their yield and quality (Moharana et al. 2021; Rangaswamy et al. 2021).

In many crops, the yield penalties linked with the increased temperature will aggravate with the rise in CO_2 concentrations where high temperatures nullify the effects of eCO_2 (Prasad et al. 2002; Reddy et al. 1995; Ahmed et al. 1993; Baker et al. 1989).

In potato, at an eCO_2 level, the biomass and area of leaves increased, but the tuber yield did not exhibit any significant change. However, an appropriate increase in CO_2 and temperature in a concurrent manner exerted balanced improvement of source and sink, and a positive effect on the rate of photosynthesis, growth, yield, and seed production can be observed (Lee et al. 2020).

Increased CO_2 and high temperature can cause alteration of sexuality in plants and also impose a breakdown of male sterility and self-incompatibility mechanisms which play crucial roles in seed production of various vegetable crops (Simmonds 1979). Male sterility reduces in carrot at 26 °C (Quagliotti 1967), breaks down above 17 °C in case of Brussels sprouts (Nieuwhof 1968), and self-incompatibility breakdown happens at 26 °C in radish (El Murabaa 1957).

9.4.2 Effect of Irregular Precipitation

Moisture stress (excess or deficit) and high temperature also affect the quality of the seeds and the resultant crop performance apart from the seed yield (Alqudah et al. 2011). Waterlogging can minimize the crop yield and is primarily influenced by the growth phase, crop type, and duration of the crop (Tian et al. 2021). Vegetables, mainly the shallow-rooted ones, are most sensitive to waterlogged conditions. Waterlogging due to heavy rainfall and failure of drainage system leads to unprecedented loss to vegetable seed production (Fig. 9.4). Water and nutrient uptake are severely affected in waterlogged conditions as the roots are under hypoxic conditions, strongly penalizing the growth and function of the plants (Fujita et al. 2021; Decoteau 1998). Heavy rainfall interferes in the natural pollination process through various means, like restricting the movement of pollinators, and is also responsible for high disease incidence in plants. The quality and shelf life of harvested vegetables are reduced due to rain at the physiological maturity stage.

Similarly, in water deficit conditions, due to untimely and insufficient rainfalls, both the vegetative and reproductive phases of vegetable crops are severely affected.

Vegetables, especially in the reproductive phase, possess high sensitivity toward drought stress due to their succulent nature. In arid and semi-arid areas, many

Fig. 9.4 Waterlogging of vegetable seed production fields due to heavy rainfall. (Source: ICAR-Indian Institute of Vegetable Research, Varanasi, U.P., India)

vegetable legumes are usually affected by water deficit conditions during their reproductive phase (Chatterjee and Solankey 2015).

Vegetables, mainly cucurbits and pod vegetables, are very sensitive to water stress conditions during flowering, fruiting, and seed development. The critical stages of water requirement are very crucial pertaining to the yield and seed production in vegetable crops (Table 9.1). Water stress during the enlargement of storage organs in tuber, root, and bulb crops significantly reduces yield and quality (Prasad and Chakravorty 2015).

Being highly succulent in nature, leafy vegetables need sufficient water for a good yield, and water stress can cause poor quality and quantity of yield as the water content in the produce is reduced (AVRDC 1990).

Gill and Singh (1973) proposed that high rainfall and snowfall in January-February are highly detrimental in the case of seed production of late cauliflower in Kullu valley and suggested the lower altitudes (1200–1450 m) for better seed production.

Floods in sweet potato field results in rotting during harvest and shrinkage in storage (Thomposon et al. 1992; Ton and Hernandez 1978). It also affects baking quality and reduces dry matter and carotenoid content (Constantin et al. 1974).

In pepper, prolonged flooding can cause reduced oxygen supply and nutrient uptake leading to stunted growth, yellow leaves, black root tips, and swelling at the shoot and root junction (Hasnain and Sheik 1976).

Table 9.1 Critical stages of water requirement of various vegetable crops

Vegetable crop(s)	Critical stage(s) of water requirement
Chilli	Flowering and fruit set
Tomato	Early flowering, fruit set, and enlargement
Brinjal	Flowering and fruit development
Okra	Flowering and pod development
Potato	Tuberization and tuber enlargement
Onion	Bulb formation and enlargement
Garden pea	Flowering and pod development
Cabbage, cauliflower, and broccoli	Curd/head formation and enlargement
Cucumber	Flowering and fruit development
Leafy vegetables	Throughout the crop duration
Radish, carrot, and turnip	Root enlargement
Melons	Flowering and fruit development
Lettuce	Throughout development
Sweet potato	Root enlargement
Snap bean	Flowering and pod enlargement

Source: Solankey et al. (2021), Malhotra (2016), Chatterjee and Solankey (2015), Kumar et al. (2012), Bahadur et al. (2011)

High rainfall leads to high humidity, which aggravates insect pest infestation and disease incidence.

9.4.3 *Effect of Wind and Light*

Pollen desiccation is a resultant of strong winds and high temperature, causing improper seed setting and poor seed quality. Heavy rainfall and strong wind cause high seed loss at the harvesting stage through shattering in vegetable crops like amaranthus and leads to the lodging of the plants. High wind velocity restricts the deposition of pollen on stigma and also causes pollen dispersal too far. This reduces the seed setting and ultimately hampers seed production. Pollen desiccation occurs due to dry winds leading to reduced pollen viability. High speed wind also interferes with essential agricultural operations like spraying, etc.

Low light intensity causes plants to have small leaves, spindly in nature, improper pollination, and poor fruit quality, whereas high light intensity coupled with high-temperature cause scorching of fruits (such as sun scald in tomatoes) and other plant parts. At high light intensity, photosynthesis is also hampered.

Some vegetables, like onion, demand specific photoperiod requirements for flower induction and the effective production of seeds (Welbaum 2015). Any deviation from this leads to a negative effect on seed yield. Short-day conditions accelerate the tuber formation in potatoes and root enlargement in sweet potatoes.

9.5 Approaches to Mitigate Climate Change Vis-À-Vis Vegetable Seed Production

Access to the adapted and improved seeds is one of the most basic and practical adaptation approaches in climatic hotspots. Productivity and incomes can be increased with the use of high-quality seeds, especially in the smallholder systems (Cacho et al. 2020). With the development of climate-resilient varieties with respect to different agro-climate zones, vegetable seed production can be stabilized. The traits of interest will be resistance to various abiotic (drought, waterlogging, temperature extremes, salinity, etc.) and biotic (pests and diseases) stresses (Ceccarelli et al. 2010). A cumulative approach of traditional, molecular, and transgenic methods can be imparted to exploit these traits and develop climate-smart vegetable crops. Genetic diversity for high temperature has already been demonstrated in crops like cowpea (Ahmed et al. 1993), bean (Porch and Jahn 2001), and soybean (Salem et al. 2007). The trait like photoperiod insensitivity also has to be exploited by the breeders to minimize the influence of photoperiod.

Wild species and wild relatives of a crop plant are a significant source of desired genes concerning biotic and abiotic stresses, so the plant genetic resources and diversity must be conserved and characterized to exploit climate-resilience breeding in vegetable crops.

Shifting seed production sites to the areas that became more suitable due to the change in climate patterns can be taken into consideration. In order to minimize the negative effect of environmental stress in particular regions, change in both latitude and elevation can come into play (Hampton et al. 2013; Shinohara et al. 2008; Egli et al. 2005). In areas like India, China, North America, the northern European countries, New Zealand, Australia, and the southern South American countries, moving of the seed production areas will be possible depending on the availability of water and land. However, this option may not be viable for the regions at lower latitudes like African countries compelling them to import even more to meet their seed requirements (Hampton et al. 2012, 2016).

In order to avoid the coinciding of abiotic stress like high temperature with flowering and seed setting in vegetables, a change in date of sowing can also be advocated as an excellent escape strategy considering various factors like water availability (Parry 2000), response to day length (especially premature flowering in short-day conditions) (Board and Hall 1984), the possibility of freezing or chilling temperature during flowering (Ridge and Pye 1985), effects of soil temperature on seed germination (Rahman et al. 2005), and increased prevalence of pests, diseases, and weeds (Hatfield et al. 2011; Olesen and Bindi 2002).

Techniques like seed priming can also be advocated, which ensures adaptability to various environmental stresses, especially drought. Seed priming also aids in the transfer of stress memory to the offspring (Saha et al. 2022). In order to adapt to harsh field conditions, the seeds of high-value crops can be covered with thermostable polymer along with growth supplements and pesticides (Maity et al. 2016).

In the era of climate change, to cope up with the possible natural or anthropogenic global or regional disasters, the Svalbard Global Seed Vault was built on the Norwegian island of Spitsbergen in 2008 to ensure food security.

Improved and sustainable water and land management practices like use of rainwater harvesting structures, micro-irrigation methods, proper management of crop residues, crop intensification, crop rotation, in-situ moisture conservation, etc. also ensure the maximization of seed production against various environmental changes.

The seed system of a particular region or country needs to be more refined and responsive towards the measures to be taken to tackle the climate change scenario. Seed producers should have access to the quality seeds of vegetable varieties with desired trait(s) of climate resilience. Growers should be strengthened to meet the cost involving the adaptive measures against environmental stresses, afford the insurance cost against climatic vagaries, and have the know-how and/or accessibility to experts to implement the necessary management changes (Singh et al. 2013). Community seed banks can facilitate the growers to maintain resilience to climate change at the household and community level (Vernooy et al. 2017). These measures would be difficult to achieve in underdeveloped and developing countries due to less efficient legislation, improper supporting activities from seed companies, and a meagre annual seed replacement rate from formal sources. However, these challenges have to be addressed by strengthening of the seed system and collaborative efforts leaving no one behind.

9.6 Conclusion

The seed production system is largely affected by global climate change. A sustainable vegetable seed production can be achieved through multi-dimensional interventions starting from the application of science and technology to the strengthening of the seed system up to the grassroots level. The need of the hour is the collaborative effort of plant breeders, plant physiologists, and molecular biologists to address climate change in terms of the development of climate-smart varieties and techniques. A holistic approach involving crop management, crop breeding, use of indigenous technical knowledge (ITK), investment in capacity building and research, etc. should be followed. With all the strategic approaches, constant monitoring and evaluation of adaptive measures and future concerns should be followed to achieve climate-resilient sustainable development in vegetable seed production.

References

Abewoy D (2018) Review on impacts of climate change on vegetable production and its management practices. Adv Crop Sci Technol 6(330)

Adams SR, Cockshull KE, Cave CRJ (2001) Effect of temperature on the growth and development of tomato fruits. Ann Bot 88(5):869–877

Ahmed FE, Hall AE, DeMason DA (1992) Heat injury during floral development in cowpea (Vigna unguiculata, Fabaceae). Am J Bot 79(7):784–791

Ahmed FE, Hall AE, Madore MA (1993) Interactive effects of high temperature and elevated carbon dioxide concentration on cowpea [Vigna unguiculata (L.) Walp.]. Plant Cell Environ 16(7):835–842

Al W, Orking G, Clima O (2008) Climate change and food security: a framework document. FAO, Rome

Aloni B, Pressman E, Karni L (1999) The effect of fruit load, defoliation and night temperature on the morphology of pepper flowers and on fruit shape. Ann Bot 83(5):529–534

Alqudah AM, Samarah NH, Mullen RE (2011) Drought stress effect on crop pollination, seed set, yield and quality. In: Alternative farming systems, biotechnology, drought stress and ecological fertilisation. Springer, Dordrecht, pp 193–213

Arora SK, Partap PS, Pandita ML, Jalal I (2010) Production problems and their possible remedies in vegetable crops. Indian Hortic 32:2–8

Atayee AR, Noori MS (2020) Alleviation of cold stress in vegetable crops. J Sci Agric 4:38–44

AVRDC (1990) Vegetable production training manual. Asian Vegetable Research and Training Center, Tainan, p 447

Ayyogari K, Sidhya P, Pandit MK (2014) Impact of climate change on vegetable cultivation-a review. Int J Agric Environ Biotechnol 7:145

Bahadur A, Chatterjee A, Kumar R, Singh M, Naik PS (2011) Physiological and biochemical basis of drought tolerance in vegetables. Veg Sci 38(1):1–16

Baker JT, Allen LH Jr, Boote KJ, Jones P, Jones JW (1989) Response of soybean to air temperature and carbon dioxide concentration. Crop Sci 29(1):98–105

Bita CE, Gerats T (2013) Plant tolerance to high temperature in a changing environment: scientific fundamentals and production of heat stress-tolerant crops. Front Plant Sci 4:273

Board JE, Hall W (1984) Premature flowering in soybean yield reductions at nonoptimal planting dates as influenced by temperature and photoperiod 1. Agron J 76(4):700–704

Bruins M (2009) The evolution and contribution of plant breeding to global agriculture. In: Proceedings of the second world seed conference: responding to the challenges of a changing world: the role of new plant varieties and high quality seed in agriculture, pp 18–31

Cacho OJ, Moss J, Thornton PK, Herrero M, Henderson B, Bodirsky BL, Humpenöder F, Popp A, Lipper L (2020) The value of climate-resilient seeds for small holder adaptation in sub-Saharan Africa. Clim Chang 162(3):1213–1229

Ceccarelli S, Grando S, Maatougui M, Michael M, Slash M, Haghparast R, Rahmanian M, Taheri A, Al-Yassain A, Benbelkacem A, Labdi M, Mimoun H, Nachit M (2010) Plant breeding and climate changes. J Agric Sci 148(6):627–637

Chakrabarti B, Singh SD, Nagarajan S, Aggarwal PK (2011) Impact of temperature on phenology and pollen sterility of wheat varieties. Aust J Crop Sci 5(8):1039–1043

Chatterjee A, Solankey SS (2015) Functional physiology in drought tolerance of vegetable crops: an approach to mitigate climate change impact. In: Climate dynamics in horticultural science. Routledge, London, pp 149–171

Constantin RJ, Hernandez TP, Jones LG (1974) Effect of irrigation and N fertilization on quality of sweet potatoes. J Am Soc Hortic Sci 99:308–310

Cruz-Ortega R, Ayala-Cordero G, Anaya AL (2002) Allelochemical stress produced by the aqueous leachate of Callicarpa acuminata: effects on roots of bean, maize, and tomato. Physiol Plant 116(1):20–27

Datta S (2013) Impact of climate change in Indian horticulture-a review. Int J Sci Environ Technol 2(4):661–671

Decoteau D (1998) Plant physiology: environmental factors and photosynthesis. Greenhouse glazing & solar radiation transmission workshop, Center for Controlled Environment Agriculture, Rutgers University, Cook College, pp 1–6

Dhankher OP, Foyer CH (2018) Climate resilient crops for improving global food security and safety. Plant Cell Environ 41(5):877–884

Djanaguiraman M, Prasad PV, Boyle DL, Schapaugh WT (2011) High-temperature stress and soybean leaves: leaf anatomy and photosynthesis. Crop Science 51(5):2125–2131
Egli DB, TeKrony DM, Heitholt JJ, Rupe J (2005) Air temperature during seed filling and soybean seed germination and vigor. Crop Sci 45(4):1329–1335
Eisenach C (2019) How plants respond to climate change: a new virtual special issue of plant, , Cell & Environment. Plant Cell Environ 42(9):2537–2539
El Murabaa AIM (1957) Effect of high temperature on incompatibility in radish. Euphytica 6(3):268–270
Erickson AN, Markhart AH (2002) Flower developmental stage and organ sensitivity of bell pepper (Capsicum annuum L.) to elevated temperature. Plant Cell Environ 25(1):123–130
Ewing EE (1981) Heat stress and the tuberization stimulus. Am Potato J 58(1):31–49
Fahem M, Haverkort AJ (1988) Comparison of the growth of potato crops grown in autumn and spring in North Africa. Potato Res 31(4):557–568
Farrar JF, Williams ML (1991) The effects of increased atmospheric carbon-dioxide and temperature on carbon partitioning, source-sink relations and respiration. Plant Cell Environ 14:819–830
Fujita S, Noguchi K, Tange T (2021) Different waterlogging depths affect spatial distribution of fine root growth for Pinus thunbergii seedlings. Front Plant Sci 12:614764
Gill HS, Singh JP (1973) Effect of environmental factors on seed production of late cauliflower in Kulu Valley. Indian J Agric Sci 43(3):234–236
Guttormsen G, Moe R (1985) Effects of day and night temperature on bolting in Chinese cabbage. Sci Hortic 25:217–224
Hampton JG, Rolston MP, Pyke NB, Green W (2012) Ensuring the long term viability of the New Zealand seed industry. Agron NZ 42:129–140
Hampton JG, Boelt B, Rolston MP, Chastain TG (2013) Effects of elevated CO_2 and temperature on seed quality. J Agric Sci 151(2):154–162
Hampton JG, Conner AJ, Boelt B, Chastain TG, Rolston P (2016) Climate change: seed production and options for adaptation. Agriculture 6(3):33
Hancock RD, Morris WL, Ducreux LJ, Morris JA, Usman M, Verrall SR, Fuller J, Simpson CG, Zhang R, Hedley PE, Taylor MA (2014) Physiological, biochemical and molecular responses of the potato (Solanum tuberosum L.) plant to moderately elevated temperature. Plant Cell Environ 37:439–450
Hasnain S, Sheik KH (1976) Effect of flooding and drainage on the growth of Capsicum annuum. Biologia 22:89–106
Hatfield JL, Prueger JH (2015) Temperature extremes: effect on plant growth and development. Weather Clim Extremes 10:4–10
Hatfield JL, Boote KJ, Kimball BA, Ziska LH, Izaurralde RC, Ort D, Thomson AM, Wolfe D (2011) Climate impacts on agriculture: implications for crop production. Agron J 103(2):351–370
Hazra P, Samsul HA, Sikder D, Peter KV (2007) Breeding tomato (Lycopersicon esculentum Mill) resistant to high temperature stress. Int J Plant Breed 1:31–40
He M, He CQ, Ding NZ (2018) Abiotic stresses: general defenses of land plants and chances for engineering multi-stress tolerance. Front Plant Sci 9:1771
IPCC (2013) In: Stocker TF, Qin D, Plattner G-K et al (eds) Climate change 2013: the physical science basis. Contribution of working group I to the fifth assessment report of the intergovernmental panel on climate change. Cambridge University Press, Cambridge
IPCC (2014) Summary for policymakers. In: Field CB, Barros VR, Dokken DJ et al (eds) Climate change 2014: impacts, adaptation, and vulnerability. Part a: global and sectoral aspects. Contribution of working group II to the fifth assessment report of the intergovernmental panel on climate change. Cambridge University Press, Cambridge, pp 1–32
Kumar PR, Yadav SK, Sharma SR, Lal SK, Jha DN (2009) Impact of climate change on seed production of cabbage in North Western Himalayas. World J Agric Sci 5(1):18–26
Kumar R, Solankey SS, Singh M (2012) Breeding for drought tolerance in vegetables. Veg Sci 39(1):1–15

Kurtar ES (2010) Modelling the effect of temperature on seed germination in some cucurbits. Afr J Biotechnol 9:9
Lee YH, Sang WG, Baek JK, Kim JH, Shin P, Seo MC, Cho JI (2020) The effect of concurrent elevation in CO_2 and temperature on the growth, photosynthesis, and yield of potato crops. PLoS One 15(10):e0241081
Litton MM, Beavers AW (2021) The relationship between food security status and fruit and vegetable intake during the COVID-19 pandemic. Nutrients 13(3):712
Lynch DV (1990) Chilling injury in plants: the relevance of membrane lipids. In: Katterman F (ed) Environmental injury to plants. Academic Press, San Diego, pp 17–34
Maity A, Vijay D, Mukherjee A, Lamichaney A (2016) Potential impacts of climate change on quality seed production: a perspective of hill agriculture. In: Conservation agriculture. Springer, Singapore, pp 459–485
Malhotra SK (2016) Achieving high water use efficiency in agriculture through per drop more crop. In: Proceedings of the 10th World Aqua Congress, international conference on water smart solutions for growing India. Aqua Foundation, New Delhi, pp 377–394
Mbow C., Rosenzweig C, Barioni LG, Benton TG, Herrero M, Krishnapillai M, Liwenga E, Pradhan P, Rivera-Ferre MG, Sapkota T, Tubiello FN, Xu Y (2019) Food security in climate change and land: an IPCC special report on climate change, desertification, land degradation, sustainable land management, food security, and greenhouse gas fluxes in terrestrial ecosystems. Intergovernmental Panel on Climate Change
Meena RS, Meena VS, Meena SK, Verma JP (2015a) Towards the plant stress mitigate the agricultural productivity: a book review. J Cleaner Prod 102:552–553
Meena VS, Meena SK, Verma JP, Meena RS (2015b) Current trend to mitigate climate change and sustainable agriculture: a book review. J Cleaner Prod 102:548–549
Menzel CM (1985) Tuberization in potato at high temperatures: interaction between temperature and irradiance. Ann Bot 55(1):35–39
Moharana DP, Singh RK, Kashyap SP, Rai N, Bhardwaj DR, Singh AK (2021) Response of Solanaceous vegetables to increasing temperature and atmospheric CO_2. In: Advances in research on vegetable production under a changing climate, vol 1. Springer, Cham, pp 91–111
Mondoni A, Rossi G, Orsenigo S, Probert RJ (2012) Climate warming could shift the timing of seed germination in alpine plants. Ann Bot 110(1):155–164
Nie LC, Chen YH, Liu M (2016) Effects of low temperature and chilling duration on bud break and changes of endogenous hormones of asparagus. Eur J Hortic Sci 81:22–26
Nieuwhof M (1968) Effect of temperature on the expression of male sterility in Brussels sprouts (Brassica oleracea L. var. gemmifera DC.). Euphytica 17(2):265–273
Olesen JE, Bindi M (2002) Consequences of climate change for European agricultural productivity, land use and policy. Eur J Agron 16(4):239–262
Pardossi A, Vernieri P, Tognoni F (1992) Involvement of abscisic acid in regulating water status in Phaseolus vulgaris L. during chilling. Plant Physiol 100(3):1243–1250
Parmesan C, Hanley ME (2015) Plants and climate change: complexities and surprises. Ann Bot 116(6):849–864
Parry ML (ed) (2000) Assessment of potential effects and adaptations for climate change in Europe. The Europe ACACIA Project/Jackson Environment Institute/University of East Anglia, Norwich
Porch TG, Jahn M (2001) Effects of high-temperature stress on micro-sporogenesis in heat-sensitive and heat-tolerant genotypes of Phaseolus vulgaris. Plant Cell Environ 24(7):723–731
Prasad BVG, Chakravorty S (2015) Effects of climate change on vegetable cultivation-a review. Nat Environ Pollut Technol 14(4):923
Prasad PVV, Boote KJ, Allen LH Jr, Thomas JM (2002) Effects of elevated temperature and carbon dioxide on seed-set and yield of kidney bean (Phaseolus vulgaris L.). Glob Chang Biol 8(8):710–721
Quagliotti L (1967) Effects of different temperatures on stalk development, flowering habit, and sex expression in the carrot (Daucus carota) L. Euphytica 16(1):83–103

Rahman MM, Hampton JG, Hill MJ (2005) Soybean seed yield as affected by time of sowing in a cool temperate environment. J New Seeds 7(4):1–15

Rama Rao CA, Raju BMK, Subba Rao AVM, Rao KV, Rao VUM, Kausalya R, Venkateswarlu B, Sikka AK (2013) Atlas on vulnerability of Indian agriculture to climate change. Hyderabad, ICAR-Central Research Institute for Dryland Agriculture, p 116

Ramírez F, Kallarackal J (2015) The effect of increasing temperature on phenology. In: Responses of fruit trees to global climate change. Springer, Cham, pp 11–13

Rangaswamy TC, Sridhara S, Ramesh N, Gopakkali P, El-Ansary DO, Mahmoud EA, Abdelmohsen SAM, Abdelbacki AMM, Elansary HO, Abdel-Hamid AM (2021) Assessing the impact of higher levels of CO_2 and temperature and their interactions on tomato (Solanum lycopersicum L.). Plan Theory 10(2):256

Rathcke B, Lacey EP (1985) Phenological patterns of terrestrial plants. Annu Rev Ecol Syst 16:179–214

Ray DK, Ramankutty N, Mueller ND, West PC, Foley JA (2012) Recent patterns of crop yield growth and stagnation. Nat Commun 3(1):1–7

Reddy KR, Hodges HF, McKinion JM (1995) Carbon dioxide and temperature effects on pima cotton growth. Agric Ecosyst Environ 54(1–2):17–29

Ridge PE, Pye DL (1985) The effects of temperature and frost at flowering on the yield of peas grown in a Mediterranean environment. Field Crop Res 12:339–346

Sadras VO, Villalobos FJ, Orgaz F, Fereres E (2016) Effects of water stress on crop production. In: Principles of agronomy for sustainable agriculture. Springer, Cham, pp 189–204

Saha D, Choyal P, Mishra UN, Dey P, Bose B, Prathibha MD, Gupta NK, Mehta BK, Kumar P, Pandey S, Chauhan J, Singhal RK (2022) Drought stress responses and inducing tolerance by seed priming approach in plants. Plant Stress 4:100066

Salem MA, Kakani VG, Koti S, Reddy KR (2007) Pollen-based screening of soybean genotypes for high temperatures. Crop Sci 47(1):219–231

Salman SA, Shahid S, Sharafati A, Salem GSA, Bakar AA, Farooque AA, Chung E-S, Ahmed YA, Mikhail B, Yaseen ZM (2021) Projection of agricultural water stress for climate change scenarios: a regional case study of Iraq. Agriculture 11(12):1288

Sato S, Kamiyama M, Iwata T, Makita N, Furukawa H, Ikeda H (2006) Moderate increase of mean daily temperature adversely affects fruit set of Lycopersicon esculentum by disrupting specific physiological processes in male reproductive development. Ann Bot 97(5):731–738

Schmidhuber J, Tubiello FN (2007) Global food security under climate change. Proc Natl Acad Sci 104(50):19703–19708

Shinohara T, Hampton JG, Hill MJ, Juntakool S, Suprakarn S, Sagwansupyakorn C (2008) Variations in pea (Pisum sativum L.) seed vigour among regions of production and cropping seasons are associated with temperature during reproductive growth. J Jpn Soc Agric Technol Manage 14:148–155

Simmonds NW (1979) Principles of crop improvement. Longman: London, UK

Singh RP, Prasad PVV, Reddy KR (2013) Impacts of changing climate and climate variability on seed production and seed industry. Adv Agron 118:49–110

Singh U, Patel PK, Singh AK, Tiwari V, Kumar R, Rai N, Bahadur A, Tiwari SK, Singh M, Singh B (2015) Screening of tomato genotypes under high temperature stress for reproductive traits. Veg Sci 42(2):52–55

Skendžić S, Zovko M, Živković IP, Lešić V, Lemić D (2021) The impact of climate change on agricultural insect pests. Insects 12(5):440

Solankey SS, Kumari M, Akhtar S, Singh HK, Ray PK (2021) Challenges and opportunities in vegetable production in changing climate: mitigation and adaptation strategies In: Solankey SS et al (eds) Advances in research on vegetable production under a changing climate, vol. 1. Springer, Cham, pp 13–59. ISBN 978-3-030-63497-1

Sonnewald S, van Harsselaar J, Ott K, Lorenz J, Sonnewald U (2015) How potato plants take the heat? Procedia Environ Sci 29:97

Thamburaj S, Singh N (2001) Textbook of vegetables, tuber crops and spices. Indian Council of Agricultural Research

Thomposon PG, Smittle DA, Hall MR (1992) Relationship of sweet potato yield and quality to amount of irrigation water. Hortic Sci 27(1):23–26

Tian LX, Zhang YC, Chen PL, Zhang FF, Li J, Yan F, Dong Y, Feng BL (2021) How does the waterlogging regime affect crop yield? A global meta-analysis. Front Plant Sci 12:634898

Tito R, Vasconcelos HL, Feeley KJ (2018) Global climate change increases risk of crop yield losses and food insecurity in the tropical Andes. Glob Chang Biol 24(2):e592–e602

Ton CS, Hernandez TP (1978) Wet soil stress effects on sweet potatoes. J Am Soc Hortic Sci 103:600–603

Velásquez AC, Castroverde CDM, He SY (2018) Plant-pathogen warfare under changing climate conditions. Curr Biol 28(10):R619–R634

Vema VK, Sudheer KP, Rohith AN, Chaubey I (2022) Impact of water conservation structures on the agricultural productivity in the context of climate change. Water Resour Manag 36(5):1627–1644

Vernooy R, Sthapit B, Otieno G, Shrestha P, Gupta A (2017) The roles of community seed banks in climate change adaption. Dev Pract 27(3):316–327

Welbaum GE (2015) Vegetable production and practices. CABI, Wallingford

Wolf S, Marani A, Rudich J (1991) Effect of temperature on carbohydrate metabolism in potato plants. J Exp Bot 42:619–625

World Health Organization (2021) The state of food security and nutrition in the world 2021: transforming food systems for food security, improved nutrition and affordable healthy diets for all, vol. 2021. Food & Agriculture Org

Wu J, Wang J, Hui W, Zhao F, Wang P, Su C, Gong W (2022) Physiology of plant responses to water stress and related genes: a review. Forests 13(2):324

Xin Z, Browse J (2000) Cold comfort farm: the acclimation of plants to freezing temperatures. Plant Cell Environ 23(9):893–902

Yadav SK (2010) Cold stress tolerance mechanisms in plants. A review. Agron Sustain Dev 30(3):515–527

Yamaguchi T, Maeda T (2015) Effect of high-temperature treatments on the breaking of dormancy in one-year-old asparagus (Asparagus officinalis L.). Environ Control Biol 53(1):23–26

Zhao C, Liu B, Piao S, Wang X, Lobell DB, Huang Y, Huang M, Yao Y, Bassu S, Ciais P, Durand JL, Elliott J, Ewert F, Janssens IA, Li T, Lin E, Liu Q, Martre P, Müller C, Peng S, Peñuelas J, Ruane AC, Wallach D, Wang T, Wu D, Liu Z, Zhu Y, Zhu Z, Asseng S (2017) Temperature increase reduces global yields of major crops in four independent estimates. Proc Natl Acad Sci 114(35):9326–9331

Chapter 10
Kitchen Gardening for Nutritional Security Under Changing Climate

Arindam Nag, Anirban Mukherjee, Kumari Shubha, Sangeeta Bhattacharyya, Ramnath K. Ray, Pinaki Roy, Anirudhha Roy, and Sheikh Mohammad Feroze

Abstract Climate change is a global phenomenon and posing formidable challenge in all sphere of life. Food system is no exception of it and impact of climate change is observed through direct and indirect influence on parameters related to weather, land, air, water etc. This is continuously making difficult to lead sustainable life; especially considering the situation of food (nutritional) security, poverty and livelihood situation of majority of world population. The nexus between food

A. Nag (✉)
Department of Extension Education, Dr. Kalam Agricultural College (Bihar Agricultural University, Sabour, Bhagalpur), Kishanganj, Bihar, India

A. Mukherjee
Division of Agricultural Extension, ICAR-Research Complex for Eastern Region, Patna, Bihar, India

K. Shubha
Division of Vegetable Science, ICAR-Research Complex for Eastern Region, Patna, Bihar, India

S. Bhattacharyya
Division of Agricultural Extension, ICAR-Central Citrus Research Institute, Nagpur, Maharashtra, India

R. K. Ray
ICAR-Research Complex for Eastern Region, Patna, Bihar, India

P. Roy
National Agricultural Science Fund, Krishi Anusandhan Bhawan-I, ICAR-Indian Council of Agricultural Research, New Delhi, India

A. Roy
Department of Agricultural Economics, Dr. Rajendra Prasad Central Agricultural University, Samastipur, Bihar, India

S. M. Feroze
Department of Agricultural Economics, College of Agriculture, Central Agricultural University, Imphal, Manipur, India

S. S. Solankey, M. Kumari (eds.), *Advances in Research on Vegetable Production Under a Changing Climate Vol. 2*, Advances in Olericulture, https://doi.org/10.1007/978-3-031-20840-9_10

security and climate change is aggravated by poverty, which needs to be challenged by enhancing resilience of food production system. World-wide development practitioners have tried to capture various coping mechanism and adaptation strategies followed by farming community. One such practice is kitchen gardening which enhances resilience of farm-families in climate change regime. In this chapter we will address concept of kitchen gardening, its prominent roles, experiences gathered from the kitchen gardening and way forward for development practitioners, in order to achieve nutritional security in changing climate scenario.

Keywords Kitchen gardening · Climate change · Homestead gardening · Nutritional security

10.1 Introduction

Climate change is very much intertwined with agricultural sectors (including allied sectors), food security and nutrition. Different sectors of agriculture are not only affected by climate change, but they also contribute to it through greenhouse gas emissions through deforestation, methane emissions from livestock rearing and rice farming, the use of biological and synthetic fertilizers etc. Global emissions must be drastically lowered in order to meet the Paris Agreement's goals of limiting global warming to 2 °C. However, by lowering greenhouse gas emissions and boosting carbon absorption in biomass and soils, agriculture can help to combat climate change. The conditions in which agricultural operations are carried out are changing substantially because of climate change, directly, by altering environmental physical properties *viz.* temperature, precipitation frequency, intensity, and distribution, the acidity of the marine environment etc., and indirectly, by altering ecosystems and interspecies relationships, particularly through affecting pollinators, parasites, weeds, illnesses etc. Household members and vulnerable groups in tropical developing nations are already suffering from harmful impacts. Negative impact of climate change on agricultural productivity will worsen in most of the regions beyond 2030. According to new research, global grain production would decline by 17% in 2050 as compared to a scenario where climate change is absent. The FAO estimates that agriculture would need to raise food output by 50% by 2050 owing to population growth and dietary changes (FAO 2017). Effects of climate change on agricultural sectors have economic and social ramifications for households (producer and consumer both) and nations that rely on agricultural output. As a result, climate change has an impact on all dimensions of food security, including food supply, physical and economic accessibility, and nutrient use, as well as the stability of these three variables. These repercussions are particularly severe for the poorest households, which rely on agricultural activities for 70% of their income, and for the least developed nations, whose agricultural sectors account for a large percentage of GDP. Climate change tackling has become a focus area in policy discourse also due to its holistic impact on life on earth. Changes not only in macro scale but at the very micro level, individual level are advocated. 'Think globally and act locally' has encouraged individuals to think over and reflect upon their own lifestyles at the realm of local to global context.

One such change in agricultural practice is kitchen gardening and current chapter addresses how multidimensional roles of kitchen garden can be a potential tool for enhancing resilience against climate change. In developing countries vegetables are the main source of livelihood for most of communities because vegetables are loaded with several vitamins, carbohydrate, salts and proteins. Now a day's vegetables become an integral part of average household's daily meals because of increasing awareness towards their health (Solankey et al. 2021). Initially this chapter will introduce the concept of kitchen garden and how it is relevant and important in current scenario. Various study reviewed to understand social-cultural-biological-environmental context of kitchen garden and how kitchen garden responses climate change also explained. This chapter has made an attempt to identify possible areas for future course of action.

10.2 Nutritional Security Under Changing Climatic Scenario

Despite tremendous improvements in food security, more than 2 crore people have faced micronutrient deficiencies; the number of chronically malnourished people stands more than 80 crores in India; nearly 25% of children under the age of five is stunted; and 0.34 crore people lost their lives every year as a result of overweight and obesity (FAO 2017). Climate change has multi-faceted impact on food safety (FAO 2016), community hygiene, water quality, food security, woman and child healthiness etc. Effect of climate change on food supply, accessibility, and stability can lead to significant changes in dietary choices and balance, which can have a bad impact on household health. Moreover, for the reason of its health implications, climate change has a detrimental influence on nutrition. Certainly, it can effect on increasing water, food, and vector-borne transmittable diseases, all of that increases nutritional requirements and reduces nutrient absorption. Farmers' health may be directly impacted by climate change, mainly due to excessive heat and a lack of drinking water. As a result, the poorest and most susceptible people have been and will be the first to suffer.

It is important to know various aspects of food safety while investigating the consequences of environmental change on food safety and developing strategies to mitigate these implications. Food security may be characterised in four ways, according to FAO:

- Accessibility of foods
- Availability of food (economical and physical)
- Consumption of food
- Stability in long-term (vulnerability and shocks)

Food Availability indicates the availability of adequate amounts of appropriate-quality food; domestically produced or imported. Accessibility of food indicates the individuals' access to sufficient finances for obtaining healthful diets. The legal, political, economic, and social features of the society in which they exist, entitlements are definite as a set of all product bundles over that a person can develop authority.

Food Utilization is to achieve a state of nutritional well-being in which physiological demands are fulfilled, food must be utilized through sufficient diet, clean water, sanitation, and healthcare.

Food Stability means a population, home, or person should have availability of enough food at all the times to be food secure. They should not be at risk of losing food due to unexpected occurrences (such as a financial or climate disaster) or cyclical occurrences (e.g., occasional food insecurity). As a result, the term "stability" refers to the availability, accessibility, and use aspects of food security.

Climate change has direct influence on food availability by impacting negatively on crop-animal production and health, as well as seafood supplies, particularly in Sub-Saharan Africa as well as South Asia, where the major parts of the global food-insecure present. A reliable supply of food is not just a basic requirement for impoverished agricultural producers; it is often their sole, and sometimes precarious method of making a living and sustaining their existence. The small and marginal farmers with lesser yields will find it more difficult to get food since they will have less money to spend.

Furthermore, decreases in food availability due to climate change will almost certainly result in higher food costs. This would have an effect on both urban and rural communities, as they spend considerably more of their capital on food. Poor landowner farmers, the major parts of them are net food buyer, would also be affected (World Bank 2008). Climate change might effect on how food is consumed. Myers et al. 2014 suggested that there might be negative consequences regarding food quality, nutrition, and food safety. However, the consequences of climate change on food security may be extremely place oriented, with foods growing in areas facing quick climatic change. Vitamin D insufficiency, for example, has been linked to environmental/ecosystem deterioration (Wahlqvist 2013).

Climate-related changes can hurt individuals who are not so poor but are vulnerable, such as, a flood destroys a small farm, a drought hurt a livestock herd, or polluted water causes a kid to get ill. These occurrences have the potential to wipe out decades of hard labour and wealth accumulation, as well as create lasting health harm. Climate change will influence overall socio-economic development thereby future trends in poverty and food insecurity. According to a recent World Bank finding, poverty levels in 2030 are likely based on many climate shocks and policy scenarios. Another investigation concluded, under an excess climate change effect condition, the number of people are in severe poverty rises by 12.2 crore in 2030, and in a prosperous condition, the increase is only1.6 crore. Policy decisions and focused adjustment methods will choose the future strategies of climate change on poverty (Hallegatte and Rozenberg 2016).

Ever increasing global population is likely to confront a broad hunger epidemic because of the reduced food supply induced by climate change. This would result in localized increases in food costs. Under the IPCC's A_2 scenario, the danger of hunger in certain emerging nations, such as India, is expected to stay quite high, with a regular rise in the number of people at risk of food hunger. By 2050, an extra13.2 crore people will be under danger; and by 2080, this number might increase to 26.6 crore (Parry et al. 2004). Coastal erosion and inundation are causing loss of cultivated land in low-lying Indian districts, making it worst for food poverty and property loss. In this example, climate change pressures might be alleviated by a variety of management strategies, including better stock improvement and more integrated agricultural ecosystems and improved soil conditions. There are various scientific evidence that climate change might have detrimental consequences on hunger as well as nutrition, endangering present agriculture and food systems (Nelson et al. 2009), as well as lower calorie intake due to income and non-income consequences (Nelson et al. 2009). In addition to that, the poverty nutrition trap is now a severe worry, which indicates that climatic changes may affect our agricultural production and income source, resulting in poor health and nutrition (Brown and Funk 2008). Under-nutrition is believed to be responsible for about 15% of the worldwide illness, thus making it a serious worldwide public health issue. As a result, poor health in the agricultural community might limit the ability to create and implement adequate food production strategies (Greenfacts 2008). One of the important issues is the high rate of malnutrition among the poor and marginalized population, specifically rural children and the huge number of others living below the poverty line in several countries, that reduces their capability of buying food. The high heterogeneity of protein deficient, energy poor and micronutrient deficiencies within countries continues to be a burden (Greenfacts 2008).

Vulnerable groups with existing inadequate dietary intakes (e.g., low-income individuals, migratory workers) and high nutritional density requirements may be at greater risk. It is commonly known that continuous nutritional deficient food in someone's life might be a critical risk health factor for chronic diseases, which is a huge burden in developing countries (Greenfacts 2008). Moreover, poor diet quality and low nutritional value might have been connected with global obesity as well as chronic sickness like heart related disease, hypertension, stroke as well as diabetes. The recent nutrition transition, that is the method through which globalization, urbanization, as well as changes in lifestyle are linked to high calorie obtain, poor diet, and low physical activities, has been the most remarkable illustration of raising nutritional pressures. Meals of people in poor countries are heavy in fat and energy rich (Gupta and Mishra 2014).

Micronutrient deficiency especially Iron deficiency is widespread in humans (Nag et al. 2020). Food availability is not an issue but the distribution and accessibility are. Child nutritional status in India during 2019–2020 is presented in Table 10.1. The states having a good amount of production are the states having maximum children malnourished (Table 10.1). Assam Bihar and Maharashtra are

Table 10.1 Child Nutritional Status in India 2019–2020

States/ National Territories	Children under 5 years who are stunted (%)	Children under 5 years who are wasted (%)	Children under 5 years who are underweight (%)
Andaman & Nicobar Islands	22.5	16	23.7
Andhra Pradesh	31.2	16.1	29.6
Assam	35.3	21.7	32.8
Bihar	42.9	22.9	41
Dadra & Nagar Haveli and Daman & Diu	39.4	21.6	38.7
Goa	25.8	19.1	24
Gujarat	39	25.1	39.7
Himachal Pradesh	30.8	17.4	25.5
Jammu & Kashmir	26.9	19	21
Karnataka	35.4	19.5	32.9
Kerala	23.4	15.8	19.7
Ladakh	30.5	17.5	20.4
Lakshadweep	32	17.4	25.8
Maharashtra	35.2	25.6	36.1
Manipur	23.4	9.9	13.3
Meghalaya	46.5	12.1	26.6
Mizoram	28.9	9.8	12.7
Nagaland	32.7	19.1	26.9
Sikkim	22.3	13.7	13.1
Telangana	33.1	21.7	31.8
Tripura	32.3	18.2	25.6
West Bengal	33.8	20.3	32.2

Source: National Family Health Survey – 5 Database (2020)

the states where all the parameters of child malnutrition are at an alarming stage. Meghalaya (46.5 per cent), Bihar (42.9 per cent), Dadra & Nagar Haveli, and Daman & Diu (39.4 per cent) are the leading states that reported children less than 5 years who are stunted. Likewise, Maharashtra (25.6%), Gujarat (25.1%) and Bihar (22.9%) are the top states where children's are wasted. Bihar (41 per cent), Gujarat (39.7 per cent), Dadra & Nagar Haveli and Daman & Diu (38.7 per cent) are the leading states of underweight children under 5.

Food security might be improved and assured by implementing policies and programmes to boost dietary diversity, as well as deployment of present and advanced technologies for food production, processing, preservation, and delivery. An adequate, safe, and diverse food supply would aid in the prevention of malnutrition. The incorporation of kitchen garden in households can be a possible step in this direction.

10.3 What Is a Kitchen Garden?

A kitchen garden is a type of area surrounding the house where vegetables as well as other herbs are planted for family consumption.

It can be done in little plot near the house for producing a variety of vegetables according to the season and choice of household. A kitchen garden gives a household the opportunity to cultivate its own food. The household can assure that the food it consumes is fresh and seasonal, and that it was cultivated organically (without the use of harmful chemical pesticides or fertilizers). A kitchen garden can supply fruits and vegetables for a family all over the year. If the product surpasses the family's usage, they can sell it and use the proceeds towards additional public places.

10.3.1 Why Make a Kitchen Garden

Vegetables are necessary element of a healthy meal since these provide a range of nutrients that are essential for a variety of bodily activities. Vegetables are necessary for growth, energy, and disease defense. Vegetables are especially beneficial for children, pregnant women, and nursing mothers who are more vulnerable to malnutrition. People need to have a nutritious diet to stay healthy. A nutritious diet consists of a well-balanced combination of grains, bread, legumes, vegetables, herbs, and fruits, among other things.

The purpose for developing a kitchen garden is one or more of the purposes stated below:

- Encourage the growth of nutrient-dense fruits and vegetables.
- Enable individuals to make better dietary choices by providing them with more information.
- Increase the availability of nutritious food options for children and adolescents at doorstep
- Improve food security and availability of nutritious foods, particularly fruits and vegetables.

Kitchen gardens demonstrate how to grow a diverse and nutritious vegetable for use in the kitchen with minimal input and maximum output.

10.4 Role of Kitchen Garden

The very essence of having a kitchen garden in household was to ensure food security of that particular family. Availability of fresh vegetables at one's disposal and that too at any time of the day without going anywhere to purchase; serve as the basic idea behind establishing such low maintenance gardens. Even today this serves as the prime motto behind kitchen gardens. Along with this, several other roles of kitchen gardens have been realized over time.

10.4.1 Role in Ensuring Food and Nutritional Security

Kitchen gardens are a genuine source of food and nutrition which if maintained properly provide fresh vegetables and fruits throughout year at low cost. Generally, kitchen gardens are established at backyards of houses near the water source and where there is presence of adequate sunlight. Usage of fertilizers and plant protection chemicals are minimum because the motto is not commercial but sustainable farming. Hence use of farm yard manure, natural sources of plant nutrition like biofertilizers and organic pesticides like *neem* oil etc. are more popular in backyard gardens. As a result, the vegetables and fruits which grow have minimum chemical residues, tastes better and are full of nutritional values. Arya et al. (2018) conducted a study in four districts *viz.* Bulandshahr, Muzaffarnagar, Gautam Budh Nagar and Baghpat of Western Uttar Pradesh, India where 160 families who had established kitchen gardens reported average production of vegetables to be 403.4 kg in the year 2011–2012 and 2012–2013. Kitchen gardens provided farm families fresh organic vegetables ensuring food and nutritional security at low cost. Asaduzzaman et al. (2011) reported that homestead gardens, in other words kitchen gardens, of Bangladesh provided 364.56 gm of vegetable/person/day containing 179.83 kcal while the minimum requirement is 200 gm per person/day. Red amaranth produced highest (69,116 kcal) kcal/ m^2 and spinach produced the lowest kcal/m^2 (10,378 kcal) in homestead gardening. Such gardens contributed 10 percent towards achievement of food and nutritional security. Tribal households with kitchen gardens were reported to have additional intake of Iron 32.70% and Ca 110.40% in Seoni district of Madhya Pradesh, India (Rana et al. 2021). Singh et al. (2020) had studied the role of kitchen gardens in improving the nutritional level amongst rural women of Kannauj district, Uttar Pradesh, India. After implementation of planned kitchen gardens in 60 selected families, 44 per cent increase in yield of vegetables was reported and benefit-cost ratio was found to be 6.13. Availability of vegetables per person for consumption increased from 80 gm to 200 gm and improvement in nutritional status was also observed through increase in BMI after six and twelve months trial. Even in Tanzania, the Per Capita Kilo Calorie Intake (PKCI) was reported to be reduced by 3.92% in households where there was absence of kitchen gardens (Sileshi et al. 2022).

10.4.2 Role in Generation of Income and Savings

The produce from kitchen gardens is a source of income which generates improved livelihood opportunities, imparts rural development and entrepreneurship (Trinh et al. 2003; Calvet et al. 2012). Mitchell and Hanstad (2004) reported that kitchen garden products can be sold to earn additional income by the households (Eyzaguirre and Linares 2004; Torquebiau 1992; Niñez 1985) either directly or through development of small cottage industry. Disposable income of households have increased through income generated from such activities and the savings generated by consuming home-grown produce. Income generated from sale of kitchen garden produce enabled households of Cambodia, Papua New Guinea and Nepal to save more, use the money for education and other services and also purchase different food items (Iannotti et al. 2009; Vasey 1985). Households in hills of Vietnam earned 22 percent additional income through sale of kitchen garden produce (Trinh et al. 2003). Arya et al. (2018) reported average production of vegetables in 2012–2013 by kitchen gardens of 160 farm families of Western Uttar Pradesh, India was 406.27 kg which saved Rs 8057.50 of each farm family. Kitchen gardens are widely promoted amongst subsistence families of developing countries as a mechanism to avert poverty through additional source of income. It helps resource poor families by making production cost effective (Galhena et al. 2013).

10.4.3 Role as Shock Absorber in Food System and Alternative Source of Livelihood

Kitchen gardens are often less commercial and more subsistence in nature. This very quality indicates that they are not the primary source of livelihood because of their nature of small-scale operation. But these gardens have immense potential to act as alternate sources of livelihood as well as source of ready homemade fresh food. Any situation of food crisis can be averted if vegetables and fruits are available at the disposal. Also, any shortage in income like failure of agricultural crops in farm families due to any natural vagary can be minimized if kitchen gardens have seasonal vegetables, perennial fruits, livestock, flowers or medicinal herbs. Livestock component in kitchen gardens minimize risk due to crop losses and act as an asset to the household (Devendra and Thomas 2002).

10.4.4 Role in Curing Diseases

Plants serve as important source of medicine for humans and livestock and used to prepare biological pesticides. Herbs and medicinal plants are grown in kitchen gardens globally. About 80 per cent people in developing countries use them to treat various illnesses in cost effective manner (Rao and Rao 2006). Perera and Rajapakse (1991) reported that 30 per cent of 125 plant species of Kandyan kitchen gardens of Sri Lanka were exclusively planted for medicinal uses and about 12 per cent for mixed purposes while in Yucatan, 70 per cent had medicinal uses (Rico-Gray et al. 1991). Medicinal plants are considered important in Bangladesh also (Millat-e-Mustafa et al. 2002) while kitchen gardens in Bukoba, Tanzania have plant species cultivated purely for medicinal purposes (Rugalema et al. 1994). Cinnamon, clove, nutmeg, lime, curry leaf, basil, aloe vera, neem, turmeric are popular medicinal plants in kitchen gardens of Tropical countries.

10.4.5 Role in Improvement of Nutrient Cycle, Ecological Balance and Biodiversity

Kitchen gardens use eco-friendly approaches for food production and conserve biodiversity and natural resources. Kitchen gardens usually have diverse flora and fauna (Galhena et al. 2013). Such gardens are complex agricultural production systems which promote biodiversity conservation. Buchmann (2009) reported that 25 kitchen gardens in Central Cuba contained 182 plant species. Kitchen gardens contain a wide variety of plant species which can be landraces, threatened or rare species or cultivars grown for a set of desirable traits (Watson and Eyzaguirre 2002). Thus, they become ideal sites for in situ conservation of biodiversity and genetic material (Gajaseni and Gajaseni 1999; Trinh et al. 2003). They also serve as habitats for animals and beneficial organisms. These gardens help in nutrient recycling, reduced soil erosion and improved pollination (Pushpakumara et al. 2010).

Humans of the household, animals and plants share a symbiotic relationship with each other within the home gardens. The plants and animals of kitchen garden provide food and other benefits to the family while the later takes care of the former. Plant wastes are used as fodder for the animals and animal manure serves as compost to fertilize plants (Mitchell and Hanstad 2004). Hence nutrient cycling and ecological balance of an area is improved through kitchen garden.

10.4.6 Socio-Cultural and Aesthetic Role

Kitchen gardens are valuable repositories preserving indigenous crops and livestock species and the production techniques of which are also transferred over the generations (Soemarwoto 1984; Blanckaert et al. 2004; Trinh et al. 2003). Interactions circling around home gardens create and reinforce social status and ties within the community. The habit of households having home gardens of exchanging or gifting saplings and plant produce for social, cultural and religious purposes (Soemarwoto 1984; Blanckaert et al. 2004), strengthen social integration and generate social capital. Beautifully decorated kitchen gardens with colourful seasonal flowers, vegetables, fruits can add to the beauty of a household and impart aesthetic value to the surroundings. Gardening is also a healthy hobby to follow which imparts positive vibes and good aura (https://agriculturegoods.com/).

10.4.7 Role in Empowerment of Women

It is not something new that women not only play the role of care giver of the family, but are also actively involved in food production. Women's role in home gardening varies across cultures, including land preparation, planting, weeding, harvesting of produce (Moreno-Black et al. 1996; Keys and Kaqchikel 1999; Pandey et al. 2007). To women, home gardening imparts social and economic enrichment. A woman of Achuar Indian community in the upper Amazon, maintaining a lush kitchen garden is looked upon as someone having high social status and as agronomically competent woman in society (Descola 1994). Such a woman is considered more committed to family's wellbeing and of high social eminence in Saraguro community of Andes (Finerman and Sackett 2003). For some women, sale of garden products are often the only source of income and livelihood (Marsh 1998). Kitchen gardens serve as important sources of income and in meeting daily food requirements in women-headed families (Rowe 2009).

10.5 Environmental Perspective of Kitchen Garden

10.5.1 Developing Eco-Literacy

Eco-literacy is vital for sustainable human communities and societies by understanding principles of ecosystems (Capra 1997). Eco-literacy helps to understand where food comes from and how it reaches on the table; so, it is foundational aspect of understanding health and sustainability and can have profound effects on the quality of life of individuals and the planet. So eco-literacy should start at an early age for preparing a person as an effective member of sustainable society. An eco-literate person shall develop head, heart, hands, and spirit for organic understanding of world, nature and environment (McBride et al. 2013). Worldwide various school kitchen gardening programme were successfully implemented for multifaceted development of students and deepen their awareness and understanding of food systems, local food movements, and nutrition; developing personal connections to the earth etc. The bunch of activities during planting, caring-nurturing, harvesting and then preparing and sharing foods in classroom may lead to enhanced ecological literacy (Fig. 10.1).

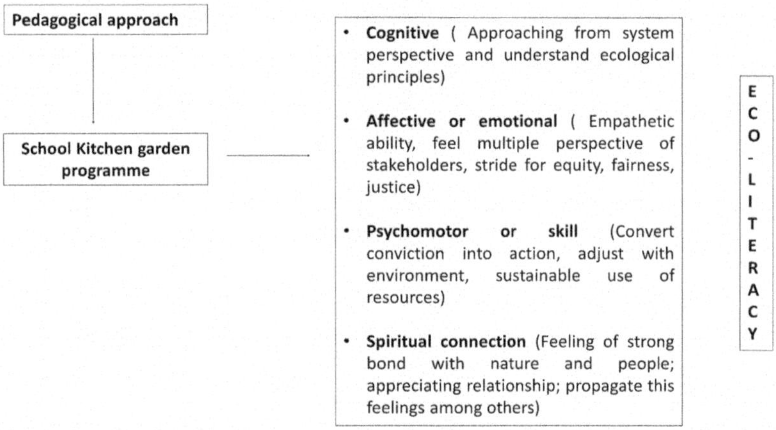

Fig. 10.1 Schematic presentation of developing eco-literacy through school kitchen garden programme

10.5.2 Environmental Behaviour

Kitchen garden or homestead garden has profound influence on individual's lifestyle. Ever-changing climatic scenario also pushes individual to 'think globally and act locally'. So, worldwide mankind is given encouragement to adopt more sustainable lifestyle by institutional bodies, groups, individuals etc. However, experience shows attitude and behaviour change is complex process specially while considering environmental behaviour. The framework suggested by Barr et al. (2001) and Barr and Gilg (2007) to understand environmental behaviour can be modified and used to understand environmental behaviour in kitchen gardening. The suggested model is mentioned below (Fig. 10.2).

10.5.3 Enhancing Resilience Against Direct and Indirect Effects of Shocks

Resilience is primarily viewed as a reactive, defensive mechanism induced by disruptive external changes (Sonnino and Griggs-Trevarthen 2013), and as a crisis-mitigating capacity (Barthel et al. 2013). Global experience highlighted resilience should be seen as proactivity, a potential for learning, as "internally produced and not just externally induced", and as a "dynamic process in which change and reinvention provide the grounds for fundamental reform" (DeVerteuil and Golubchikov 2016). Deriving lessons from Folke et al. (2003), Buchmann (2009) and Jehlička et al. (2019), following schematic representation of resilience through kitchen gardening is shown (Fig. 10.3).

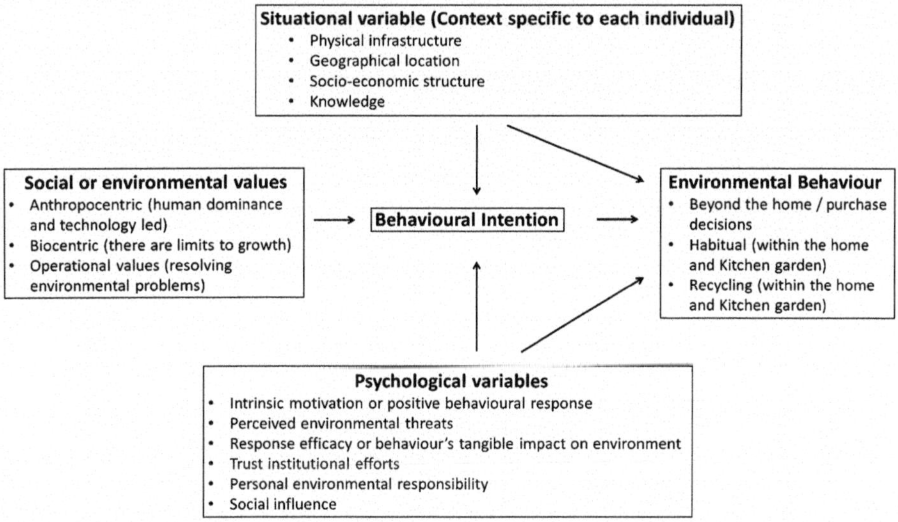

Fig. 10.2 Adapted model for Environmental behaviour in kitchen garden

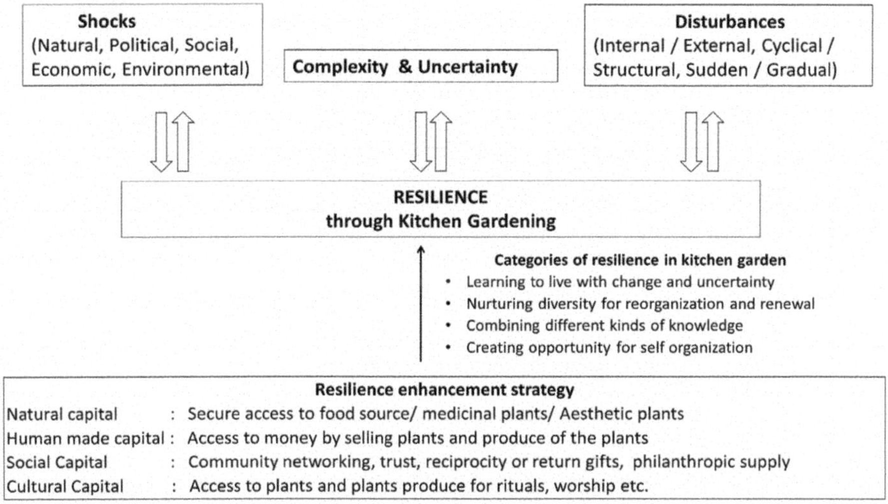

Fig. 10.3 Enhancing resilience through Kitchen Gardening

10.6 Factors Affecting Kitchen Gardening (Table 10.2)

Table 10.2 Factor and their nature of influence in Kitchen gardening

Factors	Nature of influence	Study area	Authors
Practice of school kitchen gardening programme on students and educators	This study reported children's mental, emotional and social health is promoted through activities like kitchen gardening which requires direct contact with nature.	Melbourne primary schools	Maller (2009)
	Practice of school kitchen gardening helps students in enhancing their academic performances, physical activity, language arts, healthful eating habits and acquiring nutritional knowledge.	USA	Graham and Zidenberg-Cherr (2005)
	Kitchen gardening in school involves students in practice and management of activities there by promotes physical activity	Canada	Bell and Dyment (2006)
	Study among second to fifth grade students in USA reported student's knowledge enhanced but no significant difference in attitude. However, after participating in the programme, they reported that they started eating healthier snacks.	USA	Koch et al. (2006)
	School-based Stephanie Alexander kitchen Garden program was designed for introducing children to food education through growing, caring, harvesting, preparing and sharing healthy seasonal food. Hands-on experiences positively influence children's attitude towards environmentally sustainable behaviour.	Australia	Block et al. (2009)
Motivation in volunteering	Engagement in volunteering students in school kitchen garden programme depends on their initial motivation (like desire to be more involved with children, to learn and give back to the community, and to support personal beliefs about the project's contribution to wider societal and environmental issues); benefits by actual volunteering and unexpected benefits. The challenges faced by the volunteers (like managing school community, volunteer workforce, wider community etc) did not come in the way of continuing engagement in the programme.	Australia	Henryks (2011)

(continued)

Table 10.2 (continued)

Factors	Nature of influence	Study area	Authors
Pedagogical content knowledge of teachers	Context-specific PCKG2T model was an effective professional learning tool for teacher and indicated extent to which six components (knowledge of Gardening & Kitchen Concepts; understanding of students; understanding of Whnau/ community context; knowledge of relevant curriculum areas; teacher pedagogical knowledge; knowledge of resources) were used during teaching.	New Zealand	Narayan et al. (2020)
Motivation in gardening/ kitchen gardening/ home gardening	The study suggested motivations for gardening are more related with people's way of living, keeping hobby, traditions etc. than to economic reasons. Other reasons were to continue traditional activity and also for physical exercise.	Iberian Peninsula, Europe	Reyes-García et al. (2012)
	The study conducted among 360 households with home gardens in Benin, West Africa and reported that food production (women reported) and medicinal plant production (men reported) were the main motivations of home gardens in Benin.	Benin, West Africa	Gbedomon et al. (2016)
	The study conducted in Omani home-gardens reported top gardening motives were aesthetic, hobby, source for food, physical exercise and protecting environment. The top reasons identified for non-gardening were pavement space, lack of free land, adverse weather, high water bill and lack of knowledge about gardening.	Muscat, Oman	Al-Mayahi et al. (2019)
	Relationship and appreciation for nature, apart from social issues and uses was significant motivation for gardening.	USA	Clayton (2007)
Attitude of women	Majority of the women in the Botswana study reported they were useful to the economy if they grow fruits and vegetables. Moreover, women should establish gardens to tackle food shortage due to unreliable rainfall for field crops.	Mochudi village of Botswana	Subair and Siyana (2003)

(continued)

Table 10.2 (continued)

Factors	Nature of influence	Study area	Authors
Nutritional linkage	CARE's experience in Bangladesh shows if homestead gardens are well planned for year-round production, then it can certainly increase the availability, consumption, and sales of vegetables for land-poor rural households, ultimately resulting into improved nutritional status. However, approach combining homestead production and nutrition education is required for behavioural change. Schoolchildren of grades 7 and 8 proved to be an effective vehicle for raising awareness about benefits (nutritional and economical) of homestead gardening.	Bangladesh	Khan and Begum (2002)
	Market access for rural development (MARD) project reported nutritional knowledge, feeding complementary foods to infants and preservation of foods, and consumption of 16 types of home-produced micronutrient-rich vegetables and fruits, vitamin A–rich plant products etc.	Nepal	Jones et al. (2005)
	Study on household diet diversity in Melghat concluded that increase dietary diversity might be achieved by adopting perennial kitchen garden and also through imparting adequate knowledge.	India	Birdi and Shah (2016)
Environmental behaviour	Environmentalism is related with following behavioural characteristics 1. Alternative agricultural methods like organic production shares goal of environment and food production. The pro environmental attitude which is seen through local food movement, also influences individual to produce and consume in and from kitchen garden. 2. Association of outdoor activities with environmental behaviour 3. Environmental behaviour is associated with recycling and composting activities	USA	Schupp and Sharp (2012)

(continued)

Table 10.2 (continued)

Factors	Nature of influence	Study area	Authors
Gender dimension	Experience gained from the home garden project led by Bioversity International, Nepalese NGO LI-BIRD, and the Swiss Agency for Development cooperation (SDC) a project during 2002–2012, showed that project contributed in driving change in gender norms, such as formal education of girls and the feminization of agriculture. Women from participant households have increased their decision making capacity in their own households and community. The effect was prominent across different caste categories.	Nepal	Elias et al. (2017)
	Classified gardens as men's garden, women's garden, shared garden, and separated garden. They identified gender dimensions in gardening activities and noted men's gardens tended to be "larger, more distant but women's gardens had more biodiversity.	Iberian Peninsula of Europe	Reyes-García et al. (2010)
	British women were more interested than Menin organic gardening	England	Bhatti and Church (2000)
	The way of gardening has profound influence on social relations (class, culture, power, gender etc) and environmental consequences. He argued to explore gendered division of labour in household and its impact on women's gardening activities.	USA	Hondagneu-Sotelo (2010)
Rural-urban continuum	Reciprocity economy (sharing of goods between individual members of a community) was evident in high-poverty rural areas and redistributive economy (reallocation of resources within a collective social unit by formal governmental or charitable units) was found in low-income urban areas. Part of reciprocity economy is made up by home garden produce. They also found that rural areas had three times of gardens as the urban areas.	USA	Morton et al. (2008)

10.7 Case Study on Kitchen Gardening

10.7.1 Designing and Establishment of 200m² Nutri-Garden Model

Indian Council Agricultural Research –Research Complex for Eastern Region (ICAR-RCER), Patna, India has started establishing and standardizing Nutri-garden Model. In October 2019, two nutri-garden models of 100m^2 and 200m^2 were established. Figure 10.4 shows the design of a nutri-garden model (rabi). The 100m^2 model was created for a family of 4–5 members, while the 200 m^2 model was created for a larger family of 7–8 members. The model was standardized and validated to meet the Indian Council of Medical Research (ICMR) recommendations for vegetable intake (200 g fruit vegetables, 50 g leafy vegetables, and 50 g root vegetables). The 200 m^2 models were designed with nine fruit vegetables (tomato, brinjal, cabbage, cauliflower, broccoli, dolichos (sem), pea, capsicum, and broad bean) six leafy vegetables (palak, mustard green, coriander, methi, bathua and lafasaag), three root vegetables (carrot, radish, beet) in mind (Table 10.3).

An adult man/woman requires at least 200 g of vegetables per day, whereas people in eastern India consume only 55–85 g/head/day, except potatoes and sweet potatoes. Several studies have found that the problem of malnutrition is worsening due to rural people's lack of nutritional knowledge, resulting in a low intake of balanced foods, including green vegetables. As a result, a scientific model of

Table 10.3 *Rabi* season vegetable crop, plot size, planting space and seed rate for 200 m^2 Nutri-garden Model

Vegetables (Plot size m)	Seed rate (g)	Spacing (cm)	Plant per plot
Palak (4 × 3)	180	15 × 5	1855
Coriander (4 × 3)	120	30 × 10	437
Mustard green (4 × 3)	35	15 × 5	1832
Radish (4 × 3)	45	20 × 10	785
Methi (3 × 1)	70	15 × 10	815
Beet (4 × 3)	50	20 × 15	430
Carrot (4 × 3)	35	20 × 10	635
Broccoli (4 × 3)	10	45 × 30	45
Tomato (4 × 3)	15	60 × 45	50
Capsucum (3 × 2)	7	60 × 45	18
Cauliflower (4 × 3)	10	45 × 30	45
Brinjal (4 × 3)	15	60 × 45	45
Sem (2 × 3)	150	45 × 30	24
Cabbage (4 × 3)	10	45 × 30	45
BathuaSaag (3 × 1)	2	60 × 45	15
Pea (4 × 3)	300	25 × 15	323
Lafasaag (3 × 1)	5	25 × 15	55

nutri-garden is required to provide fresh and diverse vegetables while also meeting the nutritional needs of ultra-poor farmers (Ferdous et al. 2016). It also aids in the reduction of malnutrition, particularly among children and women, as well as improvement of nutritional issues such as anaemia, which are common among rural residents.

To address these concerns, nutri-garden model was created to make available vegetables at home through year-round vegetable production in the homesteads. This model contributed to household food security through increased nutritious food intake, increased consumption of home-grown vegetables helps in reducing consumption expenditure of poor farm families, and generation of additional income for farmers through the sale of excess vegetables. While nutri-garden has many advantages for developing countries, some key constraints include limited access to agricultural inputs such as availability of quality planting material or seeds, damages from insect pests, diseases, animals, and theft, poor soil fertility etc. Despite its limitations, the nutri-garden model is an environmentally friendly sustainable agricultural practice that can improve food security and economic growth. Farmers were able to meet their daily vegetable requirements by following this model (Fig. 10.4).

Fig. 10.4 200 m^2Nutri-garden Model (Shubha et al. 2019)

10.7.2 Kitchen Gardening in Uttarakhand, India

Kitchen gardens and their remunerative aspect has been explored through case study conducted under DST-SARTHI project "Ensuring sustainable livelihood security" during 2016–2018 in Lower Shivalik Hills of Uttarakhand, India. Vegetable crops find an important place in kitchen gardens of India. Farmers aiming to maximize their income from relatively small landholding and uses their resources to high profit crops like vegetables. It was found that the whole family is engaged in intensive but small-scale horticulture.

A number of improved vegetable seeds were demonstrated for kitchen garden at farmers' field to improve nutritional security and enhancing farm income in Lower Shivalik Hills of Uttarakhand. Farmers used the seed in the backyard or free space in their house. Farmers of this region grew these vegetables without any inorganic fertilizer. The field data has been collected during Rabi 2017 from demonstrated field at Narsan, Laksar and Gaindikhata blocks of Haridwar district of Lower Shivalik Hills and it accounted average saving of Rs.120.00 per week per family from kitchen garden (Tables 10.4, 10.5 and Fig. 10.5).

Further to promote vegetable cultivation, in Rabi 2017, the following vegetables were demonstrated at farmers' field. The economic analysis was based on primary date collected from farmer's field through personal interview followed by focus group discussion and revealed kitchen gardens not only provide nutritional security of farmers, but additionally provide a handsome return also (Table 10.6).

Table 10.4 Details of kitchen gardening

Crop	Variety	No of demonstration
Palak	All green	100
Fenugreek	PEB	100
Radish	Pusa Chetki	100
Carrot	Pusa Rudhira	100
Garden pea	Pusa Pragati	100
Mustard sag	Pusa sag 1	100

Table 10.5 Economic analysis of kitchen gardening

Before	After	
Expenditure on purchasing vegetable (Rs/ week)	Expenditure on purchasing vegetable per week (Rs/week)	Saving (Rs/per week)
300.00	180.00	120.00

Source: Authors calculation

Fig. 10.5 Kitchen gardening through distributed seeds

Table 10.6 Details of vegetable demonstrations

Crop	Variety	Seed rate (acre)	Area (acre)	Yield (q/acre)	Cost of cultivation (acre)	Gross income (Rs/acre)	Net return (Rs)
Brinjal	Pusa Shyamla	42 gm	0.75	15–18	600–700	14,400–18,000	13,800–17,300
Palak	All green	1.67–2 kg	1	9–12	600–700	6000	5400
Fenugreek	PEB	1.67 kg	1	9–12	600–700	6000	5400
Carrot	Pusa Rudhira	5–6 kg	0.67	120	600–700	30,000	29,400
Garden pea	Pusa Pragati	100 kg	0.5	54–60	600–700	108,000	107,400
Mustard sag	Pusa sag- 1	3–4 kg	0.17	9–12	600–700	7200	6600

Source: Authors calculation based on field data during 2016–2018 under DST SARTHI project "Ensuring Sustainable Livelihood Security" in Lower Shivalik Hills of Uttarakhand

10.8 Constraints Faced During Kitchen Gardening Practices

Constraints or barriers impedes in developing intention to participate in kitchen gardening, in actual participation in kitchen gardening as well as it decreases the effectiveness of kitchen gardening among the practitioners. Study across globes found constraints or barriers in kitchen gardening that can be taken into consideration in time of intervention planning by the development practitioners (Table 10.7).

Table 10.7 Constraints of Kitchen gardening

Author	Constraints of kitchen garden
Sethy et al. (2010)	Following constraints were identified in kitchen garden of Burdwan District, West Bengal, India. Input constraints (unavailability of quality planting materials of fruits and HYVs seeds of vegetables); technical constraints (lack of knowledge about improved varieties, seed rate and sowing time); socio-cultural constraints (fear of theft of the farm produce); post harvest constraints (difficulties in selling for small amount of surplus produce); general constraints (high monkey menace)
Lake et al. (2011)	The study conducted in edible gardens of Eastbourne residents, New Zealand reported perceived barriers (which factored into the intention to participate) and actual barriers (plays after the intention was formed) existed in edible gardening. Perceived barriers (having sufficient time, practical skills, physical ability, access to edible gardeners for support, knowledge of good types to grow, lack of wind, sufficient space, and sun to participate) differentiated between individuals intending to garden and those who did not intend to garden. Some of the barriers also acted as actual barriers which differentiated participants to non-participants in gardening. These are sufficient time, knowledge of growing of food, practical skills, space and sun.
Kortright and Wakefield (2011)	They identified most important barrier in gardening is gardening skills, however other barriers mentioned by the participants of the study are space, lack of sun, and soil quality etc.

10.9 Conclusion

Climate change has made formidable challenges in every sphere of human life, food system is also one of them. Effect of climate change in food production-distribution-consumption has made all the value chain actors to reorient or bring changes in their usual activities. Kitchen gardening is one such practice which enhances households' resilience. The importance of eco-literacy through incorporation of suitable kitchen gardening programme, suitable competency-based skilling during volunteering etc. have immense role in developing ecological stewardship as well as imbibing sustainable lifestyle from the very early age. The practice of kitchen gardening is not new, but to enhance effectiveness of kitchen gardening simultaneous efforts in nutritional education, development and demonstration of suitable kitchen garden options; strategy to focus women from poor household etcare very much important to escape from Poverty-Nutrition Trap (PNT).

References

Al-Mayahi A, Al-Ismaily S, Gibreel T, Kacimov A, Al-Maktoumi A (2019) Home gardening in Muscat, Oman: Gardeners' practices, perceptions and motivations. Urban For Urban Green 38:286–294

Arya S, Prakash S, Joshi S, Tripathi KM, Singh V (2018) Household food security through kitchen gardening in rural areas of Western Uttar Pradesh, India. Int J Curr Microbiol App Sci 7(2):468–474

Asaduzzaman MD, Naseem A, Singla R (2011) Benefit-cost assessment of different homestead vegetable gardening on improving household food and nutrition security in rural Bangladesh. Selected Paper prepared for presentation at the Agricultural & Applied Economics Association's 2011 AAEA & NAREA Joint Annual Meeting, Pittsburgh, Pennsylvania, July 24–26, 2011

Barr S, Gilg AW (2007) A conceptual framework for understanding and analyzing attitudes towards environmental behaviour. Geografiska Annaler Ser B Human Geogr 89(4):361–379

Barr S, Gilg AW, Ford NJ (2001) A conceptual framework for understanding and analysing attitudes towards household waste-management. Environ Plan A33(11):2025–2048

Barthel S, Parker J, Ernstson H (2013) Food and green space in cities: a resilience lens on gardens and urban environmental movements. Urban Stud 50(1):1–18

Bell AC, Dyment JE (2006) Grounds for action: promoting physical activity through school greening in Canada. Evergreen, Toronto

Bhatti M, Church A (2000) 'I never promised you a rose garden': Gender, leisure and home-making. Leisure Stud 19(3):183–197

Birdi TJ, Shah SU (2016) Implementing perennial kitchen garden model to improve diet diversity in Melghat, India. Glob J Health Sci 8(4):10

Blanckaert I, Swennen RL, Paredes Flores M, Rosas López R, Lira SR (2004) Floristic composition, plant uses and management practices in home gardens of San Rafael Coxcatlán. Valle de Tehuacán-Cuicatlán, México. J Arid Environ 57:39–62

Block K, Gibbs L, Johnson B, Staiger P, Townsend M, Macfarlane S, Gold L (2009) Evaluation of the Stephanie Alexander kitchen garden program - final report to the Stephanie Alexander kitchen garden foundation, McCaughey Centre, September 2009, http://www.kitchengardenfoundation.org.au/about-the-program/proving-it-works

Brown ME, Funk CC (2008) Climate: food security under climate change. Science 319(5863):580–581

Buchmann C (2009) Cuban home gardens and their role in social–ecological resilience. Hum Ecol 37(6):705–721

Calvet ML, Gómez BE, Reyes GV (2012) Beyond food production: home gardens' ecosystem services. A case study in Vall Fosca, Catalan Pyrenees, northeastern Spain. Ecol Econ 74:153–160

Capra F (1997) The web of life: a new scientific understanding of living systems. Anchor Books, New York

Clayton S (2007) Domesticated nature: motivations for gardening and perceptions of environmental impact. J Environ Psychol 27(3):215–224

Descola P (1994) In the society of nature: a native ecology in Amazonia. Cambridge University Press, New York

Devendra C, Thomas D (2002) Smallholder farming systems in Asia. Agr Syst 71:17–25

DeVerteuil G, Golubchikov O (2016) Can resilience be redeemed? Resilience as a metaphor for change, not against change. City 20(1):143–151

Elias M, Gauchan D, Gotor E, Sthapit B (2017) Promoting gender equality, Social Inclusion and Biodiversity Conservation in Nepal's Home Gardens

Eyzaguirre PB, Linares OF (2004) Home gardens and agrobiodiversity

FAO (2016) The state of food and agriculture. Climate Change, Agriculture and Food Security. SOFA, Rome

FAO (2017) Food and agriculture organization of the United Nations Rome.

Ferdous Z, Datta A, Anal AK, Anwar M, Khan AMR (2016) Development of home garden model for year round production and consumption for improving resource-poor household food security in Bangladesh. NJAS-WAGEN J LIFE SC 78:103–110

Finerman R, Sackett R (2003) Using home gardens to decipher health and healing in the Andes. Med Anthropol Q 17(4):459–482

Folke C, Colding J, Berkes F (2003) Synthesis: building resilience and adaptive capacity in social–ecological systems. In: Berkes F, Colding J, Folke C (eds) Navigating social– ecological systems: building resilience for complexity and change. Cambridge University Press, Cambridge, pp 352–387

Gajaseni J, Gajaseni N (1999) Ecological rationalities of the traditional homegarden system in the Chao Phraya Basin, Thailand. Agrofor Syst 46:3–23
Galhena DH, Fred R, Maredia KM (2013) Home gardens: a promising approach to enhance household food security and wellbeing. Agric Food Sec 2(8):1–13
Gbedomon RC, Salako VK, Chadare FJ, Kakaï RG, Assogbadjo AE (2016) Gendered motivation for home gardening and maintenance of agrobiodiversity: a case study in Benin, West Africa. Ann Sci Agron 20(2):91–104
Graham H, Zidenberg-Cherr S (2005) California teachers perceive school gardens as an effective nutritional tool to promote healthful eating habits. J Am Diet Assoc 105(11):1797–1800
Greenfacts (2008). http://www.greenfacts.org/en/agriculture-iaastd/index.htm#5
Gupta A, Mishra DK (2014) Food consumption pattern in rural India: a regional perspective. J Econ Soc Dev X(1):1–16
Hallegatte S, Rozenberg J (2016) Climate change through a poverty lens. Nat Clim Chang 7:250–256
Henryks J (2011) Changing the menu: rediscovering ingredients for a successful volunteer experience in school kitchen gardens. Local Environ 16(6):569–583
Hondagneu-Sotelo P (2010) Cultivating questions for a sociology of gardens. J Contemp Ethnogr 39(5):498–516
Iannotti L, Cunningham K, Ruel M (2009) Improving diet quality and micronutrient nutrition: homestead food production in Bangladesh. Discussion Paper 00928, International Food Policy Research Institute, Washington DC, USA
Jehlička P, Daněk P, Vávra J (2019) Rethinking resilience: home gardening, food sharing and everyday resistance. Can J Dev Studies/Revue canadienned'études du développement 40(4):511–527
Jones KM, Specio SE, Shrestha P, Brown KH, Allen LH (2005) Nutrition knowledge and practices, and consumption of vitamin A–rich plants by rural Nepali participants and nonparticipants in a kitchen-garden program. Food Nutr Bull 26(2):198–208
Keys E, Kaqchikel G (1999) Women, children, and multiple roles of gardens among the Maya of Highland Guatemala. Yearbook, conference of Latin Americanist geographers. Austin: Paper presented at Conference of Latin Americanist Geographers
Khan MI and Begum S (2002) Addressing nutritional problems with homestead gardening: CARE's experience in Bangladesh. Alleviating Malnutrition through Agriculture in Bangladesh: Biofortification and Diversification as Sustainable Solutions P.61
Koch S, Waliczek TM, Zajicek JM (2006) The effect of a summer garden program on the nutritional knowledge, attitudes, and behaviors of children. Hort Technology 16(4):620–625
Kortright R, Wakefield S (2011) Edible backyards: a qualitative study of household food growing and its contributions to food security. Agric Human Values 28(1):39–53
Lake B, Milfont TL, Gavin M (2011) The relative influence of psycho-social factors on urban edible gardening. NZ J Psychol 40(3):50–59
Maller CJ (2009) Promoting children's mental, emotional and social health through contact with nature: a model. Health Educ 109(6):522–543
Marsh R (1998) Building on traditional gardening to improve household food security. Food Nutr Agr 22:4–14
McBride BB, Brewer CA, Berkowitz AR, Borrie WT (2013) Environmental literacy, ecological literacy, ecoliteracy: what do we mean and how did we get here? Ecosphere 4(5):1–20
Millat-e-Mustafa M, Teklehaimanot Z, Haruni AKO (2002) Traditional uses of perennial homestead garden plants in Bangladesh. For Trees Livelihoods 12:235–256
Mitchell R, Hanstad T (2004) Small homegarden plots and sustainable livelihoods for the poor. LSP Working Paper 11, Rome, Italy
Moreno-Black G, Somansang P, Thamathawan S (1996) Cultivating continuity and creating change: woman's home garden practices in northeastern Thailand. Agr Hum Val 13(3):3–11
Morton LW, Bitto EA, Oakland MJ, Sand M (2008) Accessing food resources: rural and urban patterns of giving and getting food. Agric Human Values 25:107–119

Myers SS, Zanobetti A, Kloog I, Huybers P, Leakey ADB, Bloom AJ, Carlisle E, Dietterich LH, Fitzgerald G, Hasegawa T, Holbrook NM (2014) Increasing CO2 threatens human nutrition. Nature 510(7503):139–142

Nag A, Ray P, Rakshit R (2020) Reflection on iron: strategy for nutritional security in human. Natl Acad Sci Lett 43:395–398. https://doi.org/10.1007/s40009-019-00847-5

Narayan E, Birdsall S, Lee K (2020) Developing a context specific PCK model for kitchen-garden learning programmes. ASIA-PAC J Teach Edu 48(2):112–131

Nelson GC, Rosegrant MW, Koo J, Robertson R, Sulser T, Zhu T, Ringler C, Palazzo A, Batka M, Magalhaes M Lee D (2009) Climate change: impact on agriculture and costs of adaptation. Food Policy Report, International Food Policy Research Institute, Washington, DC

NFHS-5 (2020) National family health survey –5 Database http://rchiips.org/NFHS/NFHS-5_FCTS/NFHS-5%20State%20Factsheet%20Compendium_Phase-I.pdf

Niñez VK (1985) Working at half-potential: constructive analysis of home garden programme in the Lima slums with suggestions for an alternative approach. Food Nutr Bull 7(3):6–13

Pandey CB, Rai RB, Singh L, Singh AK (2007) Home gardens of Andaman and Nicobar, India. Agr Syst 92:1–22

Parry ML, Rosenzweig C, Iglesias A, Livermore M, Fischer G (2004) Effects of climate change on global food production under SRES emissions and socio-economic scenarios. Glob Environ Change 14(1):53–67

Perera AH, Rajapakse RMN (1991) A baseline study of Kandyan forest gardens of Sri Lanka: structure, composition and utilization. For Ecol Manag 45:269–280

Pushpakumara DKNG, Wijesekara A, Hunter DG (2010) Kandyan home gardens: a promising land management system in Sri Lanka. In: Bélair C, Ichikawa K, BYL W, Mulongoy KJ (eds) Sustainable use of biological diversity in socio-ecological production landscapes. Background to the 'Satoyama initiative for the benefit of biodiversity and human Well-being'. The Secretariat of the Convention on Biological Diversity, Montreal

Rana GK, Singh NK, Deshmukh KK, Mishra SP, Saini KPS (2021) Kitchen garden: an ideal approach to enhance household nutritional security in rural areas of Seoni district (MP). Pharma Innov J 10(5):254–258

Rao MR, Rajeswara Rao BR (2006) Medicinal plants in tropical homegardens. In Tropical Homegardens: A Time-tested Example of Sustainable Agroforestry. Edited by Kumar BM, Nair PKR. Springer Science, Dordrecht,The Netherlands

Reyes-García V, Vila S, Aceituno-Mata L, Calvet-mir L, Garnatje T, Jesch A et al (2010) Gendered homegardens: a study in three mountain areas of the Iberian Peninsula. Econ Bot 64(3):235–247

Reyes-García V, Aceituno L, Vila S, Calvet-Mir L, Garnatje T, Jesch A, Lastra JJ, Parada M, Rigat M, Vallès J, Pardo-De-Santayana M (2012) Home gardens in three mountain regions of the Iberian Peninsula: description, motivation for gardening, and gross financial benefits. J Sustain Agric 36(2):249–270

Rico-Gray V, Chemas A, Mandujano S (1991) Use of tropical deciduous forest species by the Yucatecan Maya. Agrofor Syst 14:149–161

Rowe WC (2009) "kitchen gardens" in Tajikistan: the economic and cultural importance of small-scale private property in a post-soviet society. Hum Ecol 37(6):691–703

Rugalema GH, Okting'ati A, Johnsen FH (1994) Thehomegarden agroforestry system of Bukoba district, Northwestern Tanzania. 1. Farming system analysis. Agrofor Syst 26:53–64

Schupp JL, Sharp JS (2012) Exploring the social bases of home gardening. Agric Human Values 29(1):93–105

Sethy S, Sarkar S, Kumar M (2010) Constraints in adoption of improved techniques of kitchen gardening. Ind Res J Ext Edu 10(2):89–92

Shubha K, Koley TK, Mukherjee A, Kumar U (2019) Nutri-garden: a promising approach to enhance rural household food and nutritional security. Food and Sci Rep 1(A):1–4

Sileshi M, Sieber S, Friedrichs K, Rybak C, Feyisa BW, Lana MA (2022) Adoption and impact of kitchen garden on food and nutritional security of farming households in Tanzania. Ecol Food Nutr:1–18

Singh P, Kanaujia VK, Dubey SK, Pandey S (2020) Backyard kitchen garden as the viable tool for nutritional security of rural families: a study from Kannauj district (U.P.). J Commun Mobilizat Sustain Dev 15(3):643–648

Soemarwoto O (1984) The Javanese rural ecosystem. In an introduction to human ecology research on agricultural Systems in Southeast Asia. Edited by Rambo AT, Sajise PE. University of Philippines, Los Banos, The Philippines

Solankey SS, Kumari M, Kumar M, Silvana N (2021) The role of research for vegetable production under a changing climate future trends and goals. In: Solankey SS et al (eds) Advances in research on vegetable production under a changing climate Vol. 1. Springer, Cham., ISBN 978-3-030-63497-1, pp 1–12

Sonnino R, Griggs-Trevarthen C (2013) A resilient social economy? Insights from the community food sector in the UK. Entrepreneurship Reg Dev 25(3–4):272–292

Subair SK, Siyana M (2003) Attitude T oward backyard gardening in Botswana. In proceedings of the 19th Conference of the Association for International Agricultural and Extension Education. pp 612–622

Torquebiau E (1992) Are tropical agroforestry homegarden sustainable? Agric Ecosyst Environ 41:189–207

Trinh LN, Watson JW, Hue NN, De NN, Minh NV, Chu P, Sthapit BR, Eyzaguirre PB (2003) Agrobiodiversity conservation and development in Vietnamese home gardens. Agric Ecosyst Environ 97:317–344

Vasey DE (1985) Household gardens and their niche in Port Moresby, Papua New Guinea. Food Nutr Bull 7(3):37–43

Wahlqvist ML (2013) Vitamin D status and food security in North-East Asia. Asia Pac J Clin Nutr 22(1):1–5

Watson JW, Eyzaguirre PB (2002) Homegardens and in-situ conservation of plant genetic resources in farming systems. Paper presented at 2nd International Home Garden Workshop, Witzenhausen, Germany

World Bank (2008) World development report 2008: agriculture for development. Washington D.C.

Chapter 11
Protected Cultivation of High-Value Vegetable Crops Under Changing Climate

Rajiv and Meenakshi Kumari

Abstract Vegetables provide all the nutrients ingredients *viz.*, vitamins, minerals, protein and nutraceutical properties that are essential for balanced diet, which makes vegetables protective food. In order to enhance the quality production and productivity per unit area of vegetable crops, protected cultivation technologies may be opted. Protected cultivation is a cropping technique wherein microclimate surrounding the plants modified partially or fully to suite them for better production. It offers several advantages to produce vegetables of high quality and yields, thus using the land and other resources more efficiently. There are various protected structures/methods *viz.*, plastic mulching, plastic low tunnel, walk-in-tunnel, high roof tunnel with or without ventilation, insect proof nethouse, shade nethouse, naturally ventilated polyhouse, hi-tech or climate controlled greenhouse, retractable top greenhouse, rain shelters, etc. for large scale vegetable production. In hi-tech or climate controlled greenhouse, most of the parameters *viz.*, temperature, light, humidity, fertilizer, irrigation is sensed and corrected as per programme through auto control systems. Soilless cultivation technology is system of plants growing in which solid rooting growing media are used instead of soil. The solid materials of soilless growing media in alone or mixtures may provide superior environment for plants in comparison to agricultural soil. Hydroponics and aeroponics technologies are also the predominant growing systems used under protected structures. These technologies offer numerous advantages such as saves water, increases crop production, environment friendly, food can be grown round the year, and provides jobs for residents. Parthenocarpic cucumber, tomato, capsicums (coloured), and lettuce are well known for protected cultivation.

Rajiv (✉)
Department of Vegetable Science, Chandra Shekhar Azad University of Agriculture and Technology, Kanpur, Uttar Pradesh, India

M. Kumari
Department of Vegetable Science, Shree Guru Gobind Singh Tricentenary University, Gurugram, Haryana, India

S. S. Solankey, M. Kumari (eds.), *Advances in Research on Vegetable Production Under a Changing Climate Vol. 2*, Advances in Olericulture,
https://doi.org/10.1007/978-3-031-20840-9_11

Keywords Controlled environment · Crop advancement · Protected cultivation technologies · Vegetable crops

11.1 Introduction

India is the second largest producer of vegetables after China in the world with a production of 184.40 million tonnes from 10.26 million hectares area (DAC&FW 2018). Vegetables provide all the nutrients ingredients *viz.*, vitamins, minerals, protein and nutraceutical properties that are essential for balanced diet, which makes vegetables protective food. According to the study of the Indian Council of Medical Research (ICMR), New Delhi and the National Institute of Nutrition (NIN), Hyderabad, changing food habit, limited availability of vegetables and high prices of pulses are considered responsible for malnutrition in India. Generally, vegetables are eating for nutrition, health maintenance and prevention of diseases. In developing countries vegetables are the main source of livelihood for most of communities because vegetables are loaded with several vitamins, carbohydrate, salts and proteins (Solankey et al., 2021). To ensure good health it is recommended that a person should consume at least 300 g of vegetable as a part of balanced diet, comprising 125 g leafy vegetables, 75 g other vegetables and 100 g root vegetables every day. The per capita land resources are shrinking due to the tremendous pressure of the population growth; therefore, it is very imperative to ensure the higher production and productivity per unit area. Vegetable crops are more productive than other crops, which have potential of providing more food per unit time and land area (Tomar et al. 2019). Thus, in order to enhance the quality production and productivity per unit area of vegetable crops, protected cultivation technologies may be opted. Protected cultivation offers several advantages to produce vegetables of high quality (Fig. 11.1) and yields, thus using the land and other resources more efficiently. Protected cultivation is more sustainable as the effect of climate is minimized (Pachiyappan et al. 2022). Protected cultivation of high-value crops offers higher productivity which in turn increases the profitability of the farm (Prakash et al. 2022). Therefore, growers would also be interested in this technology. However,

Fig. 11.1 Cherry tomato: A quality produce of protected cultivation

protected cultivation of vegetables has need of comparatively good management practices (Kashyap et al. 2019). In addition to nutritional security, vegetables production enhances the economy because it is an excellent source of income and employment as well. The share of vegetables is maximum (59–61%) in horticulture crop productions over the last five years in India (DAC&FW 2018).

11.1.1 Protected Cultivation

It is a cropping technique wherein microclimate surrounding the plants modified partially or fully to suite them for better production. In this technology land and water requirement is minimized for plants to yield better. Protected cultivation technologies cover climate control or hi-tech greenhouses, use of sensors, naturally-ventilated green/ polyhouses, hydroponics, aeroponics, plasticulture, drip irrigation, fertigation, mulching, integrated greenhouse pest and nutrient management, low cost protected structures like insect-proof and shade nethouses, soilless cultivation and vertical farms with good agricultural practices (Singh 2018; Singh 2019). It is system of controlled environmental agriculture in which all characteristics of the real environment are altered for highest economic return (Kashyap et al. 2019).

11.1.2 Scenario of Protected Cultivation in India and World

In India, protected cultivation is become popular because of its potentiality and profitability. This hi-tech protected farming of vegetables and high-value horticultural crops came through the Indo-Israel project, initiated at Indian Agricultural Research Institute (IARI), New Delhi in 1998. By the end of the twentieth century area under the protected cultivation was about 110 ha in India and the world over 275,000 hectares. Presently, total space coated under protected cultivation in India is approx. 30,000 ha (Pattnaik and Mohanty 2021). In the last decade, India has greatly accepted this new technique, and today this farming method done by almost every Indian state.

Greenhouse crop production is currently a growing reality throughout the planet with associate calculable 405,000 hectares of greenhouses contact all the continents (Pattnaik and Mohanty 2021). China is the largest users of greenhouses. The development of greenhouse technology in China has been faster than in any other country in the world. In China, alongside an unpretentious inception in late seventies, the area under greenhouses was expanded extensively (Kacira 2011).

11.1.3 Need to Go for Protected Cultivation

There is currently a high demand from consumers for a year-round supply of quality vegetables. The challenge of supplying high quality vegetables all year round can be met by adopting one of two basic strategies such as (i) growing in high-tech greenhouses, avoiding strong dependence on the outdoor climate or (ii) growing in two or more locations with complementary harvesting periods, enabling a regular and synchronized year-round supply to markets (Castilla and Hernandez, 2007).

The United Nations (UN) population branch has estimated that the world population could reach 9.15 billion by 2050, thus the expected rate of increase in world population will be 2.25 percent over the next forty years. 80 million peoples are added to the planet's population per annum. Food security has become a major concern at present and agriculture sector is going down because of climate changes along with food borne diseases are increasing, water distribution to agriculture farming is becoming scares in many areas. To feed countless mouth we will need to generate enough food with the restricted land, water and nutrient sources and this is accomplished through protected cultivation (Jat et al. 2019).

11.2 Major Advantages of Protected Cultivation System

According to Singh 2018 and Singh 2019, the Major advantages of Protected cultivation system are as follows:

1. Larger increase in productivity, quality and income.
2. The productivity of water is high.
3. Remarkable reduction in pesticide use.
4. Higher price because of off-season production.
5. Affordable economical greenhouses/polyhouse/poly tunnel/nethouses are fabricated locally, by small-scale entrepreneurs, which generate employment.
6. Expensive greenhouse cultivation methods such as hydroponics, vertical farming, aeroponics and hi-tech plant nursery production are pre-urban and unban farming of tomorrow.
7. It balanced light, temperature and humidity inside the structure.
8. Production cycle is longer.
9. Suitable for off-season and high value crops.
10. Generates employment.

11.3 Components of the Protected Cultivation System

Singh 2019 has proposed two main constituents in protected cultivation technology.

11.3.1 Agriculture Engineering

It deals with structures and engineering inputs.

11.3.2 Crop Production Technology

It involves development of hybrids and high-yielding varieties suitable for protected cultivation, their package of practices to manage pests, harvest preferably following good agriculture practices.

11.4 Protected Structures for Vegetable Cultivation

There are various protected structures/methods *viz.*, plastic mulching, plastic low tunnel, walk-in-tunnel, high roof tunnel with or without ventilation, insect proof nethouse, shade nethouse, naturally ventilated polyhouse, hi-tech or climate controlled greenhouse, retractable top greenhouse, rain shelters, etc. for large scale vegetable production. A brief discussion of some of these is mentioned below:

11.4.1 Plastic Mulching

The plastic sheet/film of 10–15 micron thickness is used as mulching which reduces the evaporation and acts as a barrier for emergence of weeds (Fig. 11.2). It is convenient for large-scale cultivation of vegetables (Singh and Choudhary 2019). There are different colours plastic films used as mulches such as silver-black, black, yellow, red, etc. It enhances uptake of micronutrients, sustain soil moisture and increases quality (Singh 2018; Singh 2019).

Fig. 11.2 Plastic mulching

11.4.2 Plastic Low Tunnel

Plastic low tunnel is miniature structure and facilitates entrapment of carbon di oxide thereby enhancing the photosynthetic activity. It safeguards plants from severe climatic conditions such as rain, wind, hail, snow, etc. It is mainly used for raising nursery for early germination of seeds and production of healthy saplings. Plastic low tunnels are also helpful in round the year cultivation of certain crops. This low cost technology is helpful in raising early cucurbits in north India by protecting them from low temperature and frost and has expended in a remarkable way in India and abroad (Fig. 11.3). The cost of low tunnels is affordable by most of the farmers. Non-woven fabrics in place of low tunnel cladding-plastic are gradually becoming popular (Singh 2018; Singh 2019). Presently, the area under plastic low tunnel technology is estimated to be around 30–40 thousand hectares (Singh and Choudhary 2019).

11.4.3 Walk–In-Tunnel

This is a low-price and temporary structure assembled on half-inch GI pipes covered with transparent plastic (UV grade film) with 200 micron thickness. Generally, tunnel height is 2 or 3 meters with around 4 meters width. It is suitable for raising plant nursery and off-season cultivation of vegetables particularly cucurbitaceous crops (Fig. 11.4). However, it is also suitable for production of tomato, brinjal, capsicum, leafy vegetables and early production of certain vegetables. It should be prepared to hold out against wind up to 120 km/h and trellising burden up to 25 kg/m^2 (Singh 2018; Singh 2019; Singh and Choudhary 2019).

Fig. 11.3 Plastic low tunnels

Fig. 11.4 Cucurbitaceous crop under walk-in-tunnel

Fig. 11.5 Insect proof nethouse

11.4.4 Insect Proof Nethouse

It is a closed crop production system and structure covered with insect proof nylon net (UV stabilized) of 40 mesh (Fig. 11.5). UV stabilized insect nets have a longer life. Such type of structure is preferred for several vegetables like seedless cucumber, tomato, capsicum, brinjal, etc. Double layer nethouses are also available which lowers the temperature in summer months (Singh 2018; Singh 2019). The availability of the nets is in distinct strength of holes (25 to 60 mesh). It is also acceptable for hybrid seed production of high value vegetable crops (Jat et al. 2015; Jat et al. 2016).

11.4.5 Shade Nethouse

This structure is covered with shade net of different shading intensity (20–75%) and used to grow the crops which are affected adversely by high temperature and high radiation (Singh and Choudhary 2019). This technology is very useful in improvement of healthy grafts/seedlings as well as hardening for seedlings/saplings (Fig. 11.6). Shade nethouses are made keeping in view the requirements of kind of crop to be raised generally with 6–8 feet height (Singh 2018; Singh 2019). Leafy

Fig. 11.6 Shade nethouses

Fig. 11.7 Naturally ventilated polyhouses

vegetables are mainly recommended to be grown under shade nethouses (Sirohi 2003). In arid and semi-arid regions, this technology is advised for cultivation of different vegetable crops on large scale during severe summer months (Singh and Choudhary 2019).

11.4.6 Naturally Ventilated Green/Polyhouse

There are two types of naturally ventilated green/polyhouses *viz.*, single and multi-span. Both structures are erected on GI pipes covered with UV stabilized transparent plastic/film (Fig. 11.7) and are found suitable for round the year production of crops depending on the geographical area. Single span is made with central height of 5 meters, side ventilation of 3 meters along with roll-able poly-cover with or without roof ventilation and double door entry. Generally, multi-span naturally ventilated polyhouse with a central height of 6.5 meters, gutter height of 4.25 meters and side ventilation of 1.5 meters is considered more suitable and economical for raising vegetable crops. Naturally ventilated polyhouses have been found suitable for cultivation of large number of vegetables besides raising their nurseries (Singh 2018; Singh 2019). It should be equipped with foggers for reduce the temperature during peak summer months. These structures are also fit for seed production (hybrid) with seed productivity of two-three times more in comparison to open field condition (Singh and Tomar 2015).

11.4.7 Polyhouse with Pad and Fan System

This polyhouse unlike naturally ventilated polyhouse has exhaust fan and cellulose cooling pad system to regulate temperature and humidity (Fig. 11.8). They have generally 4.5 meters height with common side and top ventilation. The pads of 150 mm thickness are allied at the height of 1.8 m in the structures. The establishment, operation and maintenance charges are high. This house is required to be equipped with overhead sprinklers for cooling in peak summer (Singh 2018; Singh 2019). It is suitable for raising vegetable seedlings and off-season cultivation of vegetable crops.

11.4.8 Hi-Tech or Climate Controlled Greenhouse

Hi-tech greenhouse is a climate controlled structure and have a variety of applications such as off-season growing of vegetables, foliage and flower plants, planting material multiplication and acclimatization, plant breeding and new varieties and hybrids development. This is available in different sizes ranging from as small as $100m^2$ to $10,000m^2$ and more and constructed as per requirement. In this house, most of the parameters *viz.*, temperature, light, humidity, fertilizer, irrigation is sensed and corrected as per programme through auto control systems. The climatic requirements of vegetable crops under protected cultivation have been presented in Table 11.1. Hi-tech greenhouse can be fully or partially automated and mechanized. This is considered, based on experience, very good and essential structure for nursery multiplication of vegetable crops in plugs with soilless medium (Singh 2018;

Fig. 11.8 Polyhouse with pad and fan system

Table 11.1 Climatic necessity of different crops under protected cultivation

Crop	Temperature (°C)		Humidity (%)	Light intensity (lux)
	Day	Night		
Solanum lycopersicum L.	22 to 27	15 to 19	50 to 65	50,000 to 60,000
Cucumis sativus L.	24 to 27	18 to 19	60 to 65	50,000 to 60,000
Capsicum annuum L.	21 to 24	18 to 20	50 to 65	50,000 to 60,000
Seedling (nursery)	22 to 27	16 to 19	50 to 65	50,000 to 60,000

Singh 2019). Hybrid seed of different vegetable crops may be produced under semi-climate controlled greenhouses with three-four times more productivity in comparison to open filed conditions (Jat et al. 2017).

Solar radiation, air temperature and air relative humidity are important alterable of the greenhouse climate that can be controlled (Kittas et al. 2013). The quality of greenhouse produce enhance with the help of improved controlled climate (Castilla and Montero 2008).

Modified from Singh et al. (2015).

11.4.9 Retractable Roof Greenhouse

This structure protects leaf and root of the plants from extreme environmental conditions *i.e.* imprudent or deficient heat, cold, wind, rain, and disorders related to inadequate transpiration. The arrangement of roof, walls and curtain systems with automation in structure of the retractable roof greenhouse can create an outdoor greenhouse and modified greenhouse conditions for the plants (Singh 2018; Singh 2019).

11.4.10 Rain Shelter

This structure is also known as a low cost naturally ventilated polyhouse. It is found ideal for off-season vegetable cultivation in high rainfall areas (Fig. 11.9). Off-season cultivation of vegetables by utilizing low cost protected structures like rain shelter can be considered as a profitable enterprise besides protecting crop nurseries from high rain during monsoon months. Such structures are boon for vegetable production in rainy season (Singh 2018; Singh 2019).

Fig. 11.9 Rain shelters

11.5 Classification of Green/Polyhouse Based on Cost

Based on cost, the different types of green/ polyhouse are as follows (Singh et al., 2015):

11.5.1 Low-Cost Green/Polyhouse

This structure is made with bamboos, ropes and nails and covered with 200-micron (800 gauge) transparent polythene sheet. The temperature within structure increases by 6–10^0 C more than outer. The polythene sheet is reduced 30–40% solar radiation. Generally, this structure is used for safeguard the crop from heavy rainfall.

11.5.2 Medium Cost Green/Polyhouse

The structure is built up with 15 mm diameter of GI pipes and covered with UV-stabilized polythene (200-micron thickness) in single layer. Naturally ventilated with exhaust fans and naturally ventilated with fan-pad system are examples of this type of green/polyhouse. The fan-pad system lowers the temperature in polyhouse.

11.5.3 High-Cost Green/ Polyhouse

This structure is constructed with GI pipes having cone or dome shaped design. Light, humidity and temperature are controlled automatically with the help of sensors as per requirement of the crop. The cost of structure is approximately five to six times more. It desires genuine maintenance, appropriate care and trained operator.

Generally, depending on the covering material (glass or flexible plastic) of greenhouse structure, different terminologies are being used. A greenhouse with glass as the covering material is called as greenhouse whereas polyethylene as the covering material is known as polyhouse (Sirohi 2003).

11.6 Protected Cultivation Technologies

11.6.1 Low Tunnel Technology

Growing off-season crops in controlled atmosphere inside polythene tunnels is known as tunnel technology. During the period of December to February, it is not feasible to cultivate summer vegetables in open field conditions because of high frost and low temperature. However, inside polythene tunnels these vegetables can grow with their maximum growth and yield by providing genuine atmosphere to the plants (Tomar et al. 2019). Low tunnel technology is most acceptable as well as beneficial in northern plains of India (Singh and Solanki 2015).

11.6.1.1 Off-Season Cucurbits Production with Low Tunnel Technology

Low tunnels or row covers are pliable transparent plastic (23–30 micron) and creates microenvironment under the tunnel. Low tunnels effectively increase air and soil temperature thus enhancing the vegetative growth, improve water and nutrient use efficiency and increases yield. The cucurbits can be advanced by 30-40 days in comparison to the normal sowing time. Therefore, this technology can improve productivity and land use efficiency. This technology is most acceptable for off-season cultivation of cucurbits (Tomar et al. 2019).

11.6.1.2 Seedlings Raising of Cucurbits Under Low Tunnels for Off-Season

During the month of December or January, the cucurbits seedlings are raised in plastic pro-trays of 1.5″ cell size in soilless media with low tunnel technology. In northern parts of the India, at four leaves stage (28–32 days old) seedlings are transplanted under low tunnel in the open field when the night temperature is very low (from mid-January to mid-February). Even though, for complete off-season production the summer squash may transplant in December month and crop will ready to harvest in the first week of February. This crop will realize exceptionally good price in the market (Tomar et al. 2019).

11.6.1.3 Seedling Transplanting of Cucurbits Under Low Tunnel

The flexible galvanized iron hoops with a height of 40-60 cm are fixed on beds at a distance of 1.5–2.5 m before seedling transplanting. The distance between two ends of hoop is maintained 40-60 cm. Thereafter, the seedlings of cucurbits are transplanted at a spacing of 1.5–1.6 × 0.50 m in single row on each bed and cover the rows or beds with transparent plastic (30 micron) in the afternoon. Low tunnels keep higher temperature inside the structure than outside field. During peak day time of growing season when temperature increases inside the tunnels, 3–4 cm size vents are made in plastic at a distance of 2.5–3.0 m on eastern side just below the top. Further, the size of vents may enlarge with the increase in the temperature. Finally, in the month of February and March entirely plastic will take out from the plants depending on crop growth and current night temperature (Tomar et al. 2019).

11.6.1.4 Seedlings Raising of Tomato, Chilli and Brinjal under Low Tunnel

In Northern India nursery of tomato, chilli and brinjal is raised under low tunnel in the month of December-January for early transplanting of the seedlings. The seeds are sown in the nursery bed and nursery bed is covered with fabricated tunnel (size: 3.0 m long and 1.5 m wide along with 1.0 m central height). The semi-circular structure is clad with UV polythene sheet (20 Micron) with 75% transmittance. Low tunnel is also used for raising seedlings of cole crops during rainy season with side ventilation.

11.6.1.5 Fertigation in Low Tunnels

The water may be applied @ 4.0 m^3/1000 m^2 at six-seven days interval during the period of first month with fertilizer solution of N:P:K (5:3:5) @ 50-100 ppm per cubic meter of water. Thereafter, water @ 4.0 m^3 along with fertilizer solution @ 120-150 ppm/m^3 of water may apply at 4 days interval in second month to until beginning of flowering. The quantity of fertigation is expanded up to 300 ppm at the peak of fruiting period (Tomar et al. 2019)

11.6.1.6 Pollination under Low Tunnels

Majority of the cucurbits being monoecious need pollination. The main pollinating agent is honeybees (*Apis melifera*). At the time of complete flowering, one beehive/ acre area is required for effective pollination and bees will work in tunnels easily through the vents. For effective working of bees, the beehive box should keep on the north-west side of the field.

11.6.1.7 Harvesting and Crop Advancement under Low Tunnel

In northern plains of the India, low technology is fully cost effective for growing off-season vegetable crops in peri-urban areas. Different cucurbit crops may advance 30–60 days by transplanting them from December (first week) to February (first week) over their usual season of cultivation (Table 11.2).

Table 11.2 Crop advancement and transplanting time in cucurbits under low tunnel

S. No.	Name of the crop	Time of transplanting	Time of harvesting	Crop advancement (days)	Anticipated yield (q/ha)	Expected cost benefit ratio
1.	Cucumber	January (IIIrd week) – February (Ist week)	March (Ist week)	25–30	125–150	1:2.0–1:2.5
2.	Muskmelon	January (IIIrd week) – February (Ist week)	April (IInd week) – April (last week)	30–40	200–250	1:2.5–1:3.0
3.	Summer squash	December (Ist week)	February (Ist week)	40–60	400–500	1:3.0–1:3.5
4.	Bottle gourd	January (IIIrd week) – February (Ist week)	April (IInd week) – April (last week)	30–40	250–300	1:2.0–1:2.5
5.	Bitter gourd	January (IIIrd week) – February (Ist week)	April (IInd week) – April (last week)	25–30	100–150	1:2.0–1:2.5
6.	Water melon	January (IIIrd week) – February (Ist week)	April (IInd week) – April (last week)	25–30	200–250	1:2.0–1:2.5

Modified from Tomar et al. (2019)

11.6.2 Hydroponics Technology

Hydroponics is the predominant growing system used under protected structures. In this system, soil is replaced with growing medium like cocopeat, perlite, rockwool, gravel etc. Here, the plant roots are submerged in the solution of nutrients (Fig. 11.10). This technology can reduce the various soil associated issues *viz.*, bacteria, fungus, and insects which develop in soil. It requires relatively low maintenance. It is a cleaner process, which excludes use of animal excreta and provides an easier way to control nutrient levels and pH balance. In this technology, the best combination of nutrients is applied hence, yielded more with consistent produce. With the NFT system, a thin film of nutrient solution flows through plastic channels, which contain the plant roots with no solid planting media (Singh 2018; Singh 2019).

11.6.3 Aeroponics Technology

National Aeronautical and Space Administration (NASA) coined the term aeroponics in the 1990's. The aeroponics may be defined as growing plants in an air or mist environment with no soil and very little water. In aeroponics system, the plant boxes are stacked in such a manner that the bottom and top of the plants are hanged in air allowing the crown to grow upward and the roots down ward freely. The plants are fed through a fine mist containing nutrient rich and water mix solution with the help of sensors, which can be fully recycled as the system is an enclosed one. High density planting can be carried out in aeroponics making harvesting easier and providing higher yield (Jat et al. 2019). It is commonly used for growing leafy vegetables particularly lettuce and herbs in vertical system of farming. For plant factories, this is an appropriate technology.

Fig. 11.10 Hydroponics technology

Advantages of Hydroponics and Aeroponics Technology

These technologies are important for protected cultivation in following ways (Jat et al. 2019):

- Increase crop production
- Continuous crop production
- Crops are grown indoors therefore, no crop failures due weather related problems.
- Organic crop production is feasible
- Conservation and recycling of natural resources
- Hydroponic system uses 70% less water
- Mechanical equipment is not required hence, reduces the use of fossil fuel
- Environment friendly
- Sustainable ecology as the urban sewage is purified, solar panels and wind turbines as a source of energy making the system entirely sustainable
- Food can be grown round the year
- Provide jobs for residents

11.6.4 Microgreens: A Smart Food

Microgreens are young edible seedlings of vegetables, herbs and plants rich in flavour and nutrition. Microgreens are the infant version of plants (Fig. 11.11). They are more healthier (4–40 times) than their full-grown stages. Usually, they are harvested at a height of 2.5–7.6 cm (within 14 days of the seed germination) and consumed because of the highest amount of anti-oxidants as well as nutritional benefits at this stages (Table 11.3). They are grown under protected structures (Singh 2018; Singh 2019).

Seedling colour, flavour, texture and market demand are base for selection of the crop for microgreens. However, the most potential microgreens are found in vegetable crops belonging to the family amaranthaceae, chenopodiaceae, cruciferae and apiaceae. The crop for microgreens should have high nutritive values with very fast

Fig. 11.11 Vegetable microgreens

Table 11.3 Antioxidant potential of microgreens

S. No.	Microgreen	Days to harvesting	Antioxidant activity (μmol TE/g)
1.	Bottle gourd	11	8.35
2.	Cucumber	9	4.65
3.	Pumpkin	10	8.92
4.	Amaranthus red	9	4.60
5.	Amaranthus green	19	2.15
6.	Amaranthus Katwa	12	4.58
7.	Basella	12	3.90
8.	Water spinach	9	20.23
9.	Radish	9	17.77
10.	Palak	11	6.71
11.	Carrot	15	26.81
12.	Fennel	14	16.88

Fig. 11.12 Plug tray seedlings raising under protected conditions

germination and easy to grow as well. Generally, sterilized soilless media (coco peat, vermiculite and perlite in a ratio of 2:1:1) is recommended for best growth of microgreens. Around 6–8 seeds and 10–12 seeds per square inch are sown for larger and smaller seed, respectively. Microgreens are extremely perishable due to their more respiration rate. Hence, its shelf life is very short *i.e.*, 3–5 days at ambient temperature (Anonymous 2016).

Modified from Anonymous (2016).

11.6.5 Plug Tray Seedlings Raising Technology

Under this technology seedlings are raised in plastic multicell plug tray in artificial soilless media in principally designed protected structure (Fig. 11.12). By this technology, the healthy and off-season seedlings can produce and it will help in generate the additional employment in horticulture sector. This is the major and first key step for increasing yield, quality and profitability in vegetable crops through protected cultivation technology. A nursery greenhouse area of 1000 m^2 is able to develop about 15–20 lakhs of vegetable seedlings and can generate the net return of Rs. 2.5–3.0 lakhs. Therefore, it is capable to create employment for 4–5 youths throughout the year (Singh et al. 2005; Singh and Solanki 2015).

11.6.6 Soilless Cultivation Technology

It is systems of plants growing in which solid rooting growing media are used instead of soil. The solid materials of soilless growing media in alone or mixtures may provide superior environment for plants in comparison to agricultural soil. Therefore, these growing media may helpful in healthy vegetable production with higher yield (Gruda et al. 2013). Usually, the containers and sometimes prepared cubes, bags, slabs, mats, troughs, etc. are used for soilless cultivation (Fig. 11.13). The containers are filling up with required sterile mix and put down under controlled environment (Sehrawat and Malik 2015).

Soilless culture offers numerous advantages such as virtual absence of nematodes and pathogens, gives fine aeration and drainage, economizes the fertilizer use due to nutrient control, adequate oxygen exchange, anchorage or support to plant, light weight and standardization in comparison to natural soil culture and land-less growers are also able to grow vegetables. The plant growing media in soil-less cultivation technology are cocopeat, rockwool, vermiculite, perlite, bark chips, sawdust, sand, gravel, pumice, polyurethane mats, sand, rice hulls, bagasse, water, air, etc. It has become a desirable practice in to have vegetable and other crop nurseries in plug trays multiplied in soilless medium under protected conditions. Production of vegetables like leafy vegetables mainly lettuce, tomato, capsicum, cucumber is becoming popular to grow with soilless cultivation technology (Singh 2018; Singh 2019).

11.7 Vegetable Crops and their Varieties/Hybrids

In protected cultivation, the selection of varieties may remarkably affect the economic return (La Malfa and Leonardi 2001). The varieties for protected cultivation are vary from open-field vegetable productions. Parthenocarpic cucumber, tomato, capsicums (coloured), and lettuce are well known for protected cultivation (Table 11.4).

The details of major varieties/ hybrids of the high value vegetable crops suitable for protected cultivation and developed by public and private sectors are described below:

Fig. 11.13 Crops in soilless growing media

Table 11.4 Vegetables varieties suitable for protected cultivation

Sl. No.	Crops	Variety/ hybrids
1.	Tomato	ID-32, ID-37, Rakshita, Himsona, Himsikhar, Snehlata, Naveen, GS-600
2.	Coloured capsicum	Natasha, Swarna, Indra, Bombi, Orobelle, bachata, inspiration
3.	Parthenocarpic cucumber	Isatis, Kian, Hilton, sun star, multistar, Fadia, mini angel
4.	Summer squash	Pusa Alankar, Pusa Pasand, Australian green, Seoul green, Kora, yellow zucchini, Himanshu
5.	Bitter gourd	Pusa Rasdar
6.	Musk melon	Pusa Sarda

Adapted from Solankey et al. (2021)

*11.7.1 Tomato (*Solanum lycopersicum L.*)*

The major varieties/hybrids are as under:

1. Pant Polyhouse Tomato-2: Average fruit weight 100-105 g, pericarp thickness 0.9-1.0cm (better storage quality) and 5-6 fruits per cluster.
2. Pant Polyhouse Hybrid Tomato-1: Large fruit weight (130-140 g), pericarp thickness 1.0-1.25cm (better storage quality) and 7-8 fruits per cluster.
3. Arka Rakshak: This hybrid is resistant to triple disease (ToLCV+BW+EB) with the yield potential of 75-80 t/ha in 140 days. Fruits are medium (90-100 g) with superior storage quality (15-20 days).
4. Heemsohna: Fruit are medium in size (90–100 g) with outstanding serviceable life. Long duration, high productivity, and plants are tall vigorous with profuse branching.
5. Heemshikhar: Fruits are medium in size (80–90 g) and uniform with outstanding serviceable life. Long duration, high productivity, and plants are tall vigorous with profuse branching.
6. Pusa Hybrid-4: Fruits are medium sized (70–80 g), round, attractive, smooth with thick pericarp and uniform ripening. Plants are compact, determinate with dark green leaves and field resistance to root knot nematode. Average yield is 54.8 t/ha.
7. Pusa Hybrid-8: Fruits are medium (75–80 g), round with uniform ripening. Plants are compact, determinate and massive fruit bearer. Average yield is 43–45 t/ha.
8. Punjab Sartaj: Plants are indeterminate. Fruits are round, medium (average fruit weight 85 g) and firm. Average yield 90 t/acre under polyhouse (Dhaliwal et al. 2016).
9. Punjab Gaurav: Plants are indeterminate. Fruits are oval, medium (average fruit weight 90 g), very firm with pointed tip and excellent shelf life. Average yield 93.5 t/acre under polyhouse (Dhaliwal et al. 2016).

Some of the indigenous and multinational seed companies' varieties/hybrids are Rakshita, NS-4266 (Fig. 11.14), Snehlata, Avinash-2, GS-600, Shreshtha, Avtar, Indam Hybrid, All Rounder, Arka Samrat, Arka Meghali, Arka Surabhi, Tropic, Tolstoi, Dombello and Tuckcross-520 (Singh 2018; Singh 2019; Singh et al. 2015; Anonymous 2021).

11.7.2 *Cherry Tomato (Solanum lycopersicum var. cerasiforme)*

The major released and/or popular varieties are as under (Fig. 11.15):

1. Pusa Cherry Tomato 1: Fruits are deep red, uniform ripening, round with average fruit weight of 13 g. Average fruit weight is 22 kg/plant with yield potential of 4–5 tonnes/1000 m^2. Vine length is 9–12 m and crop duration is about 9–10 months (https://ztmbpd.iari.res.in/technologies/varietieshybrids/vegetables/cherry-tomato).
2. Punjab Red Cherry: Red fruited variety, fruit weight 8–14 g, plant having the capacity to yield 4–4.5 kg of fruit, tolerant to leaf curl virus, total soluble sugars (TSS) 8–9 per cent, 18–20 fruits per cluster and yields 110 t/ha.
3. Punjab Sona: Fruits are yellow with 7.5% TSS and 13 mg carotene/100 g fruits. Bears 20–25 fruits per cluster and yields 105 t/ha.

Fig. 11.14 Regular tomato under protected condition

Fig. 11.15 Cherry tomato under protected condition

4. Punjab Kesar: Fruits are orange with 7.6% TSS and 13 mg carotene/100 g fruits. Bears 1–23 fruits per cluster and yields 100 t/ha (Malik et al. 2017).
5. Nagmoti: Early maturing, vigorous plant with good foliage, fruits are small, round, shining deep red (ripe), firm and uniform in shape and very good flavour and sweetness.

Some of the other varieties/hybrids available in the market are BR-124 (Holland), Olle, Sairan, NS Cherry-1, Pant Cherry tomato, Pairesco, Rosa (Singh 2018), Roma and Super Sweet-100.

11.7.3 Capsicum (Capsicum annuum var. grossum L.)

The major released and/or popular varieties are as under (Fig. 11.16):

11.7.3.1 Yellow Fruited

1. Orobelle: Almost square fruits (10x9 cm) with a medium-thick wall, turning from green to yellow at maturity and average fruit weight are 150 g.
2. Yellow Wonder: Light green to golden-yellow fruit colour at maturity, upright plants that set continuously, plants are vigorous and high-yielding, ready to harvest in 62–73 days from transplanting and can be sown directly from seed in February when the night temperature is 20-300C.
3. Golden Summer: Sweet and mild flavour, fruits are lime green ripening to golden yellow and is 4-lobed, plant provides good foliage cover for the sweet bell peppers that are excellent stuffed or added to salads and pinch off early flowers to encourage plant growth.
4. Swarna: Blocky to long fruits with appealing dark green colour and thick and firm wall. Plants are vigorous and strong. Fruit weight is 200–250 g.

Fig. 11.16 Coloured and green capsicum under protected conditions

11.7.3.2 Red Fruited

1. NS 280: Fruits are firm, blocky (3–4 lobed) and good size (10x8 cm) with average fruit weight of 220–230 g. Plants are tall vigorous and strong with broad leaves.
2. Bomby: Fruit are glossy and dark green with average weight of 130–150 g. Plants are tall, strong, early branching with dense foliage and requires staking.
3. Tanvi: Plant medium tall and medium spreading, fruit weight is 150-180 g, turn yellow after maturity, high yielding and excellent keeping quality.

11.7.3.3 Green Fruited

1. California wonder: Open pollinated variety, large bell shaped fruits, ideal for gardens and large containers, relative days to maturity is 70, fruit weight is 180-200 g and fruit shape is blocky 3–4 lobed.
2. Indra: Fruits are glossy, thick-walled, dark green, 10–12 cm length, 10 cm girth and having 3–4 lobes with average weight of 170 g. Plants are bushy, medium tall and having vigorous growth with dark green leaves and dense foliage.

Pusa Deepti (green), Green Gold (green), Bharat (green), Mahabharat (green), Natasha (red), DARL-71 (yellow), Super Gold (yellow), NS-285 (yellow), Torkel (red) and Tarquino (green) are also some of the popular varieties/hybrids of the capsicum (Singh et al. 2015; Singh 2018).

11.7.4 Cucumber (Cucumis sativus *L.*)

Generally, parthenocarpic cucumber is preferred under protected conditions (Fig. 11.17). The following hybrids/varieties can be successfully grown under protected cultivation.

Fig. 11.17 Cucumber under protected condition

1. Pusa Seedless Cucumber- 6: It is parthenocarpic gynoecious cucumber and extra early (40–45 days for first fruit harvest). Fruits are uniform, glossy, dark green, appealing, cylindrical, straight and non-hairy with 14.24 cm average length, 3.45 cm average width and 105 g average weight. Average fruit productivity is 1260 q/ha (1260 kg/100 m^2) in the time of winter season.
2. Pant Parthenocarpic Cucumber- 2: Bears only female flowers (gynoecious) and fruits are ready to first harvest at 30 DAS. Average fruit weight is 400-450 g and plant produces seed less fruits.
3. Pant Parthenocarpic Cucumber- 3: Bears only female flowers (gynoecious) and fruits are ready to first harvest at 32 DAS. Single fruit weight is about 350-400 g and plant produces seed less fruits.
4. Hilton: Vigour and strong plant, uniform green lustrous fruits and high fruit setting ability with earliness and maturity days (first harvest after sowing) 38–40.
5. Isatis: Good plant vigour, good adaptability, fruit length 18–20 cm, good fruit setting, high yielding and crispy and bitter free with uniform fruits.
6. Kian: Cylindrical, glossy, medium green colour fruits, semi-multi fruited on main stem and bears 2–3 fruits per node.
7. Multistar: Fruit are about 16–18 cm length with dark green, uniform, cylindrical in shape and shiny. This is perfect for slicing and salads.
8. Deltastar: Vigorous growth, suitable for autumn, early spring and summer cultivation and fruit length is 16-18 cm with dark green colour and good flavour as well as storage life.
9. Hasan: Tender long attractive fruit with shiny whitish colour, ready to first picking in 40–45 days after sowing, average fruit length 22–25 cm, fruit weight 125-130 g, high yielder, excellent taste and keeping quality.
10. Sunstar: Small leaves and multi fruit bearing with relatively open plant type. Fruits are 17–19 cm long slightly ribbed along with dark green shiny and cylindrical in shape. Mostly used for slicing and salads.
11. Kingstar: Strong plant with less side shoots, 1–2 fruits per node, average fruit length 16–18 cm and suitable for late autumn and early winter.

Some of the popular parthenocarpic varieties from multinational seed companies are Kalunga, Bellissma, Millagon, Discover, Marianna, Fitness, Aramon, Fidelio, 90–0048, Futurea, E-1828, B-1157, Country Fair, Bush Crop, Space Master, Patio Pick, Bush Whopper, Bush Champion, Bush Pickler, Euro-American, Adrian, AI Rashid F1, Mustang, Brucona, Fanspot, and Toska-70 (Hochmuth 2012; Singh 2019; Anonymous 2021).

11.8 Crop Management and Operations

11.8.1 Nursery Raising

Seedlings of the vegetable crops are raised in multi-cell protray in soil-less growing media such as a mixture of vermicompost + sand + sterilized cocopeat (1:1:1) or cocopeat + vermiculite + perlite (2:1:1) under protected structure. The seeds are treated with carbendazim @ 2.5 g/kg seed before sowing and single seed is sown in each cell. After germination, the seedlings should be drenched with solution of copper oxy chloride (@ 3 g/litre of water). At the stage of 13–15 days old seedlings, the nutrition is applied through the drenching/foliar application of 0.2% solution of 19:19:19 (N:P:K) along with micronutrients. The seedlings can protect form insect infestation with spray of imidacloprid (0.03 ml/litre) or acephate (0.75 g/litre). The seedlings get ready for transplanting after 21–35 days of sowing (about 20-25 cm height) depending upon crop. The hardening of the seedlings is required for better establishment and it can be done by gradually reduction in frequency of irrigation and exposing them to sunlight. For better establishment and prevent from damping off in main field, the seedlings should be drenched with carbendazim (0.1%) or ridomil (0.1%) before planting.

11.8.2 Preparation and Solarization of Bed

About 15-20 cm raised beds are prepared with the width of 90-100 cm and leave the path of 50 cm in between beds. After bed preparation and application of organic manures *viz.*, vermicompost and FYM etc. it should be disinfected through fumigation with formalin 4% solution @ 4 litres per m^2 in the month of May–July. Thereafter, the beds should be covered with white and transparent polythene sheet of 100 micron (400 gauge) thickness. The doors and ventilators of the protected structure should also be closed. Polyethylene cover is removed after 15 days of formaldehyde treatment and the doors and ventilators are also opened. To remove the trapped formaldehyde fumes completely, the beds are hoed again and again. Temperature rises up to 60-70 °C during solarization process and it process kills harmful organisms, bacteria, fungal spores, nematodes and weeds.

11.8.3 Fertilizer Application and Fertigation

Before fumigation, organic manures *viz.*, vermicompost or FYM etc. are applied @ 10–15 kg/m^2 area of the growing bed depending on fertility status. The organic manure should be well decomposed and mixed thoroughly in beds. Inorganic fertilizer *i.e.* 19:19:19 (N:P:K) is added after fumigation @ 7 g/m2 in furrows close to the growing beds (Singh et al. 2015). Fertigation allows the adjustment of the amount and concentration of the applied nutrients according to the crop's needs throughout the growing season. External supply of nutrients has become important because of poor fertility status of the soil which is not able to meet the entire nutrient requirement of the crop (Rajiv and Tomar 2022).

The crop is fertigated with water soluble fertilizer (WSF) through drip fertigation as per dosage and schedule given in Table 11.5. The plants are also sprayed with mixture of micronutrient solution @ 0.3% starting from 60 DAT. Total two-three spray of micronutrients are to be given at an interval of 30 days in tomato. Whereas, in case of capsicum and cucumber, the micronutrient foliar spray can be started earlier. If required, the crop may be fertigated with calcium nitrate at 15 days interval in deficiency symptoms of calcium. The deficiency of micronutrients adversely affects the production as well as quality of vegetables (Singh et al. 2022).

Modified from Singh et al. (2015).

Table 11.5 Fertigation schedule for different crops under protected cultivation

Crop	Crop stage	NPK (water soluble)	Dose (g/500 m2)	Fertigation schedule
Tomato	Planting to first flowering	19:19:19	250	Twice a week, starting from 25 days after transplanting
	First flowering to fruit set	19:19:19	100	
		46:0:0	175	
		0:0:50	275	
	Fruit set to harvesting peak	19:19:19	100	
		46:0:0	250	
		0:0:50	275	
	Topping until crop end	19:19:19	50	
		46:0:0	125	
		0:0:50	150	
Capsicum	Planting till fruit setting	19:19:19	500	Twice a week
		0: 0: 50	25	
	Fruit set until first picking	19:19:19	500	
		46:0:0	100	
		0:0:50	250	
	Afterwards first picking and up to season end	19:19:19	4500	
		46:0:0	500	
		0:0:50	250	
Cucumber	0–14 days after transplanting	19:19:19	500	Twice a week
	14–35 days after transplanting	13:0:45	200	
		46:0:0	100	
	35 - till the end of crop	13:0:45	500	
		46:0:0	150	

11.8.4 Mulching

The growing beds are covered with silver/black polyethylene sheet of 400 gauge (100 micron) having 1.2 meter width before transplanting the plants. Edges of the sheet are buried with the help of soil. Thereafter, holes of 5 cm diameter are made on the mulching sheet using a mulching hole machine or sharp pipe at recommended spacing for the crop.

11.8.5 Spacing

The seedlings are transplanted in two rows per bed in zigzag/triangular manner. Usually, the spacing in tomato, capsicum and parthenocarpic cucumber is kept 60 x 45 cm, 45 x 30 cm and 60 x 60 cm, respectively.

11.8.6 Plant Canopy Architecture Management

The vapour concentration, temperature and radiation regime in the plant environment is significantly influenced by architecture of canopy. Interception and transmission of soil temperature and soil heat flow are also affected. It indirectly affects the plant physiological processes, transpiration, photosynthesis, cell enlargement, growth and multiplication of insects and photo-morphogenesis. The productivity and quality of produce is higher under greenhouse conditions, since near to optimum growing conditions are maintained as per the requirement of the crops unlike the open field conditions. The diffused light passing inside the greenhouse through transparent cladding materials tends the crop plants to grow upwards and utilize the vertical space inside the structure. The number of plants per unit area could be increased leading to get more productivity (Patil et al. 2019). Management of plant structure is significant activity during production of greenhouse crops. Desirable changes in plant structure can be achieved as details given below:

11.8.6.1 Capsicum

Pruning/pinching starts from 15–20 days after transplanting and it should be done at weekly intervals. Initially get 2–3 branches at 5–6 node. Remove weaker one again allow 2 stems per basal stem that is total 4 stems. Remove all flowers initially for one month and side shoots at weekly intervals. Plants training start 30 days after transplanting through a plastic twine or tressiling yarn. Each branch should have separate twine/yarn. Tie the twine/yarn to the GI wire provided at truss over the bed (Patil et al. 2019).

According to ICAR-Indian Institute of Vegetable Research (IIVR), Varanasi studies, the cv. Indra was able to produce most of 'A' grade fruits (3–4 lobes weighing 150 g and more) under two-stem canopy management practices in comparison to three-stem and unpruned plants systems (Table 11.6). Studies revealed that choice of training system in capsicum might depend upon market aimed (Anonymous 2019).

11.8.6.2 Tomato

Usually, pruning starts 20–30 DAT and it should be done at weekly intervals. In general, there are two systems (one-stem and two-stem) of management of canopy architecture in tomato (Fig. 11.18). In one-stem system, the sprouted side shoots are removed when they attain 5–10 cm length and a single main stem is leaved on each plant. Whereas, in case of two-stem management system, the side shoot is permitted to grow as a second stem just below the first floral truss and remaining side shoots on both stems are removed periodically. Studies conducted by ICAR-Indian Institute of vegetable Research (IIVR), Varanasi revealed that two-stem training system results in higher no. of clusters, no. of fruits per cluster and fruit yield/plant (Table 11.7) in cv. NS-4266 (Krishna et al. 2020). Similarly, the higher productivity was found in Heemsohna when the plants are maintained with two-stem system (Anonymous 2019). Remove the older leaves continuously (Fig. 11.19).

Table 11.6 Influence of canopy architecture management on fruit size in capsicum

	Canopy architecture management		
Variety	Two-stem	Three-stem	Unpruned (control)
Indra	200 g	110 g	70 g
Orobelle	110 g	90 g	60 g
Indus-1201	90 g	80 g	60 g

Modified from Anonymous (2019)

Fig. 11.18 Single stem *v/s* Double stem training system in tomato

Table 11.7 Effect of training system on tomato performance under polyhouse conditions

Training system	No. of clusters per plant	No. of fruits per cluster	Fruit weight (g)	Fruit yield (kg/plant)
One-stem	13.89	6.54	68.73	7.07
Two-stem	18.94	6.90	62.85	10.36
Unpruned (control)	11.83	5.56	55.61	5.26
CD 0.05	2.11	0.87	5.62	0.67

Source: Krishna et al. (2020)

Fig. 11.19 Deleafing in tomato

11.8.6.3 Cucumber

Plants produce abundant leaves and vines and direction less vine growth during the initial growth. Therefore, train lengthy vines for productivity and canopy maintenance and prune to maintain predictable growth and development. Usually, there are three training systems *viz.*, umbrella, v-system and single stem to achieve certain architecture or structure.

In case of umbrella training system, the main stem along with supporting string is permitted to develop vertically up to the overhead wire (2 m above the ground level). The apical bud is removed after producing two leaves above the overhead wire. Two healthy and vigourous lateral branches at the top of the vine are allowed to grow along the wire for about 15 cm and trained to grow down wards. All other laterals are removed. Whereas, in v-system, the main stem is permitted to grow along with supporting string and the growing point is removed at the sixth leaf stage (45-60 cm plant height). The two emerging lateral branches are then trained into a ‘v-shape’ on to the overhead wire. In single stem system, the main stem is allowed to grow as in case of the umbrella system and when the plant reaches the overhead wire, whole vine is lowered and trained to moved own ward. This system can accommodate more plants at the spacing of 60x45cm. For pruning, as the plant grows up the string, remove all the lateral buds up to the sixth node (a node being where a leaf joins the stem). In addition to the lateral buds, all the fruits should also be removed up to this point (Patil et al. 2019).

11.8.7 Harvesting and Yield

Usually, harvesting is depending on variety, distance from market and type of protected structure, however, fruit starts at 70–80 DAT and continues up to 170–180 days in tomato. Whereas, parthenocarpic cucumber cultivated under protected cultivation gives first fruit harvest at 35–40 days after transplanting. Fruit yield of tomato, capsicum and parthenocarpic cucumber under protected cultivation can be achieved up to 170–180 t/ha (17–18 kg/m^2 or 5.7–6.0 kg/plant), 100–120 t/ha (10–12 kg/m^2 or 4–5 kg/plant) and 300–400 t/ ha (30–40 kg/m^2), respectively depending upon management of crop and type of protected structure. In tomato, the average fruit weight varies from 100 g during initial harvesting to 60 g during last harvesting. Harvesting of fruits may be done usually at weekly interval or earlier (Singh et al. 2015).

11.8.8 Diseases Management

11.8.8.1 Downy Mildew

The symptoms are angular lesions that are limited by the leaf veins. Early lesions are light green in appearance and become chlorotic and finally necrotic as host plant cells die.

Management The crop should be sprayed with mancozeb @ 2 g/litre of water twice at 10 days interval.

11.8.8.2 Powdery Mildew (*Erysiphe polygoni* D.C.)

The white powdery spots are appearance on upper surface of leaves. Affected leaves lose their chlorophyll and dry up.

Management: Spray the crop with carbendazim @ 2g/litre of water. Fungicidal spray of kerathane is also effective. It can be controlled by avoid the late sown crop, foliar spray of onion extract @ 5% and spray the crop with wettable sulphur @ 1.0 kg/ha.

11.8.8.3 Wilt (*Fusarium oxysporum*)

It is a serious disease in vegetable crops and affected plants show yellowing, drooping of leaves, thereafter drying and finally plant dies.

Management: It can be effectively controlled through integrated approaches of soil solarization, healthy seed, crop rotation and treatment of seeds with *Trichoderma viridae* @ 4 g/kg seed subsequently use of *Trichoderma viridae* @ 4 kg/ha + 80 kg FYM in soil or treatment of seed with *Pseudomonas fluorescens* @ 10 g/kg seed subsequently soil application of *Pseudomonas fluorescens* @ 5 kg/ha + 100 kg FYM or treatment of seed with *Pseudomonas fluorescens* @ 10 g/kg seed subsequently use of neem cake @ 150 kg/ha in soil.

11.8.8.4 Mosaic Virus

Firstly, the greenish yellow to dark green mottling is developed in youngest leaves. Leaves are often stunted, crinkled, distorted, and curled downward. In severe cases, all except the youngest leaves at the runner tips (rosettes) may rapidly turn brown and die.

Management: Use insect proof nethouse for seedlings growing, eradication of early infected plants, two border rows with sorghum, pearl millet or maize can reduce the disease spread. Alternate hosts should remove. Spray seedlings with acephate (0.15%) prior to transplanting. Spray the crop with imidacloprid @ 1ml/3 litre of water or acephate (0.15%) at 15 days intervals after transplanting till flowering stage. Chemical spray followed by neem seed kernel extract (2%) is also effective in rotation with insecticides.

11.8.8.5 Some Approaches for IDM Practices

11.8.8.5.1 Soil Solarization

It is a technique of hydro/thermal soil heating in which moist soil is covered with polyethylene sheet for the period of 4–6 weeks during summer months. It should be adopted mandatory in protected cultivation.

11.8.8.5.2 Resistant or Tolerant Cultivars

Use of resistant varieties for plant disease management is considered as a 'painless method' because it does not require any extra expenditure for disease control. It can avoid the health hazards caused by the indiscriminate application of pesticides. It also checks environmental pollution caused by these agrichemicals. Tomato hybrid Arka Rakshak is resistant/tolerant to bacterial wilt and early blight.

11.8.9 Pest Management

The natural enemies, which manage the pest outsides conditions, are not available under protected conditions. Hence, in the greenhouse pest situations often develop quickly and some time may be more serious in comparison to outsides conditions. In India, the insect-pests scenario under protected conditions has been presented in Table 11.8.

11.8.9.1 Fruit Fly

Females fly lay eggs below epidermis of young fruits. Later on maggots feed on pulp afterward fruits starts rotting and get drop.

Management Neem oil @ 3.0% should be applied. Pheromone traps are hanged to trap and kill the insects during adults starts appear. Collect all infested fruits regularly and destroy them. Crop should be sprayed with fenthion (0.05%) or fenitrothion (0.05%).

11.8.9.2 White Fly

It acts as a vector and transmits the leaf curl virus from infected plant to healthy plant.

Management Seeds should be treated with imadacloprid 70 WS @ 3–5 g/kg of seed and nursery should be raised under insect proof nethouse (50-60mesh).

Table 11.8 Pests scenario under protected conditions

Category	Pest	Keeper
Aphids	*Aphis gossypii*	Capsicum
	Myzus persicae	Capsicum
Caterpillars	*Helicoverpa armigera*	Capsicum, tomato
	Spodoptera litura	Tomato, capsicum, cucumber
Leaf-miner	*Liriomyza trifolii*	Tomato, cucumber
Mites	*Polyphagotarsonemus latus (yellow mite)*	Capsicum
	Tetranychus urticae (spider mite)	Tomato, capsicum, cucumber
	Tetranychus neocalidonicus	Cucumber
White flies	*Bemisia tabaci*	Capsicum
	Trialeurodes vaporariorum	Tomato, cucumber, capsicum

Modified from Saha et al. (2015)

11.8.9.3 Nematode

It is a serious pest in protected cultivation. In case of heavy infestation, the roots shows welling's close to the root tips (Fig. 11.20).

Management Nematode can be managed through the soil solarization and application of multiplex niyantran (*Poaecilomyces lilacinus*), a promising bio-control agent, @ 5 kg/ha + 250 kg FYM or neemcake @ 250–400 kg/ha.

AICRP on Nematodes has recommended that the fumigation with metham sodium @ 30 ml/m^2 with polythene mulch for 15 days + neem cake 200 g/m^2 + *Pseudomonas fluorescens* 50 g/m^2 mixed 15 days prior to transplanting tomato reduced nematodes by 80%, gall index by 51% and increased yield by 49% in polyhouses. Combined application of neem cake @ 200 g/m^2 and *Paecilomyces lilacinus* @ 50 g/m^2 or *Trichoderma harzianum* @ 50 g/m^2 at 15 days before transplanting is an effective bio-management practice for root-knot nematode infesting cucumber under polyhouse conditions.

11.8.9.4 Cutworms

The caterpillar of this insect attacks on young plant/seedlings and cut them at ground level.

Management Soil should be drenched with chlorpyriphos 20% EC (1–2 ml/litre of water) at the time of preparation of field for transplanting.

11.8.9.5 Some Approaches for IPM Practices

11.8.9.5.1 Sticky Traps

Sticky traps can be incorporated in management practices of various pests like whiteflies, leaf miners and thrips. Blue and yellow coloured traps are used to monitor thrips and white fly - leaf minor infestation, respectively.

Fig. 11.20 Nematode infected cucumber (Source: AICRP on Nematodes)

11.8.9.5.2 Pheromones

There are three different types of pheromones *viz.*, aggregation (attract many individuals together), sex pheromones (attract one sex of a species to the other sex), and trail pheromones (attract pests into traps and interrupt mating). Pheromone is logical chemicals produced by animals to signal each other.

11.8.9.5.3 Bio-Pesticides

Bio-pesticides are products of microbial and plant origin based pesticides. Aphids can be controlled by clip off the heavily infested parts, pressurized water spray and foliar spray the crop with extract of neem seed kernel @ 3% or neem oil @ 1% or *verticillium lecanii* @ 3 ml/litre of water, which may be repeated three times at 15 days interval. Leaf-eating caterpillars can be controlled by foliar spray of *Bacillus thuringienses* (B.T.) @ 1 kg/ha or NPV @ 250 LE/ha. Mites (*Petrobia latens*) can be controlled by foliar spray of neem oil @ 1%. All soil insects can be managed by soil application of *Beauveria bassiana* @ 4 kg/ha + 80 kg FYM or neem cake @ 400 kg/ha.

11.8.9.5.4 Biological Control

In this method, natural enemies like predators, parasitoids and pathogens managed pest population. In tomato crop, egg parasitoid i.e., *Trichogramma chilonis* @ 50,000/ha in 6 releases starting from 45 days after transplanting or larval parasitoid i.e. *Campoletis chlorideae* @ 15,000 adults/ha or nuclear polyhedrosis virus of *H. armigera* (HaNPV) @ 250–300 larval equivalent (LE)/ha can be used for effective management of tomato fruit borer.

11.8.9.5.5 Trap Cropping

It is a specific companion planting approach of vegetative diversification. This technique has an enormous capability to attract and safeguard natural enemies. In tomato crop, the marigold and arugula are used as trap crop for management of *Helicoverpa armigera* and *Lygus* spp., respectively. The sunflower and sorghum crops can be grown as trap crop in capsicum for management of *Halyomorpha halys* and brown marmorated stink bug (Panwar et al. 2021).

11.8.10 CO_2 Enrichment

Periodical or continuous enhancement of CO_2 inside the greenhouses can lead to quality of produce and production (Shanchez-Guerrero et al. 2005). The CO_2 concentration may drops below the atmospheric level inside the greenhouses if CO_2 utilization through photosynthesis is greater than the provide amount by the vents. Improved ventilation system or CO_2 enhancement is possible solution to overcome such type of problem (Kittas et al. 2013).

11.8.11 Fruit Setting

Productivity of vegetable crops under protected conditions is totally associated with the realization of fruit-set and fruit-set is related to pollination. Due to high humidity and limited air movement under protected structures, the assistance is needed for pollination and it may be boosted by bumblebees or mechanical vibration. Mechanical vibration is a virtuous exercise; however, it is tedious. Bumble bees are the most efficient and are used in greenhouses worldwide. The advantages of pollination through bumble bees are depletion in labour expenses, greater fruit quality, higher productivity, and healthy product (Kittas et al. 2013).

11.9 Constraints in Protected Cultivation Systems

Though, numerous boom stories of protected cultivation technology have been described from various scientists and sectors as well. However, from the view point of Singh 2019, the constraints in protected cultivation systems are as follows:

1. Protected cultivation is correlated with very high beginning value.
2. Requirement of regular and uninterrupted power supply for climate-controlled greenhouses.
3. Being expensive, the cladding material is not readily available in needed standard.
4. Non-availability of tools and implements for facilitating crop production operations under greenhouse.
5. Region specific technologies are needed in absence of climate-controlled greenhouses.
6. Inadequate or scanty breeding work programs for evolution of appropriate hybrids/varieties of vegetable crops for greenhouse cultivation.
7. There is a problem of pollination in greenhouse crops in tomato and cucurbits.
8. Increasing threat of bio-stresses in greenhouse cultivation particularly root-knot nematodes and *Fusarium* wilt remain unsolved.

11.10 Opportunities in Protected Cultivation Technology

Singh 2019; Tomar et al. 2019; Singh and Choudhary 2019 have suggested the opportunities in protected cultivation technology as follow:

1. Adequate land, water and human resources are available.
2. Hard working and skilled labour force is available.
3. Less prone to main bacterial, viral and fungal diseases in arid and semi-arid regions, it becomes supportive for protected cultivation of vegetable crops.
4. An excellent network of road, rail and air is available for fast transportation of the commodities. Dedicated freight corridor will also be helpful.
5. Research and extension institutions are available to back up the protected vegetables research and development programme.
6. Protected cultivation offers very congenial environment for producing healthy, virus free, and genetically pure hybrid seed with higher seed yield per unit area.
7. Protected cultivation technology can attract to the youth towards farming sector.

11.11 Conclusion

The challenge of high demand from consumers for a year-round supply of quality vegetables can be met by growing under protected conditions or growing in two or more locations with complementary harvesting periods. Low tunnel, hydroponics, aeroponics, plug tray seedlings raising, and soilless cultivation are some important protected cultivation technologies. It offers several advantages to produce vegetables of high quality and yields, thus using the land and other resources more efficiently. Though, numerous boom stories of protected cultivation technology have been described from various scientists and sectors as well. However, very high beginning value, irregular power supply, pollination in greenhouse crops, and inadequate availability of appropriate hybrids/varieties of vegetable crops for greenhouse cultivation are also some constraints. On the other hand, the adequate land, water and human resources, hardworking and skilled labour force, excellent transportation facility, and available research and extension institutions are the opportunities in protected cultivation technology.

References

Anonymous (2016) Indian Horticulture (January – February), ICAR, Krishi Anusandhan Bhavan-I, Pusa, New Delhi, 61(1): 84

Anonymous (2019) Annual Report 2018–19. ICAR-Indian Institute of Vegetable Research, Varanasi, p 188

Anonymous (2021) Vegetable varieties for greenhouse vegetable production. Aggie Horticulture. Available from: https://aggie-horticulture.tamu.edu/greenhouse/hydroponics/varieties.html

Castilla N, Hernandez J (2007) Greenhouse technological packages for high-quality crop production. Acta Hortic 761:285–297

Castilla N, Montero JI (2008) Environmental control and crop production in Mediterranean greenhouses. Acta Hortic 797:25–36. https://doi.org/10.17660/ActaHortic.2008.797.1

DAC&FW (2018) Horticultural statistics at a glance 2018, horticulture statistics division, Department of Agriculture, Cooperation & Farmers' welfare (DAC&FW), Ministry of Agriculture & Farmers' Welfare, Government of India, New Delhi. p 458

Dhaliwal MS, Devi R, Rajinder S (2016) Vegetable varieties recommended for cultivation in Punjab. Punjab Agricultural University, Ludhiana, p 47. isbn 978-93-83092-52-9

Gruda N, Qaryouti MM, Cherubino L (2013) Growing media. In: Good agricultural practices for greenhouse vegetable crops. Food and Agriculture Organization of the United Nations, Rome, pp 271–301

Hochmuth RC (2012) Greenhouse cucumber production. Florida greenhouse vegetable production. Handbook, vol. 3, Available from: https://edis.ifas.ufl.edu/publication/CV268

Jat GS, Singh B, Tomar BS, Ram H (2015) Comparative performance of growth, flowering behaviour and fruit development traits in parental lines of bitter gourd (Momordica charantia L.) cv. Pusa Hybrid-1 Seed Res 42(2):139–145

Jat GS, Singh B, Tomar BS, Singh J, Ram H, Kumar M (2016) Seed yield and quality as influenced by growing conditions in hybrid seed production of bitter gourd (Momordica charantia L.) cv. Pusa Hybrid-1. J Appl Nat Sci 8(4):2111–2115

Jat GS, Singh B, Tomar BS, Muthukumar P, Kumar M (2017) Hybrid seed production of bitter gourd is a remunerative venture. Indian Hortic (March-April) 62(2):34–37

Jat GS, Tomar BS,Lyngdoh YA and Saha P (2019) Vertical farming of leafy vegetables under protected conditions. Chapter 16: 82–86. In: Prakash HG, Singh DP, Rajiv, Singh BK, Jaiswal VB, Khan K and Kumari M (eds) Skill development course on protected cultivation of vegetable crops. Vedant Publications, Lucknow, p. 119. isbn: 978-9380135-53-3

Kacira M (2011) Greenhouse production in US: status, challenges, and opportunities. Presented at CIGR- 2011 conference on sustainable bio-production WEF 2011, September 19-23 at Tower Hall Funabori, Tokyo, Japan

Kashyap P, Bhagat K, Singh SK, Samant D, Meena VS, Panwar AS, Prusty AK (2019) Entrepreneurship development through protected horticulture. Chapter 8: 97–112. In: Kashyap P, Prusty AK, Panwar AS, Kumar S, Punia P, Ravisankar N, Kumar V (eds) Agri-entrepreneurship: challenges and opportunities. Today and Tomorrow's Printers and Publishers, New Delhi, p 345. isbn 81-7019- (India)

Kittas C, Katsoulas N, Bartzanas T, Bakker S (2013) Greenhouse climate control and energyuse. Chapter 4: 63-95. In: Baudoin W, Womdim RN, Lutaladio NB, Hodder A, Castilla N, Leonardi C, Pascale SD, Qaryouti M (eds) Good agricultural practices for greenhouse vegetable crops: principles for Mediterranean climate areas. Food and Agriculture Organization of the United Nations, Rome, p 616. isbn 978-92-5-107649-1

Krishna, Hare, Bahadur A, Singh J, Yadava RB, Prasad RN and Singh M (2020) Two-stem training system: bountiful harvest in polyhouse tomato. In: Mohapatra T, Alagusundaram K, Choudhary SK, Jena JK, Tripathi BN, Agrawal RC, Singh AK, Singh AK, Sharma TR, Singh SK (eds) ICAR Newsletter, Vol 26(3), July–September, Indian Council of Agricultural Research, Krishi Bhawan, New Delhi, India, p. 32 issn: 2394–3289

La Malfa G, Leonardi C (2001) Crop practices and techniques: trends and needs. Acta Hort 559:31–42

Malik G, Lubna M, Nabi SU, Anil S, Singh DB (2017) Production technology of cherry tomato in Kashmir. ICAR-Central Institute of Temperate Horticulture, Srinagar, p 12

Pachiyappan P, Kumar P, Reddy KV, Kumar KNR, Konduru S, Paramesh V, Rajanna GA, Shankarappa SK, Jaganathan D, Immanuel S, Kamble AL, Selvakumar R, Immanuelraj KT, Manogaran BR, Perumal A, Maruthanayagam U, Niranjan S (2022) Protected cultivation of horticultural crops as a livelihood opportunity in Western India: an economic assessment. Sustainability 14(12):1–17. https://doi.org/10.3390/su14127430

Panwar L, Suman D, Singh Y (2021) Insect pest management in vegetable crops through trap cropping: review. Indian J Agric Sci 91(10):174–210

Patil SS, Biradar MS, Mantur SM, Patil L and Kolur S (2019) Canopy managementin vegetable cropsunder protected cultivation. Chapter 20: 103–104. In: Prakash HG, Singh DP, Rajiv, Singh BK, Jaiswal VB, Khan K and Kumari M (eds) Skill development course on protected cultivation of vegetable crops. Vedant Publications, Lucknow, p. 119. isbn: 978-9380135-53-3

Pattnaik RK, Mohanty S (2021) Protected cultivation: importance, scope, and status. Food Sci Rep 2(3):19–21. issn 2582-5437

Prakash P, Kumar P, Kar A, Singh AK, Ankukkani P (2022) Economic analysis of carnation (Dianthus caryophyllus) under protected cultivation in Maharashtra. Ind J Agric Sci 92(4):460–463. https://doi.org/10.56093/ijas.v92i4.123966

Rajiv, Tomar S (2022) Integrated use of organic and inorganic sources of nutrients in french bean (Phaseolus vulgaris). Indian J Agric Sci 92(4):17–21

Saha T, Nithya C, Ansar M, Kuma S (2015) Integrated approaches for the management of insect pest under protected condition. Rashtriya Krishi 10(1):23–25. e-issn 2321–7987

Sanchez-Guerrero MC, Lorenzo P, Medrano E, Castilla N, Soriano T, Baille A (2005) Effect of variable CO_2 enrichment on greenhouse production in mild winter climates. Agric For Meteorol 132:244–252

Sehrawat SK, Malik A (2015) Artificial media. In: Singh A, Dhankhar SS, Dahiya KK (eds) Training manual on protected cultivation of horticultural crops. CCS Haryana Agricultural University, Hisar, pp 49–54

Singh B (2018) Elements and strategic measures in maximizing protected cultivation of vegetables towards sustainable horticulture development in asian countries. Chapter 8: 132–141. In: Nath Prem (compiled by) Sustainable horticulture development and nutrition security (Vol. III: Food and nutrition security). PNASF, Bengaluru and Scientific Publishers India (SP), Jodhpur, p 659. isbn 978-93-8665-23-86

Singh B (2019) Protected cultivation of vegetables in India: An overview. Chapter 1: 1–8. In: Prakash HG, Singh DP, Rajiv, Singh BK, Jaiswal VB, Khan K and Kumari M (eds) Skill development course on protected cultivation of vegetable crops. Vedant Publications, Lucknow, p. 119. isbn: 978-9380135-53-3

Singh DK (2019) Gene exploitation for polyhouse vegetable crop improvement in India. In: Misra AK, Awasthi OP, Shukla SK, Dwivedi DH, Gupta S, Trivedi AK, Sharma D (eds) Proceedings of the progressive horticulture conclave (PHC-2019) on futuristic technologies in horticulture, pp 85–87

Singh, Balraj and Choudhary, Santosh (2019) Prospects of protected vegetable cultivation in arid and semi-arid regions of India. Chapter 3: 18–23. In: Prakash HG, Singh DP, Rajiv, Singh BK, Jaiswal VB, Khan K and Kumari M (eds) Skill development course on protected cultivation of vegetable crops. Vedant Publications, Lucknow, p. 119. isbn: 978-9380135-53-3

Singh B, Solanki RK (2015) Protected cultivation technologies for vegetable cultivation under changing climatic conditions. Chapter 7: 105-115. In: Chaudhary ML, Patel VB, Siddiqui MW, Mahdi SS (eds) Climate Dynamics in Horticultural Science (Vol. I). Apple Academic Press. isbn: 9780429173851

Singh B, Tomar BS (2015) Vegetable seed production under protected and open field conditions in India: a review. Indian J Agric Sci 85(10):3–11

Singh B, Kumar M, Yadav HL (2005) Plug-tray nursery raising technology for vegetables. Ind Hortic 49(4):10–12

Singh VK, Ranjan S, Singh A, Soni MK (2015) ICAR- Central Institute for Subtropical Horticulture, Lucknow, Uttar Pradesh, India

Singh P, Rajiv, Upadhyay A, Maurya PK, Pandey S (2022) Effect of foliar feeding of micronutrients on growth and yield of bitter gourd (Momordica charantia). Curr Hortic 10(1):66–68. https://doi.org/10.5958/2455-7560.2022.00016.4

Sirohi NPS (2003) Protective structures for vegetable cultivation. Chapter 1: 1-13. In: Singh B, Sirohi NPS, Neubauer E, Kumar M, Vidhu KP (eds) Intensive cultivation of vegetables under different growing conditions. Indian Agricultural Research Institute, New Delhi, p 141

Solankey SS, Kumari M, Kumar M, Silvana N (2021) The role of research for vegetable production under a changing climate future trends and goals. In: Solankey SS et al (eds) Advances in research on vegetable production under a changing climate Vol. 1. Springer, Cham., ISBN 978-3-030-63497-1, pp 1–12

Tomar BS, Jogendra S, Jat GS (2019) Low tunnel technology for off-season cucurbits cultivation. Chapter 4: 24–26. In: Prakash HG, Singh DP, Rajiv, Singh BK, Jaiswal VB, Khan K, Kumari M (eds) Skill development course on protected cultivation of vegetable crops. Vedant Publications, Lucknow, p 119. isbn: 978-9380135-53-3

Chapter 12
Improvement of Vegetables Through Grafting in Changing Climate Scenario

Pankaj Kumar Ray, Hemant Kumar Singh, Shashank Shekhar Solankey, Raj Narain Singh, and Anjani Kumar

Abstract Grafting in vegetables is an ancient technique to improve the yield through climate friendly practices. This technique was introduced in USA and become very popular in organic vegetables cultivation. Vegetable grafting is popular practice in many European and North American countries, Japan, Korea, and China. Because vegetable crops are so easy to grow, they are very sensitive to climate change. Drought, floods, and salt caused by temperature and precipitation shifts have severely impacted vegetable crop productivity. Vegetable crop cultivation is difficult to say the least in the face of a rapidly changing environment. Although grafting has traditionally been employed on woody and perennial fruit trees, it is increasingly being used to herbaceous plants like vegetables and flowers. In East Asia, grafting is used as a unique method for dealing with the many threats to intensive vegetable production. The genetic and physiological complexity of abiotic stress restricts the creation of tolerable cultivars at the commercial level. In addition, many vegetable crops lack resistant crossover suitable wild resistant sources, with the exception of a few. In this case, the surgical procedure of grafting a plant has been employed successfully to reduce a wide range of different environmental

P. K. Ray (✉)
Krishi Vigyan Kendra (Bihar Agricultural University, Sabour, Bhagalpur), Saharsa, Bihar, India

H. K. Singh
Krishi Vigyan Kendra (Bihar Agricultural University, Sabour, Bhagalpur), Kishanganj, Bihar, India

S. S. Solankey
Department of Vegetables & Floriculture, Agricultural Research Institute, Patna, Bihar, India

R. N. Singh
Directorate of Extension Education, Bihar Agricultural University, Sabour, Bhagalpur, Bihar, India

A. Kumar
ICAR-Agricultural Technology Application Research Institute (ATARI), Patna, Bihar, India

S. S. Solankey, M. Kumari (eds.), *Advances in Research on Vegetable Production Under a Changing Climate Vol. 2*, Advances in Olericulture,
https://doi.org/10.1007/978-3-031-20840-9_12

stress. In this chapter, we illustrate the efficacy of this plant propagation method through a review of research results on vegetable grafting.

Keywords Climate friendly · Abiotic stress · Grafting · Resistant sources · Vegetables

12.1 Introduction

Vegetable seedlings with grafts are a unique horticultural technique that has been used for a long time. In the latter half of the twentieth century, this method along with enhanced grafting procedures ideal for commercial production and productivity of grafted vegetable seedlings—was brought to Europe and other nations. Vegetable grafting is taking a vegetable plant‘s stem at the seedling stage and affixing it to the seedling's rootstock of another vegetable plant, such as a wild brinjal or a pumpkin. The grafted seedling is developed under regulated climatic circumstances once the attachment is accomplished, and then it may be planted in the field. In general, grafting onto certain rootstocks offers resistance to biotic and abiotic stress tolerance, crop development, yield, and quality, as well as nematodes and illnesses carried by the soil. Grafting is a useful tool to utilize in conjunction with more environmentally friendly crop production methods, such as lower rates and general usage of soil fumigants in many other nations. Especially the production of vegetables is sensitive to a range of abiotic environmental stresses, such drought, salt, floods and temperatures (Moretti et al. 2010). However, the insect pest and disease burden stated above might be compounded by an increase. However, a rise in mean warmth and humidity in a changeable climatic situation might worsen the Climate change may have a direct or indirect influence on food safety.

The grafting method was brought from Europe to North America. For greenhouse and organic farms, grafting techniques for fruits and vegetables are a desirable strategy. Grafting is a distinctive horticultural technique used for many years in East Asia on herbaceous seedlings to combat problems with biotic and abiotic stressors. The oldest written accounts of self-grafting to grow larger watermelon fruits date back to China in the fifth century and Korea in the seventeenth century. A watermelon farmer in Japan developed the first record of inter-specific grafting in watermelon using a squash rootstock (*Cucurbita moschata* Duch.) to boost productivity and prevent pests and illnesses. In order to produce grafted seedlings for commercial vegetable production in Japan in the 1930s, watermelon was grafted on *Lagenaria siceraria* (Mol.) Standl (Oda. 2002). Grafting of watermelons for drought tolerance is a potential strategy under limited water condition for taking higher yield as well as quality fruits (Adarsh et al. 2020). *Cucumis sativus* L. grafting was began in the late 1920s, but it wasn't successful until the 1960s (Sakata et al. 2008). In the 1950s (Oda. 1999), grafting was first used on eggplant (*Solanum melongena* L.), a Solanaceae plant, on scarlet eggplant (*Solanum integrifolium* Poir), and it was later commercialized on tomatoes (*Lycopersicon esculentum* Mill.) in the 1960s (Lee and Oda 2003). Conventional procedures for breeding which take time and advancement are gradual. In order to deal with different environmental stressors, grafting is necessary for a treatment based on the rootstock and scion compatibilities (Nilsen et al. 2014). Grafting is now commercially utilized in the vegetables such as tomato, eggplant, and cucurbits for control of soil borne diseases (Solankey et al., 2019). Grafting is a reciprocal procedure that is ecologically favourable, efficient, quick and integrative, affects both spring and rootstock. For watermelon this approach was originally used in Japan. The findings of the grafting were published in the first scientific research. The results of grafting of water melon plants to potatoes in the first scientific journal were reported to minimize fusarium fusion. In the 1920s, however, grafting was mostly considered a woody permanent fruit technique and was not employed until last time in cucurbits for the treatment of diseases of soil transmission, notably fusarium (Bhatt et al. 2015).

Many recognized researchers are currently employing this grafting approach to enhance the environmental stress tolerance in solanaceous and cucurbitaceous vegetables (Johnson et al. 2014). However, only tomatoes and potatoes are the most recent breeding technology and completely forgotten are other crops (Kato et al. 2001a). The importance of plant grafting for the control of environmental stressors, including floods, drought, heat and salinity, under a changing climate scenario will be discussed in this review paper.

12.2 Environmental Stress

The main limitations to the sustainable development of agricultural and horticultural products are the different environmental stresses, high temperatures, drought and flooding (because of irregular precipitation), and salt (Mittler. 2006). The intensity of this stress on plants varies on the stage of growth, kind and duration of stress exposure. Many researches describe the use of grafting procedures to enhance tolerance to a wide range of environmental stress on different vegetable crops, and we have compiled the key results in this chapter.

12.2.1 Flooding

Increasing flood tolerance has shown to be the use of numerous research organizations in various grafting vegetables. Tomato is a model of a flood-prone worldwide crop. Bhatt et al. (2015) used interspecies tomato greasing to improve flood tolerance. Four eggplant plants like BPLH-1, Neelkanth, Mattu Gulla, Arka Keshav have been grafted into Arka Rakshak for commercial tomatoes. Research has shown that the grateful returns of flooded and non-inundated circumstances were considerably influenced. The unexpected and self-grafted plant was murdered after five days of inundation. After five days of inundation, Arka Rakshak/Arka Keshav and BPLH-1 perished, but two combinations performed well. This study finds that eggplants withstand flooded terrain and may provide an enough radical fat material to increase the resistance to tomato flood. The World Vegetable Center (AVRDC) was also advised to increase flood tolerance for EG195 and EG203 eggplant additions to tomatoes (Black et al. 2003).

Tomato $F_{1's}$ (Arka Rakshak and Arka Samrat) in eggplant rootstock (IC-111056, IC-354557, IC-374873 and IC-2) have been found in recent research. Observations indicated that no leaf or plant withering symptoms and reduced chlorophyll decrease occur at all stages of plant growth. In contrast, following 96 hours of water stress, ungrafted plants had a 41–100 per cent decrease in chlorophyll levels and perished 4–7 days later. In 7–10 days after exposure, the flushing plants recovered entirely from flood stress. Thus, the water logging tolerance for grafting tomatoes increased for the aubergine rootstocks of IC-354557 and IC-111056 at 72–96 hours (Bahadur et al. 2015).

In addition, wild eggplant species are also employed in grease tomato rootstocks (Petran and Hoover 2014). Yetisir et al. (2006) showed that bitter melon plants grew on rootstocks of Luffa, in flooded circumstances, were more successful than ungrafted plants. The trade variety 'Crimson Tide' was also grafted in the watermelon to *Lagenaria siceraria* SKP (landrace), with chlorosis symptoms occurring both on grafted and ungrafted plants, but the symptoms were less pronounced in flooding plants.

However, non-grafted plants were less dry than grafted plants with high humidity; in addition, the fresh weight loss of ungrafted plants was around 180 per cent and of grafted plants was 50 per cent. In non-grafted and grafted plants respectively, dry weight was reduced by around 230 and 80 per cent. These results show the development of adventitious roots and aerenchyma tissue in grafted plants after 3 days, but for non-grafted plants under flood circumstances no such observations were recorded. Kato et al. (2001b) formulated a grafting trial in water-logged cucumber in accordance with research by Keatinge et al. (2014) and established that leaf chlorophyll content was increased by grafting on squash rootstocks. The rootstock suggested lately by AVRDC in East Asia for flood tolerance for tomatoes is VI045276, VI046103, VI034845, VI046104 and VI046101 for eggplants, and for rootstocks of East Asia is VI046378 (Peng et al. 2013). Rootstocks for vegetables and other economically significant crops can be placed in flood conditions. Table 12.1 provides other examples.

Table 12.1 Top-performing combinations of rootstock and scion in vegetables under induced flood stress

S. No.	Rootstock	Scion	References
1.	*S. melongena* cv. Arka Keshav	*S. lycopersicum* cv. Arka Rakshak	Bhatt et al. (2015)
2.	*S. melongena* cv. IC- 374873	*S. lycopersicum* cv. Arka Samrat	Bahadur et al. (2015)
3.	*C. maxima* x *C. moschata*	*C. sativus*	Kato et al. (2001a, b)
4.	*L. siceraria*	*C. lanatus*	Yetisir et al. (2006)

12.2.2 Drought

Drought is yet another significant challenge of water stress for sustainable worldwide vegetable production, caused by deficiencies in water supplies. Although breeding and biotechnology interventions led to new drought-tolerant crop types, the progress was mostly restricted to cereal crops. Certainly, climate change, affecting agricultural production, in particular the availability of vegetables and overall crop failures is greatly impacted by the availability of water. The lower irrigation water supply might be the explanation for decreased precipitation coupled with higher median air temperature. In drought circumstances, an upturn in evapotranspiration is also predicted as vegetables contain around 90% water (Thomas et al. 2011).

Water scarcity and drought stress are significant environmental stressors under the global climate change scenario (Schwarz et al. 2010). Grafting may thus be employed to reduce losses in production and to improve the efficiency of water usage (WEU) in a water shortage. The grafting of sensitive commercial cultivars on the rootstocks can do this by decreasing the effect of moisture stress on the shoot. In Europe, tomato hybrids, particularly *solanum* spp., and interspecific hybrids are common rootstocks for eggplants. Simile grafting on pumpkin (*Cucurbita moschata*), rootstocks, of watermelon (*Citrullus lanatus*) contributes to the water stress of watermelon shooting (King et al. 2010).

Tolerant plant grafts for vegetable plants were carried out to regulate or improve drought tolerance, in particular solanaceous and cucurbitaceous vegetables (Sanchez-Rodriguez et al. 2016). At genetic level, microRNAs (miRNAs) monitored plant growth and growth and reacted specifically to different environmental stresses. Li et al. (2016) reports that scion of squash grafted from the bottle siceraria or from the rootstock (*Cucurbita maxima* the *Cucurbita moschata*) caused a change of over 40 mil RNI expression, according to Li et al., 2016 in a spring graph of watermelon (*Citrullus lanatus*). In addition, in recent research studies, molecular processes were investigated by grafting cucumber plants on pumpkin rootstock (*Cuurebita moschata*) in 17 chosen miRNAs in the grafted plants under drought stress. As a consequence, mini-watermelon cv. Ingrid was ungrafted or grafted using a rootstock called 'PS 1313' (*Cucurbita maxima* + *Cucurbita moschata*) and findings showed that the yield, nutritional and fruit quality associated metrics of grafted plants are greater than those of non-grafted plants. In the case of gas exchange and leaf water relations no significant differences were seen between grafted and non-grafted plants. Although moisture stress sensitivity of grafted plants and ungrafted plants was equal, the greater marketable yields were reported with grafting. The findings of this study showed the benefits of rootstock 'PS 1313.' Under particular in drought circumstances, the usage of grafted rootstock plants was advised to alleviate drought stress. Another study comparing moisture stress-tolerant rootstocks for watermelon has found that wax gourds are a superior rootstock than dry gourds (Muneer et al. 2016). Other crucial physiological responses, such as changes in stomatal behaviour that have enhanced the WUE and

Table 12.2 Examples of high-performing pairings of rootstock and scion in plants under caused dry stress

S. No.	Rootstock	Scion	References
1.	*S. lycopersicum* L. (cv. Faridah)	*S. lycopersicum* L. (cv. Unifort)	Ibrahim et al. (2014)
2.	*S. lycopersicum* L. (cv. Jjak Kkung)	*S. lycopersicum* L. (cv. BHN 602)	Nilsen et al. (2014)
3.	*S. lycopersicum* L. (cv. Unifort)	*S. lycopersicum* L. (cv. Farida)	Wahb-Allah (2014)
4.	*C. annuum* (cv. Atlante)	*C. annuum* (cv. Herminio)	Lopez-Marin et al. (2013)
5.	*S. lycopersicum* (cv. Beaufort)	*S. lycopersicum* (cv. Amelia)	Chaudhari et al. (2017)
6.	*C. maxima* x *C. moschata*	*C. lanatus*	Rouphael et al. (2008)
7.	*C. annuum* L. (cv. Verset)	*C. annuum* L. (cv. Atlante)	Penella et al. (2014)
8.	*S. lycopersicum* (cv. Zarina)	*S. lycopersicum* (cv. Josefina)	Sanchez-Rodriguez et al. (2016)

photosynthetic activity have been recorded in the stress-prone environment, apart from those instances of the molecular physiological response of grassed vegetables to drought stress. For sweet pepper grafted on the rootstock of Creonte, the overall crop output and marketable fruit growth rose by 30 and 50 per cent correspondingly under the Mediterranean climate. In addition, the rootstock kept the photosynthetic leaf activity 30–60% higher and showed much greater WUE (10%). While, Amelia enhanced photosynthesis and stomach condition for tomato cultivar. In addition, the rootstock kept the photosynthetic leaf activity 30–60% higher and showed much greater WUE (10%). While enhanced photosynthesis and stomatal conduct for tomato cultivar Amelia were seen when Maxifort cultivar was utilized as a rootstock (Chaudhari et al. 2017). Thus, grafting onto the robust and tolerant and resistant rootstock, which are susceptible to drought, of commercial plants from plants in locations that are susceptible to water stress is a feasible method. The benefits of grafting in previous research were maintaining a high water and nutrient absorption capacity together with the larger WUE under drought. Table 12.2 provides further instances of research of vegetable grafting under severe stress.

12.2.3 *Thermal Stress*

Extreme temperatures can lead to loss of vegetable output by stimulating fusion and necrosis, delaying the development of truss and affecting the time of fruit maturing. Grafting can be utilized to protect plant against heat shock and to improve production performance in plants owing to the accompanying changes in physiology in the parent plant (Rivero et al. 2003). Plants are very susceptible to low as well as high temperatures. High temperatures are generally found in rainforests, whereas the temperatures or low temperatures of vegetables, particularly for tomato, squash, peat and water melon, occur in the winter, spring and fall seasons in temperate and subtropical areas. Due to high temperature the plant's physiological, morphological, molecular and biochemical response to change significantly affects on plant growth, development and economic yield of legume (Singh et al. 2021). In addition, cold stress has an effect on germination, plant growth and plant growth leading to a loss in economic returns (Venema et al. 2008). High temperatures, for example, in tomatoes cause considerable loss of agricultural output because of reduced fruit, small size of fruit and reduced quality of fruits. Similarly, the low temperature causes permanent malfunction, cell death and eventually plant death, depending upon the severity and duration of exposure. Temperature is also impacted by early fruit production and quality attributes. Species have been developed through breeding and biotechnology with increased thermal stress tolerance, but with little market success as such stressors are genetically complicated and plant specific. In vegetable crops, the genetic basis of environmental stresses tolerance such as heat, cold, drought, flood, salinity is necessity for development of superior biotype (Solankey et al. 2021). Therefore, grafting on chosen low−/high-temperature resistant rootstocks existing commercial elite varieties might be considered as a feasible plant multiplication strategy as a quick and efficient alternative. One of the best features for greenhouse vegetable production in winter or at an early spring is the low-temperature tolerance of the rootstock. Den Nijs (1980) conducted a low temperature grafting experiment with four advanced breeding lines and the rootstock of *Cucurbita ficifolia* in the Netherlands. This test showed that the survival, lifespan and fruit quality of grafted plants were excellent compared to those not grafted in low-temperature environments. Cucumber plants greased from the rootstock, *Cucurbita ficifolia*, increased transhexadecenoic Acid in phosphatidyl glyc. Horvath et al. (1983) continued this work by observing the increase in trans-hexadecenic acid in phosphatidylglycerol, which contributes to the low-temperature tolerance in plants in cucumber plants grafted onto *Cucurbita ficifolia* rootstock. In Morocco, *Cucurbita ficifolia* is also the favorite cucumber rootstock and is a great rootstock for low tolerance for soil temperature, particularly in winter spring production. During its development and growth, the grown tomato is very sensitive to the sub-optimal and cold temperature. The grafting of high yielding commercial and sensitive cultivars on tolerant rootstock is regarded as a fast approach to enhance resistance to low-temperature tomato stress. In 1777, the *Solanum habrochaites* (LA) adhesives of the wild tomato were grafted by the tomato cultivar Moneymaker to examine the low temperature impact of Venema et al. (2008). The authors found the high relative growth rate of the shooting and the increased root mass ratio under the lowest temperatures in comparison with the self-sustaining and non-grafted plants to be present on the wild tomato rootstock. The results of this study therefore record the accession of the wild tomatoes *Solanum habrochaites* (LA) in 1777 as an additional rootstock in tomatoes as well as other solanaceous vegetables for regulating inadequate root zone temperature. Further research in tomatoes has also found the importance of the resistant rootstock in maintaining appropriate hydraulic and stomach behaviour. Moreover, the impacts in grafting compared to nongrafted plants are recognized to be far less than those in greater productivity of biomass.

The findings of this investigation indicated that grafted plants were used practically under extremely high temperature circumstances. In addition, the same study groups explained that thermal stress increased phenylalanine ammonia lyase (PAL) activity in the grafted and non-grafted tomato plants, increased total phenols and increased o-diphenols, decreased polyphenol oxidase (PPO) activities and goaiacol peroxidase (GPX) activity, and decreased dry weight. In comparison with non-grafted tomato plants, however, pressure impact was reduced in grafting (Lopez-Marin et al. 2013). Under addition to increased RuBisCO activity and more photosynthetically efficient performance of grafting plants in heat stress conditions can be ascribed to better. In the instance of pepper, the behaviour of non-grafted was tested. Under the case of pepper, tests were carried out on three rootstocks, Atlante, Create and Terrano, in shaded and non-shaded circumstances for the conduct of non grafted cultivar Herminio plants. Under both situations the grafted plants were performed better than not. In plant grafted onto the rootstock of Atlante, almost 40 percent more of the leaf area was registered as neutral for Atlante than other pairings. Grabing on Creonteeli had no significant influence on the biomass of the leaves. The total and marketable fruit yields of greased plants in non-shaded and shaded circumstances were nonetheless between 30 and 50 percent higher than that of non-grafted plants. Therefore, when greased to sweet pepper, the rootstock Creontemay is more efficient in resisting heat stress. Further investigations by Del Amor et al. (2008) and Colla et al. (2008) have also demonstrated highenetic yields of sweet pepper-grafted plants for the Mediterranean climate.

High day and night temperatures affect the setting of fruits for tomato growing and cause loss of output. A high-temperature stress-related study was carried out to test greased tomato plants under high temperature stress. 'UC 82-B' has been grafted with the heat-sensitive tomato plant 'Summerset' and with the eggplant cultivar 'Black Beauty.' 'UC 82-B' has been planted. The findings indicates that at 37/27 °C at late fruits, larger leaf area and low electrolytic leakage levels of non-greaser 'UK 82-B' displayed substantially higher chlorophyll fluorescence, but Abdelmageed & Gruda had no favourable effect on total yields than non-grafted 'UC 82-B'. The number of tomato cellular proteins that have recently been shown by Muneer et al. (2016) has been 87. Further researches are presented in Table 12.3 which evaluates the application of vegetable grafting to control thermal stress.

12.2.4 Salinity Stress

Around 7% of the global surface and almost 20% of the arable land is soil salinity impacted (Shahid et al. 2018). Climate change encourages salinization and so the quantity of salt land under climate change scenarios is projected to rise. Salinity impacts output and development of plants negatively. A number of techniques have been developed to counteract the influence of salt and saline soils on vegetable crop yield. Many of the techniques for reclamation of salt soils are a temporary remedy and quite costly. The breeding of salt-tolerant was also studied for vegetables but

Table 12.3 Examples of high-level pairings of rootstock and scion in plants under heat stress

S. No.	Rootstock	Scion	References
1 .	*S. habrochaites* LA-1777	*S. Lycopersicum* cv. Moneymaker	Venema et al. (2008)
2.	*S. lycopersicum* cv. Summerset and *S. melongena* cv. Black beauty	*S. lycopersicum* cv. UC 82 B	Abdelmageed and Gruda (2009)
3.	*C. ficifolia* and *L. cylindrica* cv. Xiangfei	*C. sativus* cv. Jinyan no.-4	Li et al. (2016)
4.	*S. lycopersicum* cv. RX 335	*S. lycopersicum* cv. Tmknvf2	Rivero et al. (2003)
5.	*S. lycopersicum* cv. LA-1778	*S. Lycopersicum* cv. T-5	Bloom et al. (2004)
6.	*C. ficifolia*	*C. sativus* cv. Jinyan no.-4	Zhou et al. (2007)

Table 12.4 Examples of high-performance pairings of rootstock and scion in plants with induced salinity stress

S. No.	Rootstock	Scion	Reference
1.	*S. lycopersicum* x *S. habrochaites* Maxifort'	*S.lycopersicum* L. cv. Cuore di Bue	Di Gioia et al. (2013)
2.	*L. siceraria*	*C. lanatus* cv. Xiuli	Yang et al. (2013)
3.	*C. maxima* x *C. moschata*	*C. melo* L. cv. Brennus	Orsini et al. (2013)
4.	*C. moschata*	*C. sativus* L.	Zhen et al. (2011)
5.	a) *C. chinense* ECU-973, *C. baccatum* L. var. pendulum BOL- 58	*C. annuum*	Penella et al. (2014)
6.	*C. maxima.* X *C. moschata.* P-360 and PS-13132	(a) *C. melo* L. cv. Cyrano (b) *C. sativus* L. cv. Akito	Rouphael et al. (2012)

multiple cycles of plant breeding were necessary because of the complicated polygenic traits which trigger salt tolerance.

The adoption of resistant genotypes as rootstocks was considered a simple and effective method for enhancing the tolerance of the crop against salt stress. Research of salt tolerances for greased vegetable crops have examined throughout the last decade and most studies suggest that grafting is a highly effective technique to enhance salt tolerance. The salt tolerance has improved with interspecific, hybrid rootstock for grafted tomato plants. When bottle gourd rootstock is used for salinity tolerance for watermelon plants then the salinity tolerance can be increased many times. Orsini et al. (2013), reported that non-grafted control plants for interspecific squash rootstock, have an increased salinity tolerance, along with plant biomass and leafy regions (*Cucurbita maxima* of *Cucurbita moschata* Duch.). Salt tolerance in many vegetable crops by using grafting methods is presented in Table 12.4.

12.3 Conclusion

Summarizing, the scientific literature documented the grafting of genetically diverse crops to reduce environmental stress in a changing global climate. New studies are needed as a source of rootstock to evaluate and test different germplasm. However, the full potential of the grafting technique depends on a number of characteristics, such as the appropriate selection of scion and rootstock, location, communication between rootstock and scion, and reciprocal effects of root systems. Further research is needed in order to create automated grafting platforms that expand this technique to make it a key aspect of modern vegetable production. This alone allows the production of grafted plant material at reduced price for vegetable growers in modern simulation and automated grafting technologies. In addition, vegetable grafting is pursued in developed countries due to up-to-date farmers' understanding of the contemporary technology of vegetable manufacturing.

Plant grafting can help producers tackle climate change and overcome unsustainable production methods of vegetables that lead to soil degradation and the rapid loss of natural resources. Research is necessary in order to make this technique easier for farmers in every area of the world, requiring the development of greasing technologies that are suitable for steady, whole year round and inexpensive production of seedlings.

References

Adarsh A, Kumar A, Pratap T, Solankey SS, Sing HK (2020) Grafting in vegetable crops towards stress tolerance. In: Singh HK, Solankey SS, Roy MK (eds) Farmers' prosperity through improved agricultural technologies. Jaya Publishing House, Delhi, pp 167–184

Bahadur A, Rai N, Kumar R, Tiwari SK, Singh AK, Rai AK, Singh U, Patel PK, Tiwari V, Rai AB, Singh M, Singh B (2015) Grafting tomato on eggplant as a potential tool to improve water logging tolerance in hybrid tomato. Vegetable Sci 42(2):82–87

Bhatt R, Upreti K, Divya MH, Bhat S, Pavithra CB, Sadashiva AT (2015) Interspecific grafting to enhance physiological resilience to flooding stress in tomato (*Solanum lycopersicum* L.). Sci Hortic 182:8–17

Black LL, Wu DL, Wang JF, Kalb T, Abbass D, Chen JH (2003) Grafting tomatoes for production in the hot-wet season. AVRDC Publication, p #03-551

Chaudhari S, Jennings KM, Monks DW, Jordan DL, Gunter CC, Louws FJ (2017) Response of drought stressed grafted and non-grafted tomato to post-emergence metribuzin. Weed Technol 31(3):447–454

Colla G, Rouphael Y, Cardarelli M, Temperini O, Rea E, Salerno A, Pierandrei F (2008) Influence of grafting on yield and fruit quality of pepper (*Capsicum annuum* L.) grown under greenhouse conditions. Acta Hortic:359–364

Costello A, Abbas M, Allen A, Ball S, Bell S, Bellamy R, Friel S, Groce N, Johnson A, Kett M, Lee M, Levy C, Maslin M, McCoy D, McGuire B, Montgomery H, Napier D, Pagel C, Patel J, Antonio J, de Oliveira P, Redclift N, Rees H, Rogger D, Scott J, Stephenson J, Twigg J, Wolff J, Patterson C (2009) Managing the health effects of climate change. Lancet 373:1693–1733

Del Amor FM, Lopez-Marin J, Gonzalez A (2008) Effect of photo selective sheet and grafting technique on growth, yield, and mineral composition of sweet pepper plants. J Plant Nutr 31(6):1108–1120

Den Nijs APM (1980) The effect of grafting on growth and early production of cucumbers at low temperature. ISHS Acta Horticulturae 118: Working-party on Greenhouse Cucumber, pp. 57–64

Horvath I, Vigh L, Hasselt PR, Woltjes J, Kuiper PJC (1983) Lipid composition in leaves of cucumber genotypes as affected by different temperature regimes and grafting. Physiol Plant 57(4):532–536

Johnson S, Inglis D, Miles C (2014) Grafting effects on eggplant growth, yield, and verticillium wilt incidence. Int J Vegetable Sci 20(1):3–20

Kato C, Ohshima N, Kamada H, Satoh S (2001a) Enhancement of the inhibitory activity for greening in xylem sap of squash root with water logging. Plant Physiol Biochem 39(6):513–519

Kato C, Ohshima N, Kamada H, Satoh S (2001b) Enhancement of the inhibitory activity for greening in xylem sap of squash root with water logging. Plant Physiol Biochem 39:513–519

Keatinge JDH, Lin LJ, Ebert AW, Chen WY, Hughes Jd, Luther GC, Wang JF, Ravishankar (2014) Overcoming biotic and abiotic stresses in the Solanaceae through grafting: current status and future perspectives. Biol Agric Hortic 30 (4), 272–287

King SR, Davis AR, Zhang X, Crosby K (2010) Genetics, breeding and selection of rootstocks for Solanaceae and Cucurbitaceae. Sci Hortic 127(2):106–111

Lee JM, Oda M (2003) Grafting of herbaceous vegetable and ornamental crops. In: Janick J (ed) Horticultural reviews, vol 28. Wiley, New York, pp 61–124

Li C, Li Y, Bai L, He C, Yu X (2016) Dynamic expression of miRNAs and their targets in the response to drought stress of grafted cucumber seedlings. Hortic Plant J 2:41–49

Lopez-Marin J, Gonzalez A, Perez-Alfocea F, Egea-Gilabert C, Fernandez JA (2013) Grafting is an efficient alternative to shading screens to alleviate thermal stress in greenhouse grown sweet pepper. Sci Hortic 149:39–46

Mittler R (2006) Abiotic stress, the field environment and stress combination. Trends Plant Sci 11(1):15–19

Moretti C, Mattos LM, Calbo AG, Sargent SA (2010) Climate changes and potential impacts on postharvest quality of fruit and vegetable crops: a review. Food Res Int 43(7):1824–1832

Muneer S, Ko CH, Wei H, Chen Y, Jeong BR (2016) Physiological and proteomic investigations to study the response of tomato graft unions under temperature stress. PLoS One 11(6):e0157439

Nilsen ET, Freeman J, Grene R, Tokuhisa J (2014) A rootstock provides water conservation for a grafted commercial tomato (*Solanum lycopersicum* L.) line in response to mild-drought conditions: a focus on vegetative growth and photosynthetic parameters. PLoS One 9(12):e115380

Oda M (1999) Grafting of vegetables to improve greenhouse production. Food Fertilizer Technol Center Extension Bull 480:1–11

Oda M (2002) Grafting of vegetable crops. Scientific report of the graduate school of agriculture and biological sciences. Osaka Prefecture Univ 54:49–72

Orsini F, Sanoubar R, Oztekin GB, Kappel N, Tepecik M, Quacquarelli C, Tuzel Y, Bona S, Gianquinto G (2013) Improved stomatal regulation and ion partitioning boosts salt tolerance in grafted melon. Funct Plant Biol 40(6):628–636

Peng Y, Dong Y, Tu B, Zhou Z, Zheng B, Luo L, Shi C, Du K (2013) Roots play a vital role in flood-tolerance of poplar demonstrated by reciprocal grafting. Flora-Morphol Distribut Funct Ecol Plants 208(8–9):479–487

Petran A, Hoover E (2014) *Solanum torvum* as a compatible rootstock in interspecific tomato grafting. J Hortic 103(1)

Rivero RM, Ruiz JM, Romero L (2003) Can grafting in tomato plants strengthen resistance to thermal stress? J Sci Food Agric 83(13):1315–1319

Sakata Y. Ohara T. Sugiyama M. (2008) The history of melon and cucumber grafting in Japan. In: Prange RK, Bishop SD (ed) Proceedings of XXVII IHC-S11 sustainability through integrated and organic horticulture. Acta Hortic, 767: 217–228

Sanchez-Rodriguez E, Romero L, Ruiz JM (2016) Accumulation on free polyamines enhanced antioxidant response in fruit of grafting tomato plants under water stress. J Plant Physiol 190:72–78

Schwarz D, Rouphael Y, Colla G, Venema JH (2010) Grafting as a tool to improve tolerance of vegetables to abiotic stresses: thermal stress, water stress and organic pollutants. Sci Hortic 127(2):162–171

Shahid SA, Zaman M, Heng L (2018) Soil salinity: historical perspectives and a world overview of the problem. In: Zaman M, Shahid SA, Heng L (eds) Guideline for salinity assessment, mitigation and adaptation using nuclear and related techniques. Springer, Cham, pp 43–53

Singh HK, Ray PK, Solankey SS, Singh RN (2021) Impact of climate change on leguminous vegetables productivity and mitigation strategies. In: Solankey SS et al (eds) Advances in research on vegetable production under a changing climate, vol 1. Springer, Cham., ISBN 978-3-030-63497-1, pp 149–161

Solankey SS, Prakash S, Pawan S, Randhir K (2019) Vulnerability of vegetable crops to climate change. In: Singh P et al (eds) National Seminar on "recent advances in agriculture for sustainable rural development (RAASRD-2019)". VKSCoA, Dumraon (BAU, Sabour), pp 64–70. ISBN: 978-93-85675-11-9

Solankey SS, Kumari M, Akhtar A, Singh HK, Ray PK (2021) Challenges and opportunities in vegetable production in changing climate: mitigation and adaptation strategies. In: Solankey SS et al (eds) Advances in research on vegetable production under a changing climate, vol 1. Springer, Cham., ISBN 978-3-030-63497-1, pp 13–59

Thomas DS, Twyman C, Osbahr H, Hewitson B (2011) Adaptation to climate change and variability: farmer responses to intra-seasonal precipitation trends in South Africa. In: Williams C, Kniveton D (eds) African climate and climate change, vol 43. Springer, Dordrecht, pp 155–178

Venema JH, Dijk BE, Bax JM, Van Hasselt PR, Elzenga JTM (2008) Grafting tomato (*Solanum lycopersicum*) onto the rootstock of a high-altitude accession of *Solanum habrochaites* improves suboptimal-temperature tolerance. Environ Exp Bot 63(1–3):359–367

Xu Z, Jiang Y, Jia B, Zhou G (2016) Elevated-CO_2 response of stomata and its dependence on environmental factors. Front Plant Sci 7

Yetisir H, Çaliskan ME, Soylu S, Sakar M (2006) Some physiological and growth responses of watermelon [*Citrullus lanatus* (Thunb.) Matsum.and Nakai] grafted onto *Lagenaria siceraria* to flooding. Environ Exp Bot 58(1–3):1–8

Chapter 13
Improvement of Vegetables Through Molecular Breeding in Changing Climate Scenario

Jyoti Prakash Sahoo, Satyaprakash Barik, Menka Pathak, Barsa Tripathy, and Madhuri Pradhan

Abstract Vegetable crops have been cultivated since they were domesticated. These are the principal source of nutrients required in the growth and development of human beings. Rapid advances in the methods of next-generation sequencing technology and high throughput genotyping protocols have resulted in the collection and publication of reference genomes of major vegetable species. The large-scale genetic resource reorganization strategy has revealed the process of vegetable crops domestication and improvement of the essential traits through breeding procedures. The utilization of genetic mapping strategies and identification of quantitative trait locus has resulted in the exploration of significant molecular markers linked to essential traits in vegetables. Furthermore, the genome-based breeding approach is employed in most important vegetable crops families, such as Solanaceae and Brassicaceae, and allowing molecular selection at the single-base level. As a result, genome-wide molecular markers are extensively used for efficient genotyping in most vegetable crops. Molecular breeding has emerged as a key method for vegetables. Besides this, genome editing technology can dramatically increase vegetable breeding efficiency. This chapter examines the current scenario of genome-based molecular breeding tactics and genome editing approaches employed in

J. P. Sahoo
Department of Agricultural Biotechnology, College of Agriculture, Odisha University of Agriculture and Technology, Bhubaneswar, Odisha, India

S. Barik (✉) · M. Pradhan
Department of Agriculture and Allied Sciences, C.V. Raman Global University, Bhubaneswar, Odisha, India

M. Pathak
Department of Vegetable Science, College of Agriculture, Odisha University of Agriculture and Technology, Bhubaneswar, Odisha, India

B. Tripathy
Institute of Agricultural Sciences, S.O.A. Deemed to be University, Bhubaneswar, Odisha, India

S. S. Solankey, M. Kumari (eds.), *Advances in Research on Vegetable Production Under a Changing Climate Vol. 2*, Advances in Olericulture,
https://doi.org/10.1007/978-3-031-20840-9_13

significant vegetable crops to give insights into next-generation molecular breeding for a growing global population.

Keywords Molecular breeding · Vegetable crops · Genome editing · Climate change

13.1 Introduction

Abiotic stresses such as extremely low or high temperature, increase in soil salinity status, drought, and floods have an impact on vegetable production. Climate change significantly impacts various developmental stages of vegetables such as vegetative growth, flowering, and fruiting (Spaldon et al. 2015; Khar et al. 2022). The effects of high temperature and unpredictable rainfall can impede average plant growth and development, affecting crop yields. Environmental stress has a significant impact on soil organic matter decomposition, the nutrient cycle, and plant nutrients as well as water availability. The intensity and duration of the extreme environment, on the other hand, determines the crop growth cycle, biomass accumulation, and ultimately the range of economic benefits. Vegetable crops yields in Asia are expected to fall by 2510 per cent between 2020 and 2050, with South and Central Asia experiencing the steepest declines (Cruz et al. 2007; Ashraf et al. 2022). Several initiatives have been launched to reduce the potential impact of climate change on vegetable production. Advanced molecular breeding approaches have been used to improve the yields and quality of important vegetable crops and combat the abiotic stresses of climate change.

NGS (next-generation sequencing) technology has considerably accelerated vegetable crops genomic sequencing and resequencing, which resulted in the publication of chromosomal sequences of some major vegetable crops. The order is determined by the specific locus of the corresponding gene to a marker. Based on genotyping techniques, these linked markers can be used to perform molecular marker-assisted breeding and identify genetic relationships between different traits of vegetable crops. Various genome sequencing methods also contribute to the development of new molecular markers (Saha et al. 2022). Resequencing techniques rely heavily on the reference genome and bring researchers together to analyse sequence diversity at the genomic level. The primary approach, known as bulked segregant analysis (BSA), was assumed to be whole-genome resequencing employing mass separation analysis. BSA is based on the gene identification and marker development procedures, tested on a variety of vegetable crops. Creating the F_2 progeny's divergent phenotypes acquires the SNP index of the two diversified gene pools and matches them to the vegetable reference genome. Besides this, QTLseq and MutMap can also be utilised. As a result, there is a lot less doubt about other genes and phenotypes (Takagi et al. 2013). MutMap was also used to build a DNA pool from a single individual with the F_2 isolation mutant phenotype and compare it to wildtype and parental reference sequences (Abe et al. 2012). It can also be used to locate the markers inside the genome. Large-scale resequencing has evolved into a more generic strategy based on improved approaches for finding markers linked with vegetable crop attributes.

Transcriptome and metabolic analyses help the researchers to domesticate genes and comprehend molecular regulatory networks at the genomic and transcriptional levels by altering gene expression patterns and analysing metabolic pathways. Existing molecular markers and reference genomes for vegetable crops can be employed to alter the vegetable genomes. CRISPR/Cas9 is a standard genome editing technique used to improve a variety of vegetable crops' undesirable traits (Khan et al. 2019; Satish Chhapekar et al. 2022). Compared to other breeding methods, this form of targeted breeding is more efficient and time-saving (Chen et al. 2019; Zahid et al. 2022). This chapter summarises the progress of research on molecular methods used to create vegetable crops resistant to abiotic stress, including molecular breeding and genome editing approaches.

13.2 Signal Transduction Mechanism Against Abiotic Stresses in Vegetable Crops

Plants respond to stress in various ways, including altered gene expression, cellular metabolism, growth rates, and vegetable crops yields. A sudden change in environmental conditions typically causes plant stress. Abiotic stress imposed on plants by environmental factors can be physical or chemical. Drought, flood, extreme temperatures, salt and mineral toxicity, are the abiotic stresses impacting vegetable crops yields and seed production. The roots are the plant's first line of defence against abiotic stress. The disruption of the Na^+/K^+ ratio in the cytoplasm (Sahoo and Mohapatra 2020) of plant cells is one of the major responses to abiotic stress, including excess salt. Signalling is divided into three stages. First, according to the information received, the cell recognises extracellular signalling molecules. When a signalling molecule binds to a receptor, the signalling mechanism modifies the receptor protein and the response when the signal elicits a specific cellular response. Sensors' perception of the signal is the first step in the signalling pathway, and the generation of secondary signalling molecules follows it. It is usually a non-protein molecule such as Ca^{2+}, IP (Inositol phosphate), or ROS (reactive oxygen species). Each secondary signalling molecule can initiate a protein phosphorylation cascade that regulates the activity of specific transcription factors. Given the abundance of stress at the cellular level and its individual effects, cells develop a sensor that can detect stress (cold, drought, salt) and selectively facilitate the transmission of specific stress signals to cellular targets. A typical two-component system includes HIK (Histidine kinase), which is located on the membrane and senses the input signal, and R_{reg} (response regulator), which mediates the output signal (Sahoo and Mohapatra 2020).

13.2.1 ROS and Calcium

ROS are formed as a by-product of respiration and photosynthesis, and they are required for average plant growth. Single-electron transfer converts the O_2 molecule to superoxide anions, then reduced to hydroxyl radicals by H_2O_2. These three molecules are reactive oxygen species (ROS), which cause oxidative damage to proteins, DNA, and lipids. Drought, low and high temperatures, and other stressors activate ROS, leading to oxidative stress (Sahoo and Mohapatra 2020). Ca^{2+} is another non-protein molecule that influences how vegetable crops respond to abiotic stress. The interaction of inflow and outflow produces and maintains Ca^{2+} spikes in the cytosol. Plants become hypersensitive to ionic and cold stress when Ca^{2+} homeostasis is disrupted due to overexpression of the Ca^{2+}/H^+ antiporter calcium exchanger 1 (CAX1), which loads Ca^{2+} into the vacuole. ABA and mechanical stress signals mobilise Ca^{2+} from intracellular storage, and extracellular Ca^{2+} influx is crucial in cold stress signal transmission (Sahoo and Mohapatra 2020).

13.2.2 Phospholipids, CDPKs and MAPKs

These are critical components of the plasma membrane that can generate signalling molecules. Phosphorylated phosphatidylinositol (PI) is a significant and complex class of signalling precursors. PIP (Phosphatidylinositol phosphate) and PIP2 (Phosphatidylinositol bisphosphate) plant PI isomers complicate signal transduction. PLC (Phospholipase C) catalyses phospholipid degradation, resulting in IP_3 (Inositol-1, 4, 5-triphosphate) and DAG (Diacyl glycerol). Individually, these serve as second messengers. IP_3 levels in the cell have a direct impact on ABA-induced gene expression (Sahoo and Mohapatra 2020). Plants that overexpress the enzyme involved in IP_3 catabolism delay ABA induction of KIN1, RD29A, and RD22. Calcium-dependent protein kinases (CDPKs) are among the most numerous plant protein kinase subfamilies. It has a Ca^{2+} sensor kinase hybrid structure, a calmodulin-like domain, and a ser / thr protein kinase domain at the N-terminus (Sahoo and Mohapatra 2020). It is crucial in the signalling of abiotic stress. MAP kinase is a key player in signalling pathways initiated by various surface receptors. The MAPK target can be found in a variety of intracellular compartments. MAPK acts as a physical link between the cytoplasmic and nuclear signalling pathways (Sahoo and Mohapatra 2020).

13.3 Salinity Tolerance in Vegetable Crops

Excessive salt in agricultural land has been a problem since agriculture's inception. Irrigation has enabled agriculture to be expanded to semi-arid and arid lands and has contributed to a significant increase in food production over the last 40 years, resulting in the accumulation of large amounts of water and high salt. Soil degradation due to increased salinity now affects about 20% of the world's irrigated areas, instead of arid areas and deserts, which currently occupy 1/4thof the planet's total land area (Yeo et al. 1999). SOS3, SOS2, and SOS1 were identified as three significant additives that maintain ion homeostasis across salinity stress by molecular and biochemical mechanisms (Yamaguchi and Blumwald 2005; Flowers 2004). The membrane-bound Ca^{2+} sensor SOS3, which recruits the ser/thr protein kinase SOS2 to the moving membrane, detects Ca^{2+} signalling induced by the salt stress. SOS3 and SOS2 activate the plasma membrane Na^+/H^+ antiporter SOS1 and restores mobile ion balance via Na^+ ion outflow (Sahoo and Mohapatra 2020). The SOS3/SOS2 complex regulates the Na^+ region of the cytosol further by activating and dispersing these ions *via* the vacuolar transporter NHX1. Access to Na^{2+} ions is further complicated by the recreational inhibition of the plasma membrane transporter HKT1. SOS2 also regulates the Ca^{2+} ion domain in the cytosol by controlling the vacuolar Ca^{2+} channel, CAX1. The entire mechanism results in vegetable crops salt tolerance.

13.4 Cold Tolerance in Vegetable Crops

Cold stress, is the most significant biological stress, reducing vegetable crops productivity by affecting plant quality and postharvest lifespan. Plants adapt to such lethal cold stresses by acquiring cold and frost resistance through a process known as acclimation. This cold stress is transmitted via multiple signalling pathways, including ROS, protein kinases, protein phosphatases, ABA, and Ca^{2+}. Moreover, ABA has proven to be the best (Sahoo and Mohapatra 2020). Plants have a cold acclimation mechanism that helps them survive below-freezing temperatures, and this mechanism is induced by the COR (Cold response) gene. COR78 (cold induction) or RD29A (dehydration reaction) are the critical COR gene for studying gene regulation in cold and osmotic stress. The RD29A promoter region contains a 9-base pair sequence known as the dehydration response element (DRE), which response to cold stress. The CBF gene is activated by ICE1 (Inducer of CBF expression). ICE1 is a constitutive transcription factor that binds to the CBF3 promoter and induces its expression when activated by cold stress. CBF binds to the CRT / DRE cis-element of the COR gene and induces gene expression to provide freeze resistance.

13.5 Drought Tolerance in Vegetable Crops

The global climate is changing due to rising temperatures and CO_2 levels in the atmosphere. Because of the severe drought, the soil moisture available to the plant is constantly increasing, causing the plants to die prematurely. Stunted growth is a plant's first reaction after a drought has been imposed on the crop. ABA is an important plant hormone that aids in responding to various stress signals during drought stress. The use of ABA on plants mimics the effects of stressful conditions. In vegetable crops, ABA-mediated gene expression can be explained by two types of signalling pathways, *i.e.,* ABA-dependent and independent signalling pathways. AREB / ABF and MYC / MYB are the ABA-dependent regulons (Sahoo and Mohapatra 2020). CBF / DREB and NAC / ZFHD are ABA-independent regulons.

13.6 Genomics of Major Vegetable Crops and Identification of Genes for Abiotic Stresses

Plant genomic information research has bolstered the field of life sciences. Whole-genome sequencing can be used to investigate a species' entire genome and determine gene function. It is a valuable tool for figuring out how plants grow, develop, and differentiate at the molecular level. The advancement of high-throughput sequencing and the reduction of sequencing costs have made genome sequencing projects on various plant species more feasible. Due to their critical economic relevance and high nutrient content, vegetable crops have become an attractive subject for genetic research. Genome studies of numerous vegetable crops have been performed, including wild relatives and current cultivars. These sequenced vegetable crops can be grouped principally into Solanaceae, Cucurbitaceae, Brassicaceae, and other families (Dias and Ortiz 2021). Many major vegetable species in the Solanaceae family, including tomato, pepper, potato, and eggplant, have their genomic sequences published (Fig. 13.1).

On the other hand, wild tomatoes have a higher stress tolerance than farmed tomatoes. This discovery suggests that some critical wild tomato genes can be adapted to survive in abiotic stress (Hao et al. 2020; Pradhan et al. 2021). The following are the primary genome-based methodologies for finding genes relevant to vegetable crops. BSAseq is a popular method for resequencing the whole genome. This entails generating a new batch of descendants and comparing the resequence results to the compiled reference genome. This technique relies mainly on the reference genome to swiftly identify potential regions. Multiple candidate genes for significant features, such as in tomato for internode length (Schrager-Lavelle et al. 2019), Grey press (Chen et al. 2017a, b) and Carcass count (Gao et al. 2016a, b; Yu et al. 2021) in cucumber and, fruit shape in watermelon (Dou et al. 2018b) etc., by using this method. Some other genes are also identified by using this method in vegetables which can be useful during the phenotypic selection of genotypes in abiotic stresses (Table 13.1).

Fig. 13.1 List of reference genome information in vegetable crops. (Adapted from Hao et al. 2020); Left side: Crop name with sequence accession; Right side: Genome size in Mega Base pair (Mb)

Table 13.1 Gene discovery by genome-based approaches in major vegetable crops for changing climate

Crop species	Morphological target trait related to abiotic stresses for gene and marker discovery	Genome-based approaches	Reference
Tomato	Early flowering	QTL-seq	Ruangrak et al. (2018)
	Internode length	BSA-seq	Schrager-Lavelle et al. (2019)
	Yellow-coloured fruit	MutMap	Garcia et al. (2016)
	Fruit weight	QTL-seq	Illa-Berenguer et al. (2015)
	Leaf mold	QTL-seq	Liu et al. (2019a, b)
Cucumber	Early flowering	QTL-seq	Lu et al. (2014)
	Carpel number	BSA-seq	Li et al. (2016)
	Golden leaf	BSA-seq	Gao et al. (2016a, b)
	Tendril-less	BSA-seq	Chen et al. (2017a, b)
	Fruit length	BSA-seq	Xin et al. (2019)
	Subgynoecy	QTL-seq	Win et al. (2019)
	Yellow-green fruit peel	MutMap	Hao et al. (2018)
	Leaf variegation	MutMap	Cao et al. (2018)
	Light green coloration	MutMap	Lun et al. (2015)
	Light green peel	MutMap	Zhou et al. (2015)
	Curly leaf	MutMap	Rong et al. (2019)
	Organ size	BSR-seq	Yang et al. (2018)
	Fruit thickness	SLAF-seq	Xu et al. (2015a, b)
	Waterlogging	SLAF-seq	Liang et al. (2016)
Melon	Striped rind	BSA-seq	Liu et al. (2019a, b)
	Flavor-related	SLAF-seq	Zhang et al. (2016)
Watermelon	Fruit shape	BSA-seq	Dou et al. (2018b)
	Fruit skin color	BSA-seq	Dou et al. (2018a)
	Seed coat color	QTL-seq	Paudel et al. (2019)

(continued)

Table 13.1 (continued)

Crop species	Morphological target trait related to abiotic stresses for gene and marker discovery	Genome-based approaches	Reference
Pepper	Male sterile	BSA-seq	Cheng et al. (2018)
	Fruit colour	BSA-seq	Borovsky et al. (2019)
Ornamental kale	Lobed leaf	QTL-seq	Ren et al. (2019)
Chinese cabbage	Golden leaf	MutMap	Fu et al. (2019)
Cabbage	Purple colour trait	BSR-seq	Yan et al. (2019)
Radish	Restorer-of-fertility	BSR-seq	Lee et al. (2014)
Broccoli & cabbage	Early flowering	QTL-seq	Shu et al. (2018)

QTL Quantitative trait locus, *BSA* Bulked segregant analysis, *BSR* Bulked segregant RNA, *SLAF* Specific-locus amplified fragment, *seq* Sequencing

Other BSAseq-derived approaches, such as the QTLseq quantitative trait and the ethyl methane sulfonate (EMS) mutant MutMap, have been used to develop abiotic stress tolerance vegetable crops. QTLseq has been frequently utilised to find genes that control the quantitative features of vegetable crops since publishing reference genomes for numerous vegetable crops. This method has successfully identified quantitative features that suggest continuous phenotypic changes in different offspring of vegetable crops. It was also used to find the linkage marker Ef2.1 (Shu et al. 2018) linked to broccoli and cabbage blooms for tolerance to abiotic stress.

Significant QTL linked to cucumber early flowering during the abiotic stresses, particularly drought, has also been identified (Lu et al. 2014). This makes it easier to choose early-blooming cucumber types and shortens the cucumber growing cycle. Fruit weight and quantity of tomato loci (Illa-Berenguer et al. 2015), as well as several genes for other qualities such as ornamental cabbage locus leaves (Illa-Berenguer et al. 2015), has also been identified. In addition, QTLseq identified watermelon seed coat colour, cucumber sub-gynecology, and cucumber sub-gynecology (Win et al. 2019). The relevance of the reference genome for rapid molecular breeding is further demonstrated by the identification of candidate genes from carcinogenic EMS vegetable variations. Several pooled sequencing approaches have also been established (Abe et al. 2012), and MutMap has been proposed as a standard gene identification approach based on NGS technology. Individuals with mutant traits in the isolated F_2 group can pool their DNA. Two light green shell genes (Lun et al. 2015; Zhou et al. 2015) are found in cucumber by using MutMap technique (Hao et al. 2018), and leaf-altered genes that control cucumber fruit or leaf colour (Cao et al. 2018) has also been identified. MutMap is also used to find genes for Chinese cabbage fruits, golden leaves, and yellow tomato fruits (Fu et al. 2019). MutMap also successfully identifies random genes for plant height and leaf morphological variants.

To construct a high-density linkage map and uncover candidate genes for the Rfd1 gene, researchers used the BSRseq approach. BSAseq and BSRseq were also employed on Bell peppers at the same time to develop chloroplast and immature Bell pepper colours (Garcia et al. 2016). SLAFseq is another method for simplifying the genome and performing whole-genome resequencing. For SLAFseq applications to succeed, it is also necessary to have a known reference genome and bioinformatics foundation. SLAFseq may analyse a species' genetic diversity by simplifying its genome, and bioinformatics opens up new paths for investigating genomic alterations in evolutionary species (Xu et al. 2014). Cucumbers, peppers, and watermelons can all benefit from SLAFseq to help them figure out their complex features against abiotic stresses. For example, genes associated with cucumber fruit thickness, aphid resistance, and flooding were discovered using SLAFseq (Liang et al. 2016; Xu et al. 2015a, b; Zhu et al. 2016). It was also utilised to create a high-density map with 2634 watermelon SNPs and significantly lower distances between linked markers (Shang et al. 2016). Furthermore, SLAFseq, in conjunction with other genome-based sequencing technologies, can produce accurate and modest linkage mapping. SLAFseq and BSAseq, for example, have been used to define genes linked to pepper's first floral segment (Xu et al. 2016) and the melon flavour gene (Zhang et al. 2016; Abe et al. 2012). These finding suggests that, a combination of BSAseq and RNAseq is required, including a reference genome to identify candidate genes for abiotic stress tolerance in vegetable crops (Table 13.1).

13.7 Genome-Wide Association Analysis in Major Vegetable Crops for Abiotic Stresses

Genome-wide association analysis (GWAS) is an effective method based on genotype-phenotype correlation in a group of individuals, which looks upon genetic variation of complex traits in vegetable crops in the reference genome sequence (Saidou et al. 2014). This can be done by collecting a variety of cultural accessions, crossing lines, sequencing, and comparing genetic diversity between them, especially for traits influenced by various factors. For example, bitterness in cucumbers is an undesirable feature, hence cultivars that are not bitter after long periods of domestication were chosen. Critical GWAS studies on cucumbers have also aided in understanding the genetic mechanism of bittern production using 115 cucumber varieties and the rapid and precise detection of bitterness using linked molecular markers (Shang et al. 2014), particularly during the employment of abiotic stress.

GWAS can also be used to uncover links between fructose and volatile organic acids, which are prevalent in tomato fruit flavours during drought and salinity stress (Zhang et al. 2015; Zhao et al. 2019). Capsaicinoid content is a crucial component of pepper flavour, and GWAS has been used in several studies to explore capsaicinoids. Many transport factors and ankyrin-like proteins may be involved in capsaicinoid synthesis, and the *C. annum* accession has been used to find markers connected to capsaicin and dihydrocapsaicin levels (Nimmakayala et al. 2016). Another GWAS sample, including the *C. annuum* accession, discovered candidate genes or QTLs that modulate capsaicinoid concentration. High-density SNP mapping identified 69 QTL regions and 5 candidate genes (Han et al. 2018) during abiotic stresses. GWAS, on the other hand, has several drawbacks. Some GWAS markers produce false-positive results, and others are not repeatable among cultivars. As a result, GWAS-identified molecular markers in vegetable crops will need to be confirmed through genotyping or polymorphism amplification while studying them in the abiotic stress environment.

13.8 Genome Based Molecular Marker Discovery in Vegetable Breeding

Marker-assisted selection (MAS) has become a popular genotype-based breeding approach for increasing vegetable crops breeding performance. MAS can preemptively eliminate undesirable genetic backgrounds or nudge alleles back to the intended genetic region (Fig. 13.2). A sufficient number of molecular markers linked with the trait and their polymorphisms are also essential considerations. Molecular markers integrating RLFPs or SSRs were previously advanced without

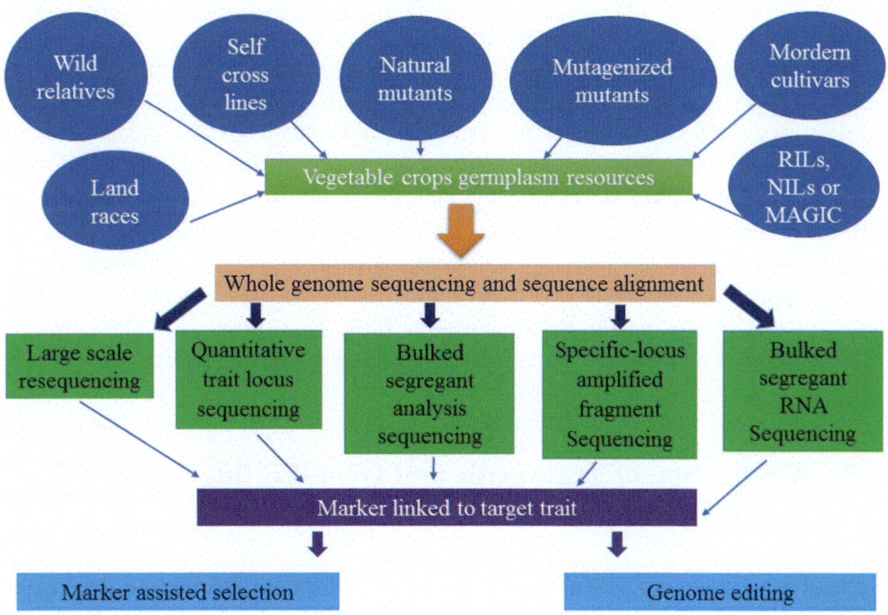

Fig. 13.2 Flowchart depicting the molecular breeding process in vegetable crops using genome-based approaches

genomic information. On the other hand, those molecular markers are far less polymorphic and have a narrow range of expression.

With the introduction of the reference genome for maximal vegetable vegetation and the advancement of the NGS era for genome sequencing and resequencing, the number of SSR markers has increased dramatically. For the black spikes of ripe cucumber fruit and the uniform colour of orange and immature fruit, near-range SSR markers were mapped (Li et al. 2013; Yang et al. 2014). In addition, in *Brassica rapa* L. var. purpurea, a high-density map using SSR markers was employed to identify the markers linked to traits related to abiotic stresses (Wang et al. 2017). In tomatoes, OVATE has been discovered to be a good indicator for fruit shape, as well as CYC B, SlMYB12, and SlMYB75 for fruit colour (Rodríguez et al. 2013; Ballester et al. 2010; Fernandez-Moreno et al. 2016; Hwang et al. 2016; Jian et al. 2019). For cucumbers, a novel targeted SSRseq approach has been devised (Jian et al. 2019). It also has higher coverage in high-throughput sequences and is less expensive than previously advanced SSR markers, allowing genotyping of many SSR markers across different accessions. This procedure can also broaden markers so that distinct cucumber ecotypes can be distinguished. SNPs, and SSR markers are important markers for vegetable plants because of the vast range and frequency of genomic modifications (Hwang et al. 2016).

As a result, SNP markers can be used to accurately genotype vegetables and plants and create accurate genetic maps for abiotic stresses (Nimmakayala et al. 2014).The molecular linkage map of tomato was also constructed in the presence of an excessive density map with more than one molecular marker (SSR, and SNP) from the preliminary 2 cM distance among the markers (Tanksley et al. 1992). As a result of the disclosure of the complete genome collection of several pepper cultivars, the genetic map has advanced dramatically (Zhang et al. 2019b). There are now 5569 SNPs in those saturation maps (Zhang et al. 2019b) available for different abiotic stresses. The essential qualities of Chinese cabbage are also linked to some molecular indications. For example, in Brassica plant breeding, flower colour is a flag function that boosts production by attracting pollinating insects. Zhang et al. (2019a) discovered markers in Chinese cabbage orange flowers and followed them through marker assisted breeding to discern different flower colour styles. These comprehensive genetic maps gave markers for inclinations in vegetable crops and increased MAS breeding for abiotic stresses. Table 13.2 summarises the genes or QTLs associated with abiotic stress tolerance in major vegetable crops.

13.9 Application of Transgenic and Gene-Editing Technology in Vegetable Breeding

Physical mutagenesis, EMS mutagenesis, and TDNA insertion can all produce a large number of mutagens, but the mechanisms behind them are yet understood (Chaudhary et al. 2019). Finding the correct strain takes a long time and many

Table 13.2 Genes identified for abiotic stress tolerance in major vegetable crops

Crop species	Name of the genes or TF or QTLs	Trait	Reference
Broccoli	BoiCesA (RNAi)	SST	Li et al. 2017
Broccoli	BoC3H	SST	Jiang et al. 2020
Broccoli	BoC3H4	SST	Jiang et al. 2019
Broccoli	BoERF1	SST	Jiang et al. 2020
Tomato	AREB cupidadehydrine	TDS	Hsieh et al. 2010
Tomato	14–3-3 genes	TDS	Xu et al. 2018
Chilli	Dreb 1A TF	TDS	Maligeppagol et al. (2016)
Tomato	bspA gene	TDS	Roy et al. (2006)
Chilli	Osmotin gene	SST	Subramanyam et al. (2011)
Tomato	HsfA1a	TDS	Wang et al. (2015)
Tomato	microRNA169	TDS	Zhang et al. (2011)
Tomato	Transcription factor SlNAC4	SST, TDS	Zhu et al. (2014)
Potato	28 drought-specific QTLs	TDS	Anithakumari et al. (2012)
Tomato	SlGATA17	TDS	Zhao et al. (2021)
Tomato	CBF/NHX1/DREB1 genes	TDS	Solankey et al. (2015)

SST Salinity stress tolerance, *TDS* Tolerance to drought stress

individuals. A genome-based gene editing approach for vegetable crops is currently being used to knock off-target genes in vegetable crops. Because of their stable gene transfer method and rich genomic information, tomatoes are an appealing food crop for proving the efficacy of gene editing. Tomato CRISPR has knocked out many genes, including regulation of the drought-tolerant gene SlMAPK3 (Wang et al. 2017). PROCERA is a gene that responds to gibberellin (Tomlinson et al. 2019) in significant vegetable crops during abiotic stresses (Soyk et al. 2017). Future studies on the domestication of wild tomatoes will also employ genome editing techniques. Wildtype tomatoes with good resistance properties can be genetically modified (Li et al. 2018; Zsögön et al. 2018). The cucumber is a model plant for squash, and its solitary nature makes reproduction difficult. As a result, selecting a female cucumber variety as a breeding target can increase cucumber yields while saving time and money. According to Hu (2017), all-female cucumber materials are conserved by boosting the cucumber genetic transformation mechanism and knocking down CsWIP, which regulates the whole female phenotype during multiple stresses. This method was also used to knock out the cucumber-specific short fruit genes sf1 and sf2 and provide the framework for molecular fruit development regulation (Xin et al. 2019; Zhang et al. 2019a). The CRISPR technique can benefit other Cucurbitaceae vegetable crops (Tian et al. 2018). For example, changing the watermelon genome revealed ClWIP1's genetic role, and female plants were kept to promote hybrid plant purity (Zhang et al. 2019a). CRISPR technology has also helped salt-tolerant squash varieties (Huang et al. 2019).

13.10 Conclusion

With the completion of the genetic sequencing of major vegetable crops, genomic information has significantly accelerated molecular breeding. By utilizing genetic data, genomic sequencing, and genotyping, several new identification approaches have been developed and applied to the majority of vegetable crops in order to identify major genes tolerance to abiotic stresses. For marker assisted breeding, several molecular markers that are directly related to essential properties of vegetable crops and linked to abiotic stress tolerance genes have also been established for major vegetable crops. Advances in gene-editing technology have also paved the way for developing new vegetables varieties with improved characters. However, in the case of vegetable farming, the breeding strategy must be improved.

Even though the genetic sequences of the majority of vegetable crops have been identified, many approaches, such as MutMap Plus and MutMap Gap, have yet to be successfully applied to vegetables for identifying genes associated with abiotic stress tolerance in order to combat climate change. These methodologies, which are reliable in other model cultures, should be used to discover target genes and related molecular markers in vegetable crops. Researchers and breeders can improve communication and information exchange by using connected data. Molecular identifiers can thus be effectively transformed into widely used markers in this manner. Superior traits in vegetable crops and cultivar genotypes can also been studied to better understand the molecular process of superior trait development and to provide a theoretical foundation for molecular breeding.

References

Abe A, Kosugi S, Yoshida K, Natsume S, Takagi H, Kanzaki H, Matsumura H, Yoshida K, Mitsuoka C, Tamiru M, Innan H (2012) Genome sequencing reveals agronomically important loci in rice using MutMap. Nat Biotechnol 30(2):174–178

Anithakumari AM, Nataraja KN, Visser RG, van der Linden CG (2012) Genetic dissection of drought tolerance and recovery potential by quantitative trait locus mapping of a diploid potato population. Mol Breed 30(3):1413–1429

Ashraf U, Mahmood S, Shahid N, Imran M, Siddique M, Abrar M (2022) Multi-omics approaches for strategic improvements of crops under changing climatic conditions. In: *Principles and practices of OMICS and genome editing for crop improvement*. Springer, Cham, pp 57–92

Ballester AR, Molthoff J, de Vos R, Hekkert BTL, Orzaez D, Fernandez-Moreno JP, Tripodi P, Grandillo S, Martin C, Heldens J, Ykema M (2010) Biochemical and molecular analysis of pink tomatoes: deregulated expression of the gene encoding transcription factor SlMYB12 leads to pink tomato fruit color. Plant Physiol 152(1):71–84

Borovsky Y, Monsonego N, Mohan V, Shabtai S, Kamara I, Faigenboim A, Hill T, Chen S, Stoffel K, Van Deynze A, Paran I (2019) The zinc-finger transcription factor cc LOL 1 controls chloroplast development and immature pepper fruit color in Capsicum chinense and its function is conserved in tomato. The Plant J 99(1):41–55

Cao W, Du Y, Wang C, Xu L, Wu T (2018) Cscs encoding chorismate synthase is a candidate gene for leaf variegation mutation in cucumber. Breed Sci:18023

Chaudhary J, Deshmukh R, Sonah H (2019) Mutagenesis approaches and their role in crop improvement. Plan Theory 8(11):467
Chen F, Fu B, Pan Y, Zhang C, Wen H, Weng Y, Chen P, Li Y (2017b) Fine-mapping identifies CsGCN5 encoding a histone acetyltransferase as putative candidate gene for tendril-less1 mutation (td-1) in cucumber. TheorAppl Genet 130(8):1549–1558
Chen K, Wang Y, Zhang R, Zhang H, Gao C (2019) CRISPR/Cas genome editing and precision plant breeding in agriculture. Annu Rev Plant Biol 70:667–697
Chen S, Zhao H, Zou C, Li Y, Chen Y, Wang Z, Jiang Y, Liu A, Zhao P, Wang M, Ahammed GJ (2017a) Combined inoculation with multiple arbuscular mycorrhizal fungi improves growth, nutrient uptake and photosynthesis in cucumber seedlings. Front Microbiol 8:2516
Cheng Q, Wang P, Liu J, Wu L, Zhang Z, Li T, Gao W, Yang W, Sun L, Shen H (2018) Identification of candidate genes underlying genic male-sterile msc-1 locus via genome resequencing in Capsicum annuum L. TheorAppl Genet 131(9):1861–1872
Cruz RV, Harasawa H, Lal M, Wu S, Anokhin Y, Punsalmaa B, Honda Y, Jafari M, Li C, Huu N (2007) Asia climate change 2007: impacts, adaptation and vulnerability. In: Parry ML, Canziani OF, Palutikof JP, van der Linden PJ, Hanson CE (eds) Contribution of working group ii to the fourth assessment report of the intergovernmental panel on climate change. Cambridge University Press, Cambridge, pp 469–506
Dias JS, Ortiz R (2021) New strategies and approaches for improving vegetable cultivars. In: The basics of human civilization. CRC Press, pp 349–381
Dou J, Lu X, Ali A, Zhao S, Zhang L, He N, Liu W (2018b) Genetic mapping reveals a marker for yellow skin in watermelon (Citrulluslanatus L.). PLoS One 13(9):e0200617
Dou J, Zhao S, Lu X, He N, Zhang L, Ali A, Kuang H, Liu W (2018a) Genetic mapping reveals a candidate gene (ClFS1) for fruit shape in watermelon (Citrulluslanatus L.). TheorAppl Genet 131(4):947–958
Fernandez-Moreno JP, Tzfadia O, Forment J, Presa S, Rogachev I, Meir S, Orzaez D, Aharoni A, Granell A (2016) Characterization of a new pink-fruited tomato mutant results in the identification of a null allele of the SlMYB12 transcription factor. Plant Physiol 171(3):1821–1836
Flowers TJ (2004) Improving crop salt tolerance. J Exp Bot 55(396):307–319
Fu W, Ye X, Ren J, Li Q, Du J, Hou A, Mei F, Feng H, Liu Z (2019) Fine mapping of lcm1, a gene conferring chlorophyll-deficient golden leaf in Chinese cabbage (Brassica rapa ssp. pekinensis). Mol Breed 39(4):1–12
Gao M, Hu L, Li Y, Weng Y (2016a) The chlorophyll-deficient golden leaf mutation in cucumber is due to a single nucleotide substitution in CsChlI for magnesium chelatase I subunit. TheorAppl Genet 129(10):1961–1197
Gao M, Hu L, Li Y, Weng Y (2016b) The chlorophyll-deficient golden leaf mutation in cucumber is due to a single nucleotide substitution in CsChlI for magnesium chelatase I subunit. TheorAppl Genet 129(10):1961–1973
Garcia V, Bres C, Just D, Fernandez L, Tai FWJ, Mauxion JP, Le Paslier MC, Berard A, Brunel D, Aoki K, Alseekh S (2016) Rapid identification of causal mutations in tomato EMS populations via mapping-by-sequencing. Nat Protoc 11(12):2401–2418
Han K, Lee HY, Ro NY, Hur OS, Lee JH, Kwon JK, Kang BC (2018) QTL mapping and GWAS reveal candidate genes controlling capsaicinoid content in capsicum. Plant Biotechnol J 16(9):1546–1558
Hao N, Du Y, Li H, Wang C, Wang C, Gong S, Zhou S, Wu T (2018) CsMYB36 is involved in the formation of yellow green peel in cucumber (Cucumissativus L.). TheorAppl Genet 131(8):1659–1669
Hao N, Han D, Huang K, Du Y, Yang J, Zhang J, Wen C, Wu T (2020) Genome-based breeding approaches in major vegetable crops. TheorAppl Genet 133(5):1739–1752
Hsieh TH, Li CW, Su RC, Cheng CP, Tsai YC, Chan MT (2010) A tomato bZIP transcription factor, SlAREB, is involved in water deficit and salt stress response. Planta 231(6):1459–1473
Hu B (2017) Engineering non-transgenic gynoecious cucumber using an improved transformation protocol and optimized CRISPR/Cas9 system. Mol Plant 10(2):1575–1578

Huang Y, Cao H, Yang L, Chen C, Shabala L, Xiong M, Niu M, Liu J, Zheng Z, Zhou L, Peng Z (2019) Tissue-specific respiratory burst oxidase homolog-dependent H2O2 signaling to the plasma membrane H+-ATPase confers potassium uptake and salinity tolerance in Cucurbitaceae. J Exp Bot 70(20):5879–5893

Hwang I, Kim Y, Han J, Nou IS (2016) Orange color is associated with CYC-B expression in tomato fleshy fruit. Mol Breed 36(4):42

Illa-Berenguer E, Van Houten J, Huang Z, van der Knaap E (2015) Rapid and reliable identification of tomato fruit weight and locule number loci by QTL-seq. TheorAppl Genet 128(7):1329–1342

Jian W, Cao H, Yuan S, Liu Y, Lu J, Lu W, Li N, Wang J, Zou J, Tang N, Xu C (2019) SlMYB75, an MYB-type transcription factor, promotes anthocyanin accumulation and enhances volatile aroma production in tomato fruits. Hort Res 6(1):1–15

Jiang M, Miao L, Zhang H, Zhu X (2020) Over-expression of a transcription factor gene BoC3H4 enhances salt stress tolerance but reduces Sclerotinia stem rot disease resistance in broccoli. J Plant Growth Regul 39(3):1162–1176

Jiang M, Ye ZH, Zhang HJ, Miao LX (2019) Broccoli plants over-expressing an ERF transcription factor gene BoERF1 facilitates both salt stress and sclerotinia stem rot resistance. J Plant Growth Regul 38(1):1–13

Khan MZ, Zaidi SSEA, Amin I, Mansoor S (2019) A CRISPR way for fast-forward crop domestication. Trends Plant Sci 24(4):293–296

Khar A, Singh H, Verma P (2022) Mitigating abiotic stresses in under changing climatic scenario. In: *Genomic designing for abiotic stress resistant vegetable crops*. Springer, Cham, pp 253–278

Lee YP, Cho Y, Kim S (2014) A high-resolution linkage map of the Rfd1, a restorer-of-fertility locus for cytoplasmic male sterility in radish (Raphanussativus L.) produced by a combination of bulked segregant analysis and RNA-Seq. TheorAppl Genet 127(10):2243–2252

Li S, Pan Y, Wen C, Li Y, Liu X, Zhang X, Behera TK, Xing G, Weng Y (2016) Integrated analysis in bi-parental and natural populations reveals CsCLAVATA3 (CsCLV3) underlying carpel number variations in cucumber. TheorAppl Genet 129(5):1007–1022

Li S, Zhang L, Wang Y, Xu F, Liu M, Lin P, Ren S, Ma R, Guo YD (2017) Knockdown of a cellulose synthase gene BoiCesA affects the leaf anatomy, cellulose content and salt tolerance in broccoli. Sci Rep 7(1):1–14

Li T, Yang X, Yu Y, Si X, Zhai X, Zhang H, Dong W, Gao C, Xu C (2018) Domestication of wild tomato is accelerated by genome editing. Nat Biotechnol 36(12):1160–1163

Li Y, Wen C, Weng Y (2013) Fine mapping of the pleiotropic locus B for black spine and orange mature fruit color in cucumber identifies a 50 kb region containing a R2R3-MYB transcription factor. TheorAppl Genet 126(8):2187–2196

Liang D, Chen M, Qi X, Xu Q, Zhou F, Chen X (2016) QTL mapping by SLAF-seq and expression analysis of candidate genes for aphid resistance in cucumber. Front Plant Sci 7:1000

Liu C, Chen L, Zhao R, Li R, Zhang S, Yu W, Sheng J, Shen L (2019a) Melatonin induces disease resistance to Botrytis cinerea in tomato fruit by activating jasmonic acid signaling pathway. J Agric Food Chem 67(22):6116–6124

Liu G, Mao S, Kim JH (2019b) A mature-tomato detection algorithm using machine learning and color analysis. Sensors 19(9):2023

Lu H, Lin T, Klein J, Wang S, Qi J, Zhou Q, Sun J, Zhang Z, Weng Y, Huang S (2014) QTL-seq identifies an early flowering QTL located near flowering locus T in cucumber. Theor Appl Genet 127(7):1491–1499

Lun Y, Wang X, Zhang C, Yang L, Gao D, Chen H, Huang S (2015) A CsYcf54 variant conferring light green coloration in cucumber. Euphytica 208(3):509–517

Maligeppagol M, Manjula R, Navale PM, Babu KP, Kumbar BM, Laxman RH (2016) Genetic transformation of chilli (Capsicum annuum L.) with Dreb1A transcription factor known to impart drought tolerance

Nimmakayala P, Abburi VL, Saminathan T, Alaparthi SB, Almeida A, Davenport B, Nadimi M, Davidson J, Tonapi K, Yadav L, Malkaram S (2016) Genome-wide diversity and association mapping for capsaicinoids and fruit weight in Capsicum annuum L. Sci Rep 6(1):1–14

Nimmakayala P, Levi A, Abburi L, Abburi VL, Tomason YR, Saminathan T, Vajja VG, Malkaram S, Reddy R, Wehner TC, Mitchell SE (2014) Single nucleotide polymorphisms generated by genotyping by sequencing to characterize genome-wide diversity, linkage disequilibrium, and selective sweeps in cultivated watermelon. BMC Genomics 15(1):1–15

Paudel L, Clevenger J, McGregor C (2019) Chromosomal locations and interactions of four loci associated with seed coat color in watermelon. Front Plant Sci 10:788

Pradhan K, Rout S, Tripathy B, Mishra UN, Sahoo G, Prusty AK, Dash L (2021) Role of biotechnology in vegetable breeding. Turkish Online J Qual Inq 12(3):5092–5102

Ren J, Liu Z, Du J, Fu W, Hou A, Feng H (2019) Fine-mapping of a gene for the lobed leaf, BoLl, in ornamental kale (Brassica oleracea L. var. acephala). Mol Breed 39(3):1–14

Rodríguez GR, Kim HJ, Van Der Knaap E (2013) Mapping of two suppressors of OVATE (sov) loci in tomato. Heredity 111(3):256–264

Rong F, Chen F, Huang L, Zhang J, Zhang C, Hou D, Cheng Z, Weng Y, Chen P, Li Y (2019) A mutation in class III homeodomain-leucine zipper (HD-ZIP III) transcription factor results in curly leaf (cul) in cucumber (Cucumissativus L.). TheorAppl Genet 132(1):113–123

Roy R, Purty RS, Agrawal V, Gupta SC (2006) Transformation of tomato cultivar 'PusaRuby'withbspA gene from Populustremula for drought tolerance. Plant Cell Tissue Organ Cult 84(1):56–68

Ruangrak E, Su X, Huang Z, Wang X, Guo Y, Du Y, Gao J (2018) Fine mapping of a major QTL controlling early flowering in tomato using QTL-seq. Can J Plant Sci 98(3):672–682

Saha P, Singh S, Bhatia R, Dey SS, Das Saha N, Ghoshal C et al (2022) Genomic designing for abiotic stress resistant brassica vegetable crops. In: *Genomic designing for abiotic stress resistant vegetable crops*. Springer, Cham, pp 153–185

Sahoo JP, Mohapatra U (2020) Plant signal transduction mechanism against abiotic and biotic stresses. https://doi.org/10.22271/ed.book.1027

Saidou AA, Clotault J, Couderc M, Mariac C, Devos KM, Thuillet AC, Amoukou IA, Vigouroux Y (2014) Association mapping, patterns of linkage disequilibrium and selection in the vicinity of the PHYTOCHROME C gene in pearl millet. TheorAppl Genet 127(1):19–32

Satish Chhapekar S, Singh S, Singh S, Ma Y, Jeevan Rameneni J, Ryun Choi S, Pyo Lim Y (2022) Genomic Design for Biotic Stress Tolerance in vegetable brassicas. In: *Genomic designing for biotic stress resistant vegetable crops*. Springer, Cham, pp 189–231

Schrager-Lavelle A, Gath NN, Devisetty UK, Carrera E, López-Díaz I, Blázquez MA, Maloof JN (2019) The role of a class III gibberellin 2-oxidase in tomato internode elongation. The Plant J 97(3):603–615

Shang J, Li N, Li N, Xu Y, Ma S, Wang J (2016) Construction of a high-density genetic map for watermelon (*Citrullus lanatus* L.) based on large-scale SNP discovery by specific length amplified fragment sequencing (SLAF-seq). Sci Hortic 203:38–46

Shang Y, Ma Y, Zhou Y, Zhang H, Duan L, Chen H, Zeng J, Zhou Q, Wang S, Gu W, Liu M (2014) Biosynthesis, regulation, and domestication of bitterness in cucumber. Sci 346(6213):1084–1088

Shu J, Liu Y, Zhang L, Li Z, Fang Z, Yang L, Zhuang M, Zhang Y, Lv H (2018) QTL-seq for rapid identification of candidate genes for flowering time in broccoli × cabbage. Theor Appl Genet 131(4):917–928

Solankey SS, Singh RK, Baranwal DK, Singh DK (2015) Genetic expression of tomato for heat and drought stress tolerance: an overview. Int J Veg Sci 21(5):496–515

Soyk S, Lemmon ZH, Oved M, Fisher J, Liberatore KL, Park SJ et al (2017) Bypassing negative epistasis on yield in tomato imposed by a domestication gene. Cell 169(6):1142–1155

Spaldon S, Samnotra RK, Chopra S (2015) Climate resilient technologies to meet the challenges in vegetable production. Int J Curr Res Aca Rev 3(2):28–47

Subramanyam K, Sailaja KV, Subramanyam K, Rao DM, Lakshmidevi K (2011) Ectopic expression of an osmotin gene leads to enhanced salt tolerance in transgenic chilli pepper (Capsicum annum L.). Plant Cell Tissue Organ Cult 105(2):181–192

Takagi H, Abe A, Yoshida K, Kosugi S, Natsume S, Mitsuoka C, Uemura A, Utsushi H, Tamiru M, Takuno S, Innan H, Cano LM, Kamoun S, Terauchi R (2013) QTL-seq: rapid mapping of

quantitative trait loci in rice by whole genome resequencing of DNA from two bulked populations. Plant J 74(1):174–183
Tanksley SD, Ganal MW, Prince JP, De Vicente MC, Bonierbale MW, Broun P, Fulton TM, Giovannoni JJ, Grandillo S, Martin GB (1992) High density molecular linkage maps of the tomato and potato genomes. Genet 132(4):1141–1160
Tian S, Jiang L, Cui X, Zhang J, Guo S, Li M, Zhang H, Ren Y, Gong G, Zong M, Liu F (2018) Engineering herbicide-resistant watermelon variety through CRISPR/Cas9-mediated base-editing. Plant Cell Rep 37(9):1353–1356
Tomlinson L, Yang Y, Emenecker R, Smoker M, Taylor J, Perkins S, Smith J, MacLean D, Olszewski NE, Jones JD (2019) Using CRISPR/Cas9 genome editing in tomato to create a gibberellin-responsive dominant dwarf DELLA allele. Plant Biotechnol J 17(1):132–140
Wang L, Chen L, Li R, Zhao R, Yang M, Sheng J, Shen L (2017) Reduced drought tolerance by CRISPR/Cas9-mediated SlMAPK3 mutagenesis in tomato plants. J Agric Food Chem 65(39):8674–8682
Wang Y, Cai S, Yin L, Shi K, Xia X, Zhou Y, Yu J, Zhou J (2015) Tomato HsfA1a plays a critical role in plant drought tolerance by activating ATG genes and inducing autophagy. Autophagy 11(11):2033–2047
Win KT, Zhang C, Silva RR, Lee JH, Kim YC, Lee S (2019) Identification of quantitative trait loci governing subgynoecy in cucumber. TheorAppl Genet 132(5):1505–1521
Xin T, Zhang Z, Li S, Zhang S, Li Q, Zhang ZH, Huang S, Yang X (2019) Genetic regulation of ethylene dosage for cucumber fruit elongation. Plant Cell 31(5):1063–1076
Xu H, Liu Q, Yao TFX (2014) Shedding light on integrative GA signaling. CurrOpin Plant Biol 21:89–95
Xu L, Li T, Wu Z, Feng H, Yu M, Zhang X, Chen B (2018) Arbuscular mycorrhiza enhances drought tolerance of tomato plants by regulating the 14-3-3 genes in the ABA signaling pathway. Appl Soil Ecol 125:213–221
Xu X, Chao J, Cheng X, Wang R, Sun B, Wang H, Luo S, Xu X, Wu T, Li Y (2016) Mapping of a novel race specific resistance gene to phytophthora root rot of pepper (Capsicum annuum) using bulked segregant analysis combined with specific length amplified fragment sequencing strategy. PLoS One 11(3):e0151401
Xu X, Lu L, Zhu B, Xu Q, Qi X, Chen X (2015a) QTL mapping of cucumber fruit flesh thickness by SLAF-seq. Sci Rep 5(1):1–9
Xu X, Xu R, Zhu B, Yu T, Qu W, Lu L, Xu Q, Qi X, Chen X (2015b) A high-density genetic map of cucumber derived from specific length amplified fragment sequencing (SLAF-seq). Front Plant Sci 5:768
Yamaguchi T, Blumwald E (2005) Developing salt-tolerant crop plants: challenges and opportunities. Trends Plant Sci 10(12):615–620
Yan C, An G, Zhu T, Zhang W, Zhang L, Peng L, Chen J, Kuang H (2019) Independent activation of the BoMYB2 gene leading to purple traits in Brassica oleracea. TheorAppl Genet 132(4):895–906
Yang L, Liu H, Zhao J, Pan Y, Cheng S, Lietzow CD, Wen C, Zhang X, Weng Y (2018) LITTLELEAF (LL) encodes a WD40 repeat domain-containing protein associated with organ size variation in cucumber. Plant J 95(5):834–847
Yang X, Li Y, Zhang W, He H, Pan J, Cai R (2014) Fine mapping of the uniform immature fruit color gene u in cucumber (*Cucumis sativus* L.). Euphytica 196(3):341–348
Yeo AR, Flowers SA, Rao G, Welfare K, Senanayake N, Flowers TJ (1999) Silicon reduces sodium uptake in rice (*Oryza sativa* L.) in saline conditions and this is accounted for by a reduction in the transpirational bypass flow. Plant Cell Environ 22(5):559–565
Yu D, Gu X, Zhang S, Dong S, Miao H, Gebretsadik K, Bo K (2021) Molecular basis of heterosis and related breeding strategies reveal its importance in vegetable breeding. Horticul Res 8(1):120–120
Zahid G, Iftikhar S, Farooq MU, Soomro SA (2022) Advances in DNA based molecular markers for the improvement of fruit cultivars in Pakistan-a review. Sarhad J Agricult 38(3):812–832

Zhang H, Yi H, Wu M, Zhang Y, Zhang X, Li M, Wang G (2016) Mapping the flavor contributing traits on" Fengwei melon" (*Cucumis melo* L.) chromosomes using parent resequencing and super bulked-segregant analysis. PLoS One 11(2):e0148150

Zhang J, Zhao J, Xu Y, Liang J, Chang P, Yan F, Li M, Liang Y, Zou Z (2015) Genome-wide association mapping for tomato volatiles positively contributing to tomato flavor. Front Plant Sci 6:1042

Zhang N, Zhang H, Ren Y, Chen L, Zhang J, Zhang L (2019a) Genetic analysis and gene mapping of the orange flower trait in Chinese cabbage (*Brassica rapa* L.). Mol Breed 39(6):1–11

Zhang X, Zou Z, Gong P, Zhang J, Ziaf K, Li H, Xiao F, Ye Z (2011) Over-expression of microRNA169 confers enhanced drought tolerance to tomato. Biotechnol Lett 33(2):403–409

Zhang XF, Wang GY, Dong TT, Chen B, Du HS, Li CB, Geng SS (2019b) High-density genetic map construction and QTL mapping of first flower node in pepper (*Capsicum annuum* L.). BMC Plant Biol 19(1):1–13

Zhao J, Sauvage C, Zhao J, Bitton F, Bauchet G, Liu D, Huang S, Tieman DM, Klee HJ, Causse M (2019) Meta-analysis of genome-wide association studies provides insights into genetic control of tomato flavor. Nat Commun 10(1):1–12

Zhao T, Wu T, Pei T, Wang Z, Yang H, Jiang J, Zhang H, Chen X, Li J, Xu X (2021) Overexpression of SlGATA17 promotes drought tolerance in transgenic tomato plants by enhancing activation of the Phenylpropanoid biosynthetic pathway. Front Plant Sci 12

Zhou Q, Wang S, Hu B, Chen H, Zhang Z, Huang S (2015) An accumulation and replication of chloroplasts 5 gene mutation confers light green peel in cucumber. J Integr Plant Biol 57(11):936–942

Zhu M, Chen G, Zhang J, Zhang Y, Xie Q, Zhao Z, Pan Y, Hu Z (2014) The abiotic stress-responsive NAC-type transcription factor SlNAC4 regulates salt and drought tolerance and stress-related genes in tomato (*Solanum lycopersicum*). Plant Cell Rep 33(11):1851–1863

Zhu WY, Huang L, Chen L, Yang JT, Wu JN, Qu ML, Yao DQ, Guo CL, Lian HL, He HL, Pan JS (2016) A high-density genetic linkage map for cucumber (*Cucumis sativus* L.): based on specific length amplified fragment (SLAF) sequencing and QTL analysis of fruit traits in cucumber. Front Plant Sci 7:437

Zsögön A, Čermák T, Naves ER, Notini MM, Edel KH, Weinl S, Freschi L, Voytas DF, Kudla J, Peres LEP (2018) De novo domestication of wild tomato using genome editing. Nat Biotechnol 36(12):1211–1216

Chapter 14
Emerging Insect-Pests of Vegetables Due to Changing Climate

M. Prashant, M. A. Waseem, Kalmesh Managanvi, Erayya, and Vijay Laxmi Rai

Abstract Climate change, an emerging global worry to farming in many countries and it is the most discussed topic in many forums as it affects the many livelihoods including human beings. Many weather affecting the different forms of life including hexapods by impacting in hexapod fecundity, habit, habitat, behavior, endurance, mobility, generation's asynchrony between plants and pests, altered inter-specific interaction, increased hexapod vectored diseases and reduce the effectiveness of natural enemies. The global warming is also threatening overall food production any country by way of shifting hosts, moving to higher elevation, minor pests becoming major pests and increased invasiveness due to trade of different agricultural commodities. Many of the insect-pests are affecting hosts plants including vegetables as most of the insect are polyphagous in nature, they easily shift from one crop to other crop irrespective of cereals, pulses, vegetables and fruit crops there by damaging host plant and in turn reducing the productivity. Before we deal with different management strategies there is wide concern to address the climate change effects on the hosts as well environment as whole.

Keywords Insect-pest · CO_2 · Temperature · Plants · Climate change

---

M. Prashant · M. A. Waseem
Department of Entomology, University of Horticultural Sciences, Bagalkot, Karnataka, India

K. Managanvi (✉)
Department of Entomology, Dr. Kalam Agricultural College (Bihar Agricultural University, Sabour, Bhagalpur), Kishanganj, Bihar, India

Erayya
Department of Plant Pathology, Dr. Kalam Agricultural College (Bihar Agricultural University, Sabour, Bhagalpur), Kishanganj, Bihar, India

V. L. Rai
College of Agriculture Campus, (Archarya Narendra Dev University of Agriculture & Technology, Ayodhya), Kotwa, Azamgarh, Uttar Pradesh, India

S. S. Solankey, M. Kumari (eds.), *Advances in Research on Vegetable Production Under a Changing Climate Vol. 2*, Advances in Olericulture,
https://doi.org/10.1007/978-3-031-20840-9_14

14.1 Introduction

A regular swelling in the average heat of the Earth's troposphere and its oceans is termed as climate change and a transformation that is thought to be altering the Earth's weather endlessly. The overall temperature of the world has been progressively escalating from 1900 along with arise of approximately 1 °C. The highest upsurge has been observed in America, but in India's the raise in temperature was between 0.2 °C and 1 °C.Likewise, the rate at which globe is heating was increasing; in the last 50 years the temperature increase was twice as compared to the last 100 years. The average temperature from July to October season (*Kharif*) and November to March season (*rabi*), in India is expected to augment up to 1.7 °C and 3.2 ° C respectively along with anticipated increase of 10% in precipitation by year 2070 (Gupta 2011).

The dialogue of the fast few decades remains the climate change. Due to variations in climatic aspects *viz.*, temperature, rainfall, relative humidity and other meteorological gears will have impact on the superiority and amount of crop produce. Climate change is intimidating worldwide food production through straight effects in agricultural productivity and insect-pest and disease associated losses of crop plants. Every notch increase in temperature might cause 10–25% crop losses from insect pests. Intergovernmental Panel on Climate Change (IPCC 2007), defined it as "Change in climate over time, either due to natural variability or as a result of human activity". Over the last 50 years the maximum of global warming reported is due to the anthropological actions, as described by IPCC. The increase in the temperature of globe is the consequence of the greater greenhouse effect, triggered due to the higher concentrations of greenhouse gasses (GHG) viz. Chlorofluorocarbon (CFC), Methane (CH_4), Carbon-dioxide (CO_2) and Nitrous oxide (N_2O) in the surrounding environment. Over last 10 decades, the carbon-dioxide concentration in the atmosphere has gone up radically from280 ppm to 370 ppm and by 2100 it is expected to be doubled. According to IPCC (2007), temperature of globe has amplified by 0.6 + 0.2 °C and is predicted to attain 1.1–5.4 °C over the end of century. IPCC explains if 2 °C temperatures increase over the subsequent 100 years, then undesirable global warming effects might start to spread to maximum areas of the globe and it might straightly disturb many of the earth's living organisms. Global warming with increase in temperature and carbon dioxide along with climate variability of the environment may have numerous consequences in agricultural sector viz. change in precipitation pattern, extended period of drought and reduced crop production due to minor insects becoming major insect.

The climatic conditions which regulate much of the natural and agro-ecological systems during the season and species diversity in that particular area. In agro-ecological system, weather disturbs quality of crop and its harvest, as the insect pests and diseases were managed by natural enemies; the management practice that mainly ignored (De Bach 1964). The preceding valuation report from the IPCC forecasts, by the year 2100 an augmentation in 1.1 to 5.4 °C mean temperature (Meehl et al. 2007).

14.2 Vegetables in Climate Change Scenario

Vegetables are considered as defensive food as they source vital vitamins, nutrients and minerals to the human lives and are the greatest source for killing micronutrient insufficiencies. The universal production of vegetables has doubled over the last 25 years and the worth of international trade in vegetables now surpasses that of cereals. In Asia, produces are maximum in the eastern side where the weather is predominantly sub-temperate and temperate. India is the second major producer after China in the world in vegetables production with17.3 t/ha and 22.5 t/ha respectively. In the last20 years, the olericulture production in India has been amplified 2.5 times. Vegetables are more succulent (have 90% water) and are usually more delicate to weather extravagances (Solankey et al. 2019), so therefore higher temperature along with inadequate moisture in soil remain the chief reasons of reduced harvests because they significantly disturb many biochemical and physiological developments like decreased photo synthesis, enzymatic and metabolic transformation, hot air wound to the vegetable plant tissues, reduction in pollination and ultimately reduced fruit set.

The implications of the changes in climate seriously hit the vegetable production. Under the shifting climatic circumstances failure of crops, reduction in yields and quality along with higher incidence of insect pest and disease complications are natural and they tend to the vegetable farming a loss-making enterprise. Thus, eventually questions the accessibility of nutrient source in man's food. The South Asian seasonal monsoon will be arriving lately or becoming uncertain and that temperature upsurges will be dominant during the winter period. The disappointment of the reduced rains fallouts in shortage of water, ensuing in lesser average harvest of the crop. It is very much predominant in the many droughts hit areas viz. Rajasthan, northern Karnataka, eastern and southern Maharashtra Andhra Pradesh and Orissa. In the states like Karnataka, Andhra Pradesh and Tamil Nadu a high temperature along with insufficient precipitation at the sowing time and severe shower at the time of crop harvesting cause high crop losses.

The proportion of increase in minimum temperature throughout the Rabi season is significantly larger than in rainy season. The maximum temperature presented growing yearly in rainy season crop and rabi crop time periods nonetheless substantial increase was detected from 2000 year onwards. Ahead and noteworthy undesirable precipitation movements were detected many states of India. In spite of having a straight consequence on rain fed vegetable production, climate change disturbs water storage and water availability for irrigation. As the accessibility of water is partial, drought will become the foremost stress issue to vegetable growing, more stressing farming systems. Climate change happens to be the utmost important reason of biodiversity damage over the next 100 years, which causes fluctuations in species phenology, distributions and ecological interactions in the environment. For instance, most vegetable crops like Cole crops, onion and root crops are fertilized by insects as changes in pollinating insect species viz. honey bees, syrphids distribution will affect the pollination there by reducing the yield. Forays of agricultural crops by weeds can be additional big problem. Post-harvest value of vegetables was disturbed by seasonal variations may be by heavy yield losses, affect safety of food storage throughout storage period, by instigating fluctuations in fungi populations which are producing aflatoxin. The recurrent extreme weather actions under climate change might harm infrastructure, with destructive effects on storage and dispersal of vegetables. Along with the physiological and biochemical changes, climate change may impact the insect pest and disease incidence, host-pathogen interactions, distribution and biology of insects, time of appearance in cropping season, movement to new locations and their hibernating capacity. Growth and expansion of plant diseases largely depends on environment prevailing about the host and pathogen and a modification in the constituents may impact host vulnerability and therefore host, parasitoid and parasite relationship. Overall, global climate change has the capacity to change plant composition, mechanism of resistance and to modify expansion of the pathogen. Many other significant factors manipulating diseases in vegetable plant are air contamination, predominantly UV-B radiation and ozone along with nutrient accessibility.

14.3 Insect-Pests and Climate Change

Insects-pests which belong to arthropods are coldblooded which means they change the temperature of their bodies is roughly similar to the surrounding environment. So, the most important ecological factor affecting insect/hexapod development, behavior, dispersal, existence, and reproduction is undoubtedly the temperature. The carbon-dioxide emitted by anthropogenic activities is almost twice more important for temperature rise than other greenhouse gases altogether. Even though, augmented CO_2 would not straightly affect hexapods, the temperature surges determined by the intensification in CO_2 by human activities will affect the hexapods in their phenology, dispersal, nourishment and as a disease carrying vectors.

14.3.1 Crop Production Influenced by Climate Change in Three Ways

1. Fluctuations in precipitation, temperature and carbon dioxide stages directly on the host plant growth and well-being.
2. Indirect consequence on plant well-being via climatic persuaded fluctuations in plant and competitor circulation along with species richness.
3. Consequence on host plant well-being indirectly via deviations in advanced food chain communications of predators, parasitoids and competitors on plants and/or competitor dispersal along with richness.

The above-mentioned fluctuations in global temperature have radical impact on the financial system of farm oriented, biodiversity prosperous country like Republic of India. The global temperature raises is projected to have remarkable implications on numerous species on the earth (Table 14.1). The sign of a reply to long-term ecological trends, directly or indirectly connected to current global warming are remarkably quick and many fold particularly in insect species. It comprises movement towards northern and altitudinal in topographical choice, increased sum of generations for insect species especially in multivoltine, improved cold persistence along with extended hibernations. The climate change estimated patterns will contain manifold collaborative undeviating properties on the functioning herbivores and plant species as well. The ecological factors manipulating crop phenology like blossom bud maturation, strength in blossoming, period of development, produce of the crop laterally with superiority, insect pests, diseases occurrence, physiological ailments that might deliver whichever destruction of conventional places or ensure additional places becoming possible for farming of these crops, in all these ways the climate change looked to have obstructed. The communications between host plant and herbivores are manipulated by climate change in so many ways. To cite a few, the expected ecological fluctuations in CO_2concentrations, raincloud cover, temperature, nutrient and water accessibility will disturb host plant vulnerability to plant feeders (Johnson and Lincoln 1991). The herbivores dietary necessities, growth time and hibernation survival will directly be influenced by climate change (Solbreck 1991). Indirectly the herbivores will involve variations in foodstuff in addition to the above impacts. There will be occasional penalty at the stage of populations, niche, ecosystem and communities due to modification in nature and strength of various plant herbivore exchanges induced by climate change. The progress might constitute the capability of numerous cultivators of temperate nuts and fruits to fruitfully yield the similar quantity as in the history.

Table 14.1 Climate change impact on insect pests

Component	Means of action	Effect (+/−)	Influence
Increased CO_2	Host plant size and canopy density increase and raise in carbon nitrogen ratio	–	Additional amino acids extraction by herbivores by increased eating
		–	Fungal spore increases and influence disease causing agents
	Decreased breakdown rate could increases the crop residue	–	Increases in inoculum concentration at the start of the cropping season as the pest can overwinter

14.3.2 Directs Effects of Climate Change on Insect-Pests

14.3.2.1 Effect on Population Growth Rate

The temperature can put for the diverse effects due to the different developmental stages of hexapods. Species with soaring thermal acceptance will carry out superior during harsh and recurrent extreme temperature events than other species (Burgi and Mills 2010). In tropical and subtropical regions, the temperature increases inside certain favorable ranges speed up the growth rates, multiplication and existence of hexapods. Thereupon, hexapods might be competent in finishing a higher cycles per year and eventually resulting in additional yield losses (Bale and Hayward 2010). In recent times, it was clearly observed in case of pests like Aphids and plant hoppers. Pandi et al. (2018) projected that for every 2 °C rise in temperature, hexapods will enhance one to five extra cycles in development per planting season. In addition to sucking pests, elevated temperature benefits some lepidopterans in number of ways by escalating their flight, thus leading to increased mating accomplishment and egg laying and eventually larger brood expansion. Elevated CO_2 and temperature will increase activities of enzymes (e.g. midgut proteases, carbohydrates, and mitochondrial enzymes) and metabolic rates which in turn may guide to increased number of generations of insect pests (Akbar et al. 2016). Shrestha (2019) reported that insect causing yield losses could increase for every additional degree of temperature increase it would rise by 10–25%additionally.

14.3.2.2 Effect on Migrating Behavior and Habitat Ranges

Hughes et al. (2003) reported that rise in temperature likely to vigor the insect species to change their dispersals by intensifying into the newer zones of climate and by vanishing from places that have developed inappropriate climatically. Increasing temperatures will allow hexapod pests to transfer from subtropical and tropical areas to colder or temperate areas at increased elevations laterally with modifications in farming places host plant. Parry and Carter 1989 predicted that for every 1 °C increase in temperature might spread dispersal of insects140 km upwards in elevation or 200 km north. This is clearly indicating that global warming will cause more crop yield losses in temperate countries due to the expansion of insect range. At the same time, it makes them as vulnerable places for introduction of new invasive pests having potential threats to their agro ecosystems. Insect pests which increase its main host range outside those of their major predators and parasitoids potentially escaping biological pest control and may cause outbreaks. In near future, it is unfortunate to note that because of short life cycle and increased fighting ability of sucking insects, viral diseases may become prevalent and cause severe crop losses to farmers (Sharma et al. 2005). Several modeling studies predicted the range expansion of forest as well as agricultural insect pests. Few examples are given below. Global warming consequential heights wise host series enlargement and augmented hibernating existence of *Heliothis zea* (maize earworms) and cob worm *Helicoverpa armigera* might result in hefty decrease in crop yield and present chief obstacle for insect pests' management in corn (Diffenbaugh et al., 2008). Increase in temperature will let the unfriendly pink bollworm, *Pectinophora gossypiella* (Lepidoptera: Gelechiidae), to increase the appetite on cotton interested in previously unreceptive parts effected by substantial ices and harm percentage might rise through-out the existing areas (Gutierrez et al. 2006). The northward expansion of southern pine beetle, *Dendroctonus frontalis*, has stayed associated to upgraded circumstances for hibernating beetles may be in larval or adults' stages (Williams and Liebhold 2002).

14.3.2.3 Effect on Overwintering or Resting Periods

Hexapods undertaking hibernation are probable to have the experience the important thermal variations in their environment (Bale and Hayward 2010). Faster biological activity at advanced temperatures reduces the time period required for hexapod diapause that may be by faster reduction of stockpiled nutrient possessions. In winter, the global warming might postponement in beginning and initial summer season may help in quicker end of overwintering in hexapods, which can at that point recommence their vigorous development. Lower winter mortality of hexapods owing to hot midwinter temperatures might be significant in raising hexapod inhabitants (Hahn and Denlinger 2007). A warm, dry winter will aid aphid survival and increase population in wheat. The pupae of *Helocoverpa armigera* could emerge up to 7 days earlier from winter diapause when temperatures increase (Ouyang et al. 2016).

14.3.2.4 Effect on Abundance and Biodiversity

The comparative richness of different hexapods species might alter quickly owing to changes in climate, and the insect species incapable of withstanding the traumas might be vanished in the coming times (Jump and Penuelas 2005). As temperature increases, the species present in mountains or high latitudes are likely to be affected more comparatively to lower altitude and they will be enforced to modify their innate places to further advanced altitudes. Nonetheless, still if they are clever to transfer, they will ultimately run out of livable ranges and may unavoidably get destroyed. In addition to temperature, significant variations in precipitation will have a main effect on the richness and of diversity of insects. Severe shower may decrease the occurrence of sucking insect pests like thrips, aphids, leaf hoppers and whiteflies. Aberrant weather events cause the outbreak of insect pests for example, red hairy caterpillar outbreak may be seen due to heavy and frequent rains and extended dry spells followed by heavy precipitation reasons the epidemic of cutworm's pests (Sardana and Bhat 2016).

14.4 Indirect Effects of Climate Change on Insects through Host Plants

14.4.1 Effect on Host Plant and Insect Pest Synchrony

Normally, the natural timing of the insect lifecycle synchronizes with the lifecycle of the plants on which they feed. However, climate change has caused mostly negative consequences from the increasing asynchrony in ecological systems, due to variation in the phenological responses of insects and their host plants (Vissers and Holleman 2001). Many insect species feeding on specific plants will face pressure to adapt as the plants on which they feed undergo changes in their growth cycle. Consequently, predictable response in hexapods could comprise an advance in period for adult and larval emergence and upsurge in span of flying time (Menéndez 2007). Asynchrony in phenology of winter moth (*Operophtera brumata*, Lepidoptera: Geometridae) with its wood host Oak tree where moth eggs gamble with their own lives by hatching early (Van Asch and Visser 2007). It was observed the egg masses laid in 23 of 30 by the bay checker spot butterfly *Euphydrya seditha* (Lepidoptera: Nymphalidae) were emerged later the plants (Plantago and Castilleja) had gone through ageing, thereby none of the larvae from these batches attained the third instar stage (Singer and Parmesan 2010).

14.4.2 Effect of Increased CO_2 on Host Plants

Lamarche et al., 1984 believed that increased atmospheric CO_2 associated with global warming could stimulate plant growth because of increased photosynthesis rate of most of C3 plants, thus increasing the amount of food available which scientist originally believed that swelling carbon dioxide concentration might be an answer for global nourishment. Regrettably, these hopeful forecasts have not established precise. One motive for this is that hexapods also plague higher when florae are raised in higher levels of CO_2 owing to poor nutritious superiority of host plant. According to the "*Nutrition compensation hypothesis*," raised CO_2concentrations can disturb indirectly the growth fitness of plants by altering the nutritive value of host plants. It had negative effects on insects by diluting the nitrogen in leaf tissues by 15–25% and increases the C: N ratio due to buildup of non-structural sugars (Lincoln et al. 1993). Insects especially chewing insects such as moths and butterflies may revealed to reply this proportion by accelerating eating in order to achieve the metabolic requirements for nitrogen to derive more amino acids called "*compensatory feeding*". However, xylem and phloem eating hexapods such as the hemipterans may be least exaggerated by raised CO_2 levels (Petzoldt and Seamann 2010). After investigating the effects of elevated CO_2 on many lepidopteran pests, several studies revealed that it lengthy larval and pupal period and reduced pupation rate and pupal mass. It also resulted in increased relative consumption rate, decreased efficiency of conversion of ingested as well as digested food and low relative growth rate. It has been designated that raised CO_2, increased 22% abbreviated tannins, 19% phenols, and 27% flavonoids whereas terpenoids nitrogen based secondary metabolites decreased by 13% and 16%, respectively (Robinson et al. 2012). Under elevated CO_2 (> 550 ppm) augmented digestibility, and reduced effectiveness of alteration of swallowed diet into body matter have been documented in four consecutive cycles of caster semilooper, *Archaea janata* (L.) under raised CO_2 (Srinivasa Rao et al. 2013). Higher activity of digestive and mitochondrial enzymes thereby increased rate of consumption of *Helicoverpa armigera* was observed by Akbar et al. (2016). Chen et al. (2004) reported that increase in population of *Sitobion avenae* (wheat aphid)*:* Prasannakumar et al. (2012) on *Nilaparvata lugens* (Brown Plant Hopper) while Sudderth et al. (2005) on *Macrosiphum euphorbiae* (Potato aphid) due to higher CO_2 levels.

14.4.3 Increased Temperature and Insect-Pests

Many effects of augmented temperature on hexapod routine have to do with the straight effects on insects due to temperature. As hexapods are exothermal, in warmer circumstances the insects incline to be more energetic. It is projected that for every 2° C temperature upsurge hexapod will have one to five added life sets in every season. The dietary quality host plants in will be increased due to higher photosynthetic activity as increase in temperature raises. It has been revealed that, to improve the plant feeding by hexapods indirectly by altering the development probability of host plants, which turn lead to the epidemic of insect pests in that area (Visser and Both 2005).

In few insects like thrips the gender percentages may be altered by temperature possibly affecting parthenogenic speed. Hexapods which devote significant time of their life cycle on the earth may be additionally exaggerated by temperature variations than individuals that are just above soil just for the reason that soil offers a protecting intermediate that determine or incline to shield temperature variations more than the air. Lesser winter death of hexapods due to warmer winter temperatures might be significant in swelling insect inhabitants. With higher latitude and altitude, the hexapods species variation per area leans towards to decline, indicating that increasing temperatures could consequence in higher hexapod species feeding a greater number of hosts plants in temperate weathers. In addition to above the production of plant secondary metabolites and some other plant distrustful characters will be affected and may become vulnerable to insect attack. Further, some other studies have revealed the contrary effects of raised temperature on plants, together with altered developing time period and enlargement in the usage of plant possessions for other plant feeders. Zhang et al. (2018) observed the noteworthy reduction in the larval period, pupal period of *Spodoptera litura* nourished on soybean raised in elevated temperature in divergence to ambient temperature. Primary start of invasion by *Helicoverpa armigera* in pulses and cotton in North areas of

India may be due to reduced host defenses as a result of temperature stress (Sharma et al. 2005).

14.4.4 Precipitation Pattern and Insect-Pests

Under the changing climatic conditions early and appropriate sowing become more ambiguous. For example, with temperature and rainfall variations can influence hexapod pest parasitoids, predators and diseases ensuing in a multifaceted effect. Spores of fungus affecting hexapods are influenced by high relative humidity and their occurrence would be augmented by changes climate which increase time of high relative humidity and decreased by those individuals that lead to drier situations. Most of the smaller sized insects are delicate to rainfall and are get murdered or detached from host plants under heavy precipitation, and the deliberation is significant during selecting proper management practices for onion thrips.

14.4.5 Combined Effect of Elevated CO_2 and Temperature on Insect-Pests

Overall, the herbivores cultivated in raised carbon dioxide are lesser nourishing to hexapod herbivores, as they might affect their behavior and performance. Physical appearance of host-plant deviates characteristically make foliage material plagued by hexapods which less healthy. As a result of this the hexapods have a more problematic time altering the nutrition they eat into biomass. To alleviate the diet which is of less nutritive, hexapod herbivores frequently eat higher quantity. Zvereva and Kozlov (2006) assessed that the level of nitrogen concentration in plants is decreased both under higher CO_2 and higher temperature conditions.

In plants the ratio of carbon and nitrogen was increased under raised temperature and CO_2 conducts, nonetheless the organic carbon related minor composites did not display an important reply to intensifications in any of the temperature and CO_2. It was also reported that plant feeder act in relations to existence, growth of rate, pupal weight and egg laying were undesirably exaggerated by higher CO_2 unaided, but then again definitely exaggerated by raised temperature; when studied concurrently, as two features had none of the evidences on the hexapod act (Cornelissen 2011). Antixenotic properties possessed by plants may also protect automatically, may be possessing hard leaf surfaces or by possessing leaf hair, scales, trichomes like structures on it.

The increased carbon dioxide led to better success of natural enemies praying upon the their pray as it is more visible. As the hexapods have more life cycle stages and typically take more time to develop, so that they are more visible in time to the attack of natural enemies. Higher leaf damage and frass production due to increased consumption rates are also the signals to natural enemies. However, recent studies observed detectable variation in insect population due to the interactive effect of elevated CO_2 and temperature. Pandi et al. (2018) observed that elevated CO_2 and temperature had showed encouraging effect on brown plant hopper reproduction, increasing its inhabitants (55.2 ± 5.7hoppers/hill) in contrast to normal carbon dioxide and temperature (25.5 ± 2.1 hoppers/hill). In a China based study, Zhang et al. (2018) observed the decrease in relative growth rate (RGR), Efficiency conversion of ingested food (ECI), Efficiency conversion of assimilated diet and increase of the actual daily when *S. litura* fed on a soybean cultivar.

Even Karthik et al. (2021) suggested that Temperature and elevated CO_2 has a strong effect on insect growth, survival and reproduction and enrolls a major role in controlling the development and growth of their host plants. In addition, the development of plant secondary chemicals as well as the structural characteristics used to protect against herbivores are influenced by temperature. Thus, for both insects and plants, temperature has potentially significant consequences (Fig. 14.1).

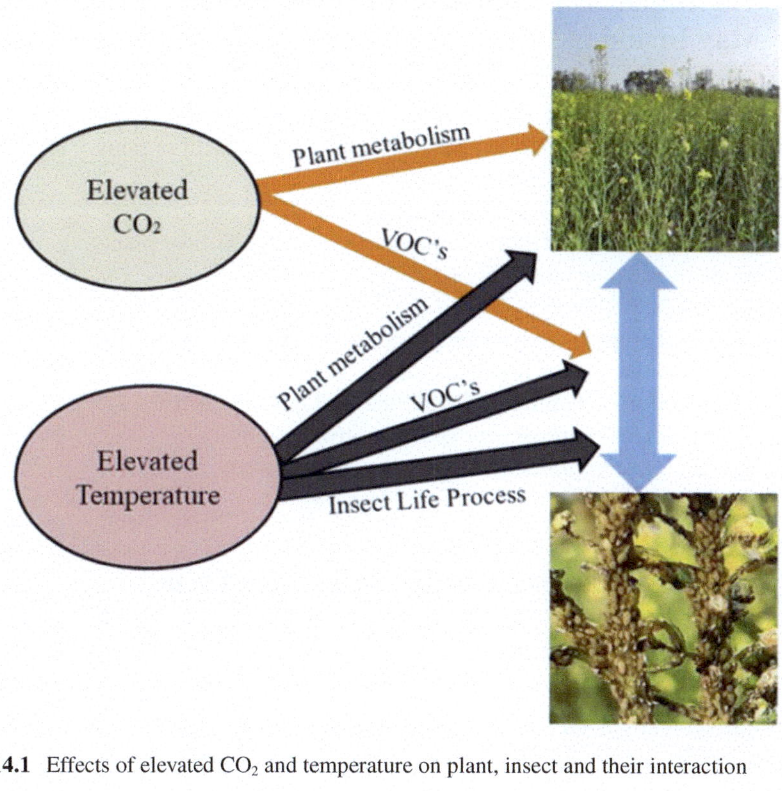

Fig. 14.1 Effects of elevated CO_2 and temperature on plant, insect and their interaction

14.4.6 Effect on Host Plant Distribution

Global warming can also influence the geological spreading of plant species and their growth outlines. There will be variations in the richness and dispersal of its grown and non-grown plants would the insect pest dispersal. Many new insect pests may notice, while some insect pests may turn out to be invasive by reaching new areas due to extended range of their host plants. Wang et al. (2017) in their modeling study on international dispersal of Colorado potato beetle revealed the risk of invasion of the beetle from its native range. Climate change indirectly can also impact biotic dispersal of vector borne viruses and diseases triggered by parasite will increase. In some part of the globe, there will be bigger disease epidemics and some other part of the world may eyewitness reduced outbreak of diseases too based on the type of disease parasite or vector accessible at a given time and space (Nwaerema 2020).

14.5 The Impacts of Climate Change on Insect-Pests May Include

- Variation in diversity and richness of hexapod pests
- Variations in pests' dispersal among different geographical regions
- Amplified hibernation in insects
- Swift population build up and number of cycles
- Fluctuations in synchronization of insect pests and their host plants
- Deviations in resistance of host plants
- Changes in hexapod biotypes
- Variations in tritrophic exchanges
- Influence on extinction of insect species
- Changes in action and comparative richness of natural enemies
- Amplified threat of invasive pest species introduction
- Effectiveness of crop protection skills may be reduced

14.6 Impact of Climate Change and Insect-Pests

14.6.1 Rising Temperature

Temperature is recognized as main abiotic factor straight affects herbivorous hexapod pests. Insects being poikilothermic, have temperature of their bodies is similar as that of the surrounding situation. So, the growth rates of the insect's life cycle phases are sturdily reliant on temperature. Many hexapods will be affected to some extend for every notch variation in temperature and so, insect life histories will be affected manifold due to this. Fleming and Volney (1995), Fye and McAda (1972), and Cammell and Knight (1991) based on many experiments models and support the notion that the ecology of hexapod pests is probable to reply to amplified temperatures. Through each notch increase in worldwide temperature, the overall life history of hexapod pests will be lesser as compared to normal temperature. The faster the insect life cycle, the more developed might be the inhabitants of insects. In colder climate, many of hexapod stake their development time throughout the hot portions of the year thereby the insect species which has a definite climatic regime whose niche area is well-defined, might reply additionally to changes in climate while individuals in which the place is unfinished by other living or nonliving factors will be fewer expectable (Bale and Hayward 2010). In the earlier situation, the overall forecast is that if temperatures of the globe upsurge, the insect species in question might swing their terrestrial ranges nearer to the poles or to advanced raises and increase their inhabitant's population (Harrington et al. 2001; Sutherst 2000; Bale and Hayward 2010; Samways 2005). How the raise in temperature affects the insect pests is depicted in Fig. 14.2 (Skendzic et al. 2021).

Due to climate change, there will be rise in global temperature which ultimately influence the hexapod pests in many ways as.

(a) terrestrial areas expansion.
(b) more hibernation and aestivation.
(c) fluctuations in inhabitants' development rate.

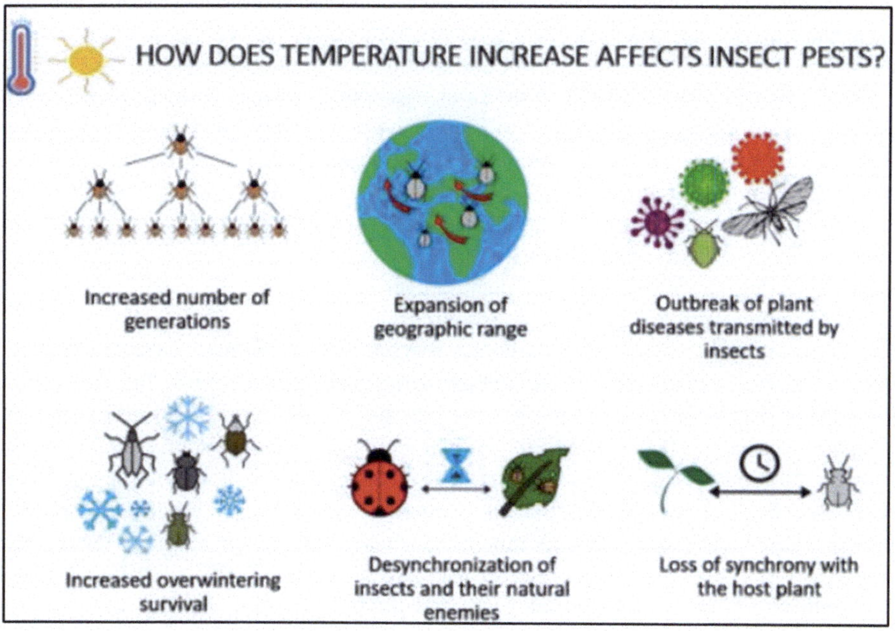

Fig. 14.2 Effect of rise in temperature to the insect pests

(d) amplified sum of life cycles.
(e) delay of expansion time.
(f) fluctuations in insect and crop synchronization.
(g) deviations in inter and intra-specific communications.
(h) augmented dangers of incursions and.
(i) entry of different hosts.

14.6.1.1 Changes in Insect-Pests Diversity

Increased realization of the deterioration of biological systems over the last couple of decades has led to a better appreciation of the loss of genes, species and ecosystems. In the overall animal kingdom insects play a major and dynamic role in providing many amenities of the ecosystem (Kannan and James 2009; Kremen et al. 1993). Hexapods multiplicity in an environment designates the healthiness of an environment as they are actual decent pointers of changes in environment (Gregory et al. 2009), which show a significant part in dietary chains and food webs. Alfred (1998) reported that in India approximately 6.83% of global hexapod species are present. The changes in climate might disturb the comparative wealth of unlike hexapod species and the incapable species to familiarize the variations may be vanished over the period of time (Thomas et al. 2004). Anand and Pereira (2010) and Hampson (1908) reported that in India western ghats is the only habitation to many species of colorful moths and butterflies which are erratic, endemic in the world. At present numerous butterfly species are under an actual danger just because of exhaustion of the normal flora for numerous man-made happenings as reported by Costanza et al. (1987), Sachs (2008), and Sidhu and Mehta (2008).

There is an essential to rise functional multiplicity in agricultural ecosystems susceptible to changes in climate to recover structure flexibility, and reduce the number of fatalities owing to hexapods (Newton et al. 2011). Still, variations in planting designs as a consequence of global warming might radically disturb the equilibrium amongst hexapods and their biocontrol agents. Foremost impact of changes in climate on hexapod pests and natural enemies 'effect in diminished richness of predators, parasitoids and decomposers and higher plant feeding, which might have undesirable significances for building and facilities of the whole ecosystems. Answers of hexapod pests and natural enemies hang on both rainfall and temperature and ecosystem inclusive antagonistic problems are probable to higher beneath projected climatic changes (Zvereva and Kozlov 2006). Significances of temperature rises of 1-2 °C might be equivalent in size to the presently seen climate change (Bokhorst et al. 2008). Intensification in rainfall will mainly disturb the insect species with meager distribution competences, which will bind their capability to enlarge their native range (Xannepuccia et al. 2009). Huge gauge of variations in precipitation due to changes in climate might have a main effect on the richness and multiplicity of both hexapod pests and natural enemies. In case of extreme drought due to climate change are expected to decline trophic variety and alter the arrangement of ecosystem (environment and organisms). Habitat disintegration consequences in the subdivision of habitats into lesser units ensuing in their amplified narrowness as well as loss of total habitat area. Such disintegration deviates the local weather at the split limits and limits effects comprise microclimate, light fluctuations, relative humidity and temperature individually these can have a noteworthy influence on the liveliness.

14.6.1.2 Extension of Geographical Areas

The topographical dispersal and richness of animal and plant kingdom in nature is identified by insect specific weather necessities vital for their development, existence and replication. Circulation in insects will rise if there is small increase in temperature. One projection state that for 1 °C increases in temperature might authorize swiftness of 200 km north or 40 m elevation in upward direction. Parts that are non-favorable at existing due to less temperature might convert favorable with increase in temperature. Minimum temperature compared to maximum temperature act as significant player in influential the worldwide spreading of hexapod species, henceforth any surge in temperature might upshot in a better capability to hibernate at higher elevations, eventually producing a change. Numerous hexapods have topographical arrays that are non-straightly restricted due to foliage, nonetheless in its place are delimited by temperature. Prior investigations have revealed that whenever there will increase in temperature, the hexapods are predictable to spread their topographical areas at higher elevations laterally with variations in farming areas of their host species (Hill and Dymock 1989; Sharma et al. 2005; Kuchlein and Ellis 1997). The heating in pleasant area may direct to reduction in comparative richness of temperature subtle hexapod inhabitants (Sharma et al. 2005, 2010; Petzoldt and Seamann 2010). Typically, the Glacial areas are unnatural since the hexapod outburst sowing to less temperature then regularly happening ices (Volney and Fleming 2000). In the upcoming period, predictable global heating (Carroll et al. 2004) and amplified famine occurrence (Logan et al. 2003) is anticipated to source additional recurrent hexapod occurrences in mild areas too. As the species prosperity of hexapods tends to surge with temperature, it is imaginary that with arise in temperature additional species will be increased than lost. Hibernating survival and timing of the instigation spring are significant at higher latitudes, leading to residents increase. Choice allowance in *Helicoverpa armigera*, a key insect instigating damage to vegetables and many other crops in India is projected with global heating (Sharma et al. 2005). Afterward, the continuing changes in hexapod circulation and array owing to altering weather might change local construction, multiplicity and working of ecosystems (IPCC 2007).

14.6.1.3 Changes in Insect Phenology

Along with the modification in planetary of insect species dispersals, changes in climate led to an environmental swing in period, by means of fluctuations in insect species phenology (timing of life history stages insects). Root et al. (2003) noted that was ace of the coolest influences of changing climatic conditions to screen and is utmost recognized in this esteem for an extensive variety of creatures. Many of the insect species might restrict their larval stages for shorter period and may emerge as adults earlier than required due to the increase in temperature. So, predictable replies in hexapods might comprise an early payment in the judgment of larval as well as adult appearance with an upsurge in the duration of the flying time (Menéndez 2007). Moths and butterflies are the finest illustrations of such seasonal changes. Roy and Sparks (2000) reported variations in lepidopteran phenology in United Kingdom, where in 26 of 35 insect have increased their primary presence. Harrington et al. (2007) described premature adult emergence and an initial influx of migrant species might similarly been described for aphid's insect pests in the United Kingdom. The influences climate changes on four hexapod species like honey bee, butterfly, fly and beetle was examined in Mediterranean and specified that entire insect species showed variations in their initial arrival day for the preceding five decades, which was associated with upsurges in spring season temperature (Gordo and Sanz 2005).

The fluctuations in insect phenology might be deliberated over continuing researches by adjustable planting times for witnessing the arrival of insects on plants. Similarly, the effectiveness of influx of hexapods can also be documented by different lures like light, suction trap or pheromone trap. Examination of continuing information on seasonal variation would disclose variations in the judgments of insect pest arrival under the changing climate scenario (Pathak et al. 2012). Equally, enduring statistics from numerous insect pest recording systems in Europe and North America have given indication for species becoming active, drifting or reproducing quicker in the year due to rises in temperatures that lead directly to augmented development rates or earlier appearance from winter hibernation (Roy and Sparks 2000). Swelling temperatures have also acceptable a many insect of species to persist dynamic for a lengthier time throughout the year or to rise fecundity. The nodal agency ICAR through its All India Coordinated Rice Improvement Programme (AICRIP) have also extensive system of the collecting data from each center on insect light trap throughout the year. Investigation of past trap data compared with present facts can deliver significant evidence on the ill effects of changes in climate on paddy insect.

14.6.1.4 Increased Hibernating Existence

Being ectotherms, hexapods have restricted capability of self-regulation with outside changes in temperature. Bale and Hayward (2010) revealed that a variety of insect established ways such as interactive evasion over relocation and adaptations in physiology like overwintering to back life span under thermally traumatic situations. The diapause is a deferred evolving period happening, the display of which is administered by ecological issues like relative humidity, photoperiod and temperature. The hibernation as an adaptive characteristic, plays vigorous role in periodic instruction of hexapod life cycles because of which the hexapod stake healthier benefit to live countless deal of ecological hardships. According to Chapman (1998), there are two key types of diapause in hexapods. One is aestivation and another is hibernation to with stand high and low temperature respectively during the life cycle. The researchers (IPCC 2007; IMD 2010) have revealed that increase in temperature is happening highest at high latitudes particularly in winter season than in summer. Observing at the historical 10 decades weather outline of India, global heating was noticeable higher all through midwinter and it was the maximum temperature not the least temperature where noteworthy upsurge was detected (IMD 2010). Bale and Hayward (2010) thus observed the hexapods experiencing hibernation are expected to practice the utmost noteworthy variations in their thermal situation.

The diapause time of insects might be increased at higher temperature and due to high metabolism leads to quicker exhaustion of deposited food capitals (Hahn and Denlinger 2007). Increase in temperature in cold season might lead to postponement in beginning and initial summer might lead to quicker relieving from diapause in hexapods, which can then recommence their energetic development and expansion (Harrington et al. 2001).

14.6.1.5 Increase in Number of Generations

Yamamura and Kiritani (1998) observed that with a temperature rise 2 °C, hexapods may have the capacity to increase life cycles to the tune of one to five per year supplementary to the existing. Some insects advance additional quickly through ages of period with appropriate temperatures. Augmented temperatures will quicken the growth of these kinds of hexapods may be ensuing in additional cycles per year (Awmack et al. 1997).

Yamamura and Kiritani (1998); Petzoldt and Seamann (2010) observed that the universal problem of global warming is within convinced promising range might hasten the rates of expansion, multiplication and existence in insect pests of tropical and subtropical regions. Thus, hexapods will be skilled of implementation a greater cycle within same period of time. Increase in temperature might reduce the incidence of heavy cold measures, which may in turn enlarge the hibernating area for hexapods (Patterson et al. 1999).

14.6.1.6 Introduction of Invasive Alien Species

The changes in climate can as well encourage introduction and establishment of the exotic insect species. Danger of introduction invasive alien hexapod species, upsurge with worldwide climate change. Dukes and Mooney (1999), IPCC (2007) revealed that for the biological incursions many reasons may exists and they are many-sided fluctuations in abiotic, biotic machineries of the atmosphere and are named as chief influencers of insect species assault. World agricultural trade globalization and liberalization joined with the speedy transportation and communication means these days have considerably and reasonably increased the probabilities of exotic introductions. The CBD i.e., Convention on Biological Diversity, is of the view that introduction of new species is the utmost danger to damage of diversity of that biosphere (Mooney and Hobbs 2000) and levy huge prices to farming and marine ecologies by changing their local construction, variety and working (Timoney 2003; Sutherst 2000).

14.6.1.7 Outbreaks and Population Dynamics of Insect Pests

The population dynamics is the feature of population ecology dealing with aspects distressing changes in population concentrations. The seasonal effects of climate and enduring fluctuations in climatical circumstances resolve straightly paves way for alterations in spreading and expansion of hexapod pests. According to Bale and Hayward (2010) the fluctuations in neighboring temperature surely include modifications in voltinism, expansion speed and existence of hexapods and later turn upon size, compactness and hereditary alignment of inhabitants and also on the degree of plant manipulation.

According to IPCC (2007) it might lead in distressing environmental equilibrium since random deviations in the inhabitants of insects along with natural enemies like predators and parasitoids. Variations in climatic factors might lead to augmented occurrence and increase in strength of epidemics hexapod pests (Rao et al. 2009). Epidemic of sugarcane aphid *Ceratovacuna lanigera* in cane growing areas of Karnataka and Maharashtra during 2002–2003 lead to 30% crop decrease (Joshi and Viraktamath 2004; Srikanth 2007). These circumstances of augmented and recurrent nuisance to the plants have made additional giant hole in the bags of previously upset growers by enhancing the input charge of crop defense and plummeting the percentage of return.

14.6.1.8 Crop-Pest Exchanges

The capacity of a plant feeding hexapods to finish its growth hinge on the variation equally, the host plant and the ecological environments. The global change of climate has been initiated to apply both the effects from bottom to top and from top to bottom on the tri-tropic communications amongst host crops, hexapods and their predators and parasitoids by the way of convinced biological variations particularly connected to host-suitability and dietary standing (Coviella and Trumble 1999; Gutierrez et al. 2008). This has been observed in gypsy moth feeding on sugar maple and red maple which had condensed larval weight, augmented eating time and lengthy development (Williams et al. 2000). The big epidemics seen in development parts on the novel host's plants might be clarified whichever by the more vulnerability of the host's plants or by the helplessness of predators and parasitoids to find the larvae of moth on a rare host (Battisti et al. 2006). Cleland et al. (2007) reported that photoperiod and temperature have remained to disturb intensely the serious measures viz., shoot extension, blossoming and fruit setting in the life sequence of plants.

14.6.1.9 Augmented Occurrence of Hexapod Vectored Diseases of Plant

Changes in climate might give way to increased occurrence of hexapods transmitted many plant viruses through host range extension and quicker production of hexapod vectors (Sharma et al. 2005; Petzoldt and Seamann 2010). Robert et al. (2000) in his findings reported that amplified temperatures, mostly in the initial season found to increase the frequency in potato due to initial establishment of aphids which contains virus, as it is the chief potato viruses vector potato growing areas. Amplified concentrations of CO_2, one of the more evaluated features of global warming is the consequence of swelling meditations of carbon dioxide on florae. Florae basically made up of carbon and raised carbon dioxide concentration permit them to produce additionally quickly as they might adapt carbon very fast.

14.7 Tactics to Alleviate the Bad Effects of Climate Change

Cultural practices in the field, predator and parasitoid, plant resistance mechanisms, insecticides of plant and animal origin and organic insecticides are now in existence which are extensively used for management of pest and disease. Though, most of the approaches of insect pest management are extremely delicate to the surrounding weather. So, keeping in mind the changes in climate there is need to plan the suitable pest's management methods for effective control of pests and diseases.

- the plants existential struggle to hexapods is utmost the ecological approachable mechanism for the management. It is significant to categorize and advance in more varietal release which are steady in appearance of confrontation to the marked pests.
- Recently developed genetically modified plants have been governing around many of the stubborn insect pest infestations.

- Crop modification is best approaches of growing the action profusion of predators and parasitoids. It is the requirement of the hour to progress plant varieties that are welcoming to the predators and parasitoids.
- It is also need of the hour to study and formulate the synthetic organic insecticides based on the changing climatic factors which are most effective and least affecting the environment.

Lastly, there is a necessity to use the newer methodologies in pests' management like IPM practices that look into deliberation the alteration in insect pest range, patterns of florae and efficiency of diverse mechanisms of insect supervision for justifiable plant manufacture.

14.8 Conclusion

Hexapod has a marvelous influence on vegetable plants. The modification in circumstances of climate is a acknowledge threatening many for fruit and vegetable crops. Predictable changes in climate may also consequence in increasing the number of pesticide application as there will be increase in number of insect cycles per season. Evolutionary variations may ambiguous our capability to distinguish species reply to global warming- consequently, insects retort contrarily to fluctuations in weather. As many communications mechanisms exists and it is very difficult to forecast the influence on insect pests due to changes in climate in the future. Existing stride of ecological modification stresses novel methods to hasten the acceptance of innovative IPM practices. Plant and insect pest controllers might have to develop 'nimbler' in demand to manage with the quickening vitality of changes in climate.

It is an essential of the hour for policy makers in administration, business and other strategy architects to recognize the broader problems in upholding actual IPM systems in the aspect of modification and acknowledgement assumed to the kind of data desirable. Improved appraisal of existing hereditary capitals to recognize auspicious agreements for such insect is a decision for justifying harm by such deviations in concentration of precise insect. There is a necessity to emphasis on calibration of insect control strategies for usage, in circumstance of altered climatical situations. Specified the magnitude of the encounters we are facing to alleviate the influences of changes in climate, the period is accurate to strengthen goal slanted towards exploration within many disciplines in the direction to deliver the immediately required and lucrative practical answers for maintainable hexapod control. The greatest economic approach for crop cultivators to track is to use of IPM practices to carefully observe hexapods and viral incidence.

References

Akbar SM, Pavani T, Nagaraja T, Sharma HC (2016) Influence of CO_2 and temperature on metabolism and development of *Helicoverpa armigera* (Noctuidae: Lepidoptera). Environ Entomol 45:229–236

Alfred JRB (1998) Faunal diversity in India: an overview. In: Alfred JRB (ed) Faunal diversity in India. ENVIS Centre, zoological survey of India, Calcutta, pp 1–495

Anand T, Pereira GN (2010) Butterflies-the flying jewels of the Western Ghats. http://www.daijiworld.com/chan/exclusive_arch.asp?ex_id=1003. 2008, Accessed on 28 Dec 2010

Awmack CS, Woodcock CM, Harrington R (1997) Climate change may increase vulnerability of aphids to natural enemies. Ecol Entomol 22:366–368

Bale JS, Hayward SAL (2010) Insect overwintering in a changing climate. J Exp Biol 213:980–994

Battisti A, Stastny M, Buffo E, Larsson S (2006) A rapid altitudinal range expansion in the pine processionary moth produced by the 2003 climatic anomaly. Glob Chang Biol 12:662–671

Bokhorst S, Huiskes A, Convey P, van Bodegom PM, Aerts R (2008) Climate change effects on soil arthropod communities from the Falkland Islands and the maritime Antarctic. Soil Biol Biochem 40:1547–1556

Burgi LP, Mills NJ (2010) Cold tolerance of the overwintering larval instars of light brown apple moth *Epiphyas postvittana*. J Insect Physiol **56**:1645–1650

Cammell ME, Knight JD (1991) Effects of climate change on the population dynamics of crop pests. Adv Ecol Res 22:117–162

Carroll AL, Taylor SW, Regniere J, Safranyik DL (2004) Effects of climate change on range expansion by the mountain pine beetle in British Columbia. In: Shore TL, Brooks JE, Stone JE (eds) Mountain pine beetle symposium: challenges and solutions. Natural Resources Canada, Canadian Forest Service, Pacific Forestry Centre, Victoria, British Columbia, pp 223–232

Chapman RF (1998) The insects-structure and function, 4th edn. Cambridge university press, p 788

Chen FJ, Wu G, Ge F (2004) Impacts of elevated CO_2 on the population abundance and reproductive activity of aphid, *Sitobion avenae* Fabricius feeding on spring wheat. J Appl Entomol 128:723–730

Cleland EE, Chuine I, Menzel A, Mooney HA, Schwartz MD (2007) Shifting plant phenology in response to global change. TREE 22:357–365

Cornelissen T (2011) Climate change and its effects on terrestrial insects and herbivory patterns. Neotrop Entomol 40:155–163

Costanza R, d'Arge R, de Groot R, Farber S, Grasso M, Hannon B, Limburg K, Naeem S, O'Neill RV, Paruelo J, Raskin RG, Sutton P, van den Belt M (1987) The value of the World's ecosystem services and natural capital. Nature 387:253–260

Coviella CE, Trumble JT (1999) Effects of elevated atmospheric carbon dioxide on insect-plant interactions. Conserv Biol 13:700–712

De Bach P (1964) Biological control of insect pests and weeds, p 350. Reinhold Publishing Corporation, New York

Diffenbaugh NS, Giorgi F, Pal JS (2008) Climate change hotspots in the United States. Geophys Res Lett 35:L16709. https://doi.org/10.1029/2008GL035075

Dukes JS, Mooney HA (1999) Does global change increase the success of biological invaders? TREE 14:135–139

Fleming RA, Volney WJ (1995) Effects of climate change on insect defoliator population processes in Canada's boreal forests: some plausible scenarios. Water Air Soil Pollution 82:445–454

Fye RE, McAda WC (1972) Laboratory studies on the development, longevity, and fecundity of six Lepidopterous pests of cotton in Arizona. USDA, Washington, DC

Gordo O, Sanz JJ (2005) Phenology and climate change: a long-term study in a Mediterranean locality. Oecologia 146:484–495

Gregory PJ, Johnson SN, Newton AC, Ingram JSI (2009) Integrating pests and pathogens into the climate change/food security debate. J Exp Bot 60:2827–2838

Gupta HS (2011) Climate change and Indian Agriculture: Impacts, mitigation and adaptation. In: Proceedings of Xth Agricultural Science Congress on Soil, Plant and Animal Health for Enhanced and Sustained Agricultural Productivity. ICAR-NBFGR, Lucknow, India, pp 73–81

Gutierrez AP, D'Oultremont T, Ellis CK, Ponti L (2006) Climatic limits of pink bollworm in Arizona and California: effects of climate warming. Acta Oecol 30:353–364

Gutierrez AP, Ponti L, d'Oultremont T, Ellis CK (2008) Climate change effects on poikilotherm tritrophic interactions. Clim Chang 87:S167–S192

Hahn DA, Denlinger DL (2007) Meeting the energetic demands of insect diapause: nutrient storage and utilization. J Insect Physiol 53:760–773

Hampson GF (1908) The moths of India- supplementary papers to the volume in the Fauna of British India. Ser.III, Pt.X. J Bombay Nat Hist Soc 18:257–271

Harrington R, Fleming R, Woiwod I (2001) Climate change impacts on insect management and conservation in temperate regions: can they be predicted? Agric For Entomol 3:233–240

Harrington R, Clark SJ, Weltham SJ, Virrier PJ, Denhol CH, Hullé M, Maurice D, Roun-sevell MD, Cocu N (2007) Environmental change and the phenology of European aphids. Global Change Biol **13**:1556–1565

Hill MG, Dymock JJ (1989) Impact of climate change: agricultural/horticultural systems. DSIR Entomology Division Submission to the New Zealand Climate Change Program. Department of Scientific and Industrial Research, Auckland, New Zealand, p 16

Hughes CL, Hill JK, Dytham C (2003) Evolutionary trade-offs between reproduction and dispersal in populations in expanding range boundaries. Proceedings of the Royal Society of London. Series B: Biolog Sci 270(2):147–150

IMD (2010) Annual climate summary. India meteorological department. Pune. Government of India, Ministry of Earth Sciences, p 27

IPCC (2007) Impacts, adaptation and vulnerability: scientific-technical analyses -contribution of working group II to the fourth assessment report of the intergovernmental panel on climate change, p 976. Cambridge University Press, Cambridge, UK

Johnson RH, Lincoln DE (1991) Sagebrush carbon allocation and grasshopper nutrition: the influence of CO_2 enrichment and soil mineral nutrition. Oecologia 84:103–110

Joshi S, Viraktamath CA (2004) The sugarcane woolly aphid, *Ceratovacuna lanigera* Zehntner (Hemiptera: Aphididae): its biology, pest status and control. Curr Sci 87:307–316

Jump AS, Penuelas J (2005) Running to stand still: adaptation and the response of plants to rapid climate change. Ecol Lett 8:1010–1020

Kannan R, James DA (2009) Effects of climate change on global diversity: a review of key literature. Tropical Ecol 50:31–39

Karthik S, Sai Reddy MS, Yashaswini G (2021) In: Stuart A (ed) Climate change and its potential impacts on insect-plant interactions. The Nature, Causes, Effects and Mitigation of Climate Change on the Environment. Global Warming and Climate Change, Harris, pp 01–23

Kremen C, Colwell RK, Erwin TL, Murphy DD, Noss RF, Sanjayan MA (1993) Terrestrial arthropod assemblages: their use in conservation planning. Conserv Biol 7:796–808

Kuchlein JH, Ellis WN (1997) Climate-induced changes in the micro lepidoptera fauna of the Netherlands and the implications for nature conservation. J Insect Conserv **1**:73–80

Lamarche VC, Graybill DA, Fritts HC, Rose MR (1984) Increasing atmospheric carbon dioxide: tree ring evidence for growth enhancement in natural vegetation. Science 225(4666):1019–1021. https://doi.org/10.1126/science.225.4666.1019

Lincoln DE, Fajer ED, Johnson RH (1993) Plant-insect herbivore interactions in elevated CO_2 environments. Trends Ecol Evolut **8**(2):64–68

Logan JA, Regniere J, Powell JA (2003) Assessing the impacts of global warming on forest pest dynamics. Frontiers in Ecol Env 1:130–137

Meehl GA, Stocker TF, Collins WD, Friedlingstein P, Gaye AT, Gregory JM, Kitoh A, Knutti R, Murphy JM, Noda A, Raper SCB, Watterson IG, Weaver AJ, Zhao ZC (2007) Global climate projections. In: Solomon S, Qin D, Manning M, Chen Z, Marquis M, Averyt KB, Tignor M, Miller HL (eds) Climate change 2007: the physical science basis. Contribution of working group I to the fourth assessment report of the intergovernmental panel on climate

change. Cambridge University Press, Cambridge, United Kingdom and New York, NY, USA, pp 747–845
Menéndez R (2007) How are insects responding to global warming? Tijdschrift voor Entomologie 150:355–365
Mooney HA, Hobbs RJ (2000) Invasive species in a changing world. Island Press, Washington, DC
Newton AC, Johnson SN, Gregory PJ (2011) Implications of climate change for diseases, crop yields and food security. Euphytica 179:3–18
Nwaerema P (2020) Impact of climate change on insects, pest, diseases and animal biodiversity. Inter J Environment Sci Nat Res 23(5):151–153
Ouyang F, Hui C, Men X, Zhang Y, Fan L, Shi P (2016) Early eclosion of overwintering cotton bollworm moths from warming temperatures accentuates yield loss in wheat. Agricult Ecosyst Environ 217:89–98
Pandi GG, Chander S, Singh MP, Pathak H (2018) Impact of elevated CO_2 and temperature on brown planthopper population in Rice ecosystem. Nat Acad Sci India Section B Biolog Sci 88:57–64
Parry ML, Carter TR (1989) An assessment of the effects of climatic change on agriculture. Clim Chang 15(2):95–116
Pathak H, Aggarwal PK, Singh SD (2012) Climate change impact, adaptation and mitigation in griculture: methodology for assessment and application division of environmental sciences. Indian Agricultural Research Institute, New Delhi, pp 111–130
Patterson DT, Westbrook JK, Joyce RJV, Lingren PD, Rogasik J (1999) Weeds, insects and disease. Clim Chang 43:711–727
Petzoldt C, Seamann A (2010) Climate change effects on insects and pathogens. Climate change and agriculture: promoting practical and profitable responses 6–16; Accessed online at http://www.climateandfarming.org/pdfs/FactSheets/III.2Insects.Pathogens.pdf
Prasannakumar N, Chander S, Pal M (2012) Assessment of impact of climate change with reference to elevated CO_2 on rice brown planthopper, *Nilaparvata lugens* (Stal.) and crop yield. Curr Sci 103:1201–1205
Rao GGSN, Rao AVMS, Rao VUM (2009) Trends in rainfall and temperature in rainfed India in previous century. In: Aggarwal PK (ed) Global climate change and Indian agriculture case studies from ICAR network project. ICAR Publication, New Delhi, pp 71–73
Robert Y, Woodford JAT, Ducray-Bourdin DG (2000) Some epidemiological approaches to the control of aphid-borne virus diseases in seed potato crops in northern Europe. Virus Res 71:33–47
Robinson EA, Ryan GD, Newman JA (2012) A meta-analytical review of the effects of elevated CO_2 on plant–arthropod interactions highlights the importance of interacting environmental and biological variables. New Phytol 194:321–336
Root TL, Price JT, Hall KR, Schneider SH, Rosenzweig C, Pounds JA (2003) Fingerprints of global warming on wild animals and plants. Nature 421:57–60
Roy DB, Sparks TH (2000) Phenology of British butterflies and climate change. Global Change Biol **6**:407–416
Sachs J (2008) Common wealth: economics for a crowded planet. Allen Lane, London
Samways M (2005) Insect diversity conservation. Cambridge University Press, Cambridge, p 342
Sardana HR, Bhat MN (2016) Pest scenario, plant protection approaches in the current context of changing climate. In: Chattopadhyay C, Prasad D (eds) Dynamics of crop protection and climate change. Studera press, New Delhi, pp 167–186
Sharma HC, Dhillon MK, Kibuka J, Mukuru SZ (2005) Plant defense responses to sorghum spotted stem borer, Chilo partellus under irrigated and drought conditions. Inter Sorghum Millets Newsletter 46:49–52
Sharma HC, Srivastava CP, Durairaj C, Gowda CLL (2010) Pest management in grain legumes and climate change. In climate change and Management of Cool Season Grain Legume Crops. In: Yadav SS, McNeil DL, Redden R, Patil SA (eds). Business Media, Springer Science, Dordrecht, The Netherlands, pp 115–140
Shrestha S (2019) Effects of climate change in agricultural insect pest. Acta Scientific Agricult 3:74–80

Sidhu AK, Mehta HS (2008) Role of butterflies in the natural ecosystem with special reference to high altitude (Pangi Valley, Himachal Pradesh). In: Proceedings of international conference on climate change, biodiversity and food security in the South Asian Region, 3–4 November, 2008. Punjab State Council for Science and Technology, Chandigarh and United Nations Educational, Scientific and Cultural Organization, New Delhi, p 36

Singer MC, Parmesan C (2010) Phenological asynchrony between herbivorous insects and their hosts: signal of climate change or pre-existing adaptive strategy? Philosoph Transact Royal Society B: Biolog Sci 365:3161–3176

Skendzic S, Zovko M, Živković IP, Lesic V, Lemic D (2021) The impact of climate change on agricultural insect pests. Insects 12(5):440

Solankey SS, Prakash S, Pawan S, Randhir K (2019) Vulnerability of vegetable crops to climate change. In: Singh P (ed) National Seminar on "recent advances in agriculture for sustainable rural development (RAASRD-2019)". VKSCoA, Dumraon (BAU, Sabour), Bihar, pp 64–70

Solbreck C (1991) Unusual weather and insect population dynamics: *Lygaeus Equestris* during an extinction and recovery period. Oikos 60:343–350

Srikanth J (2007) World and Indian scenario of sugarcane woolly aphid. In: Mukunthan N, Srikanth J, Singaravelu B, Rajula Shanthy T, Thiagarajan R, Puthira Prathap D (eds) Woolly aphid management in sugarcane, vol 154. Extension Publication, Sugarcane Breeding Institute, Coimbatore, pp 1–12

Srinivasa Rao M, Srinivas K, Vanaja M, Manimanjari D, Rao CAR, Venkateswarlu B (2013) Response of multiple generations of semilooper, *Achaea janata* feeding on castor to elevated CO_2. J Environ Biol 34:877–883

Sudderth EA, Stinson KA, Bazzaz FA (2005) Host-specific aphid population responses to elevated CO_2 and increased N availability. Glob Chang Biol 11:1997–2008

Sutherst R (2000) Climate change and invasive species: a conceptual framework. In: Mooney HA, Hobbs RJ (eds) Invasive species in a changing world. Island Press, Washington, DC, pp 211–240

Thomas CD, Cameron A, Green RE (2004) Extinction risk from climate change. Nature 427:145–148

Timoney KP (2003) The changing disturbance regime of the boreal forest of the Canadian prairie provinces. Forest Chronicle 79:502–516

Van Asch M, Visser ME (2007) Phenology of forest caterpillars and their host trees: the importance of synchrony. Annu Rev Entomol 52:37–55

Visser ME, Both C (2005) Shifts in phenology due to global climate change: the need for a yardstick. Proceed Royal Society London Series B Biolog Sci 272:2561–2569

Vissers ME, Holleman LJM (2001) Warmer springs disrupt the synchrony of oak and winter moth phenology. Proceed Royal Society London Series B Biolog Sci 268:289–294

Volney WJA, Fleming RA (2000) Climate change and impacts of boreal forest insects. Agric Ecosyst Env 82:283–294

Wang C, Hawthorne D, Qin Y, Pan X, Li Z, Zhu S (2017) Impact of climate and host availability on future distribution of Colorado potato beetle. Sci Rep 7:4489

Williams DW, Liebhold AM (2002) Climate change and the outbreak ranges of two north American bark beetles. Agricult Entomol 4:87–99

Williams R, Norby RJ, Lioncln DE (2000) Effects of elevated CO_2 and temperature grown red and sugar maple on gypsy moth performance. Global Change Biol **6**:685–695

Xannepuccia AD, Ciccihino A, Escalante A, Navaro A, Isaach JP (2009) Differential responses of marsh arthropods to rainfall-induced habitat loss. Zool Stud 48:173–183

Yamamura K, Kiritani K (1998) A simple method to estimate the potential increase in the number of generations under global warming in temperate zones. Appl Entomol Zool 33:289–298

Zhang YF, Wan GJ, Liu B, Zhang XG, Xing GN, Chen FJ (2018) Elevated CO_2 and temperature alter development and food utilization of *Spodoptera litura* fed on resistant soybean. J Appl Entomol 42:250–262

Zvereva EL, Kozlov MV (2006) Consequences of simultaneous elevation of carbon dioxide and temperature for plant-herbivore interactions: a meta-analysis. Glob Chang Biol 12:27–41

Chapter 15
Emerging Diseases of Vegetables Due to Changing Climate

Erayya, Subhashish Sarkhel, Kalmesh Managanvi, Santosh Kumar, and Ayon Alipatra

Abstract Global climate has been adversely affected due to rapid increase in concentration of atmospheric greenhouse, increasing population, globalization, increasing temperature, soil erosion, atmospheric pollution, deforestation, etc. The various human activities during past five decades is being the primary cause for the climate change. Global temperature is increasing by 0.8 °C during the past century and in the next century (2100) it is expected to be raise between 0.9 to 3.5 °C that is mainly due to increasing level of CO_2. The changing climate may cause adverse effect on cultivation and production of agricultural crops and also on occurrence, severity and spread of plant diseases. Emerging crop diseases poses a serious threat to crop production and productivity. The adaption potential of pest/ pathogen may be one of the key indicators of impact of climate change. Few crops' diseases/ potential pathogen strains have led adverse changes in agricultural and horticultural cropping system. Emergence of new pathogen races has not only caused crop losses, but also have adverse effect on vegetable production. Many pathogens have also crossed national boundaries naturally or through the recent trade practices. Many viral diseases in vegetables, rise in bacterial and nematode diseases of vegetable crops and late blight of potato are some of the recent examples of occurrence of new epidemics. Suitable actions on research priorities to focus on quarantine pest/diseases and international

Erayya (✉) · S. Sarkhel
Department of Plant Pathology, Dr. Kalam Agricultural College (Bihar Agricultural University, Sabour, Bhagalpur), Kishanganj, Bihar, India

K. Managanvi
Department of Entomology, Dr. Kalam Agricultural College (Bihar Agricultural University, Sabour, Bhagalpur), Kishanganj, Bihar, India

S. Kumar
Department of Plant Pathology, Mandan Bharti Agricultural College (Bihar Agricultural University, Sabour, Bhagalpur), Agwanpur, Bihar, India

A. Alipatra
Department of Agronomy, Dr. Kalam Agricultural College (Bihar Agricultural University, Sabour, Bhagalpur), Kishanganj, Bihar, India

S. S. Solankey, M. Kumari (eds.), *Advances in Research on Vegetable Production Under a Changing Climate Vol. 2*, Advances in Olericulture,
https://doi.org/10.1007/978-3-031-20840-9_15

regulations may help to prevent the epidemics caused by plant pathogens. The review briefs about impact of climate change on occurrence and emergence of pest and diseases, emergence, dispersion/ spread and possible threats due to emergence of new strains pathogens, identification of priority research areas, and encouragement of research works which should focus on quarantine pest/pathogens to address the problems and risks associated with the emergence of new pest and diseases.

Keywords CO_2 · Climate change · Temperature · Greenhouse gases · Vegetable diseases

15.1 Introduction

Increasing global temperature and CO_2 concentration and global temperatures indicates the rise of temperature by 2–4 °C in the next 100 years. Extreme weather events such as Erratic rainfall, heat waves and drought like extreme weather events are likely become more common. Climate change will affect agriculture in many ways. Rise of temperatures in tropical areas may detrimental to crop cultivation and production. Crop production may not be possible in semi-arid areas of Asia and Africa due to extreme drought and heat. Concurrently, changes in climatic pattern may increase the risk of crop disease epidemics which may cause heavy crop losses. Vegetables are usually succulent and sensitive plants therefore, severely affected by minor changes in the climate (Solankey et al. 2021).

Vegetables are rich in carbohydrate, vitamins and protein. Vegetable cultivation is highly profitable and provides more job per acre to the farmers with than any other crops. India stands second in vegetables vegetable production. Since, vegetables crops are more sensitive to environmental factors like temperature rainfall, humidity, leaf wetness period etc. During the recent years drastic increase in the volume of literature on impact of changing climate on diseases of vegetable crops (Pautasso et al. 2011). Many recent references indicate the chances of emergence of many new pathogen strains and frequent occurrence of epidemics. The present review briefs the challenges of agricultural research to respond to changing climate scenario and also warns the uncertainties associated with emerging plant diseases, heavy crop losses and threats to future vegetable production (Chakraborty and Newton 2011). There is much more need of research and funding emerging strains of pathogens and occurrence of new crop diseases. Decades of studies have been generated greater understanding of the uncertainties of humidity, rainfall and temperature and their effect on diseases of major crops. However, data on long-term effect of climate change on crop disease are still limited. Many studies also indicates that plant disease responses differently to changing climatic parameters (Eastburn et al. 2011). The impact of changing climate on any host-parasite interaction may not be readily predictable. Another uncertainty is the limitation of multiple-climate-change weather parameters, as these parameters may interact significantly and exhibits varying impact on crop diseases. Many parameters determine plant-pathogen interaction under a varying climatic condition, their interactions and feed-back loops raise the issue of uncertainties to predict this complex system (Garrett et al. 2011). There is a need to develop models which predicts the uncertainties of plant pathogen interactions under changing climate scenario. Climate change has been causing direct and indirect implications on disease incidence, severity, spread, epidemics and special and temporal distribution of vegetable diseases (Burdon and Zhon 2020).

A recent study in Europe on wheat rust illustrates the uncertainty of leaf rust caused by *Puccinia triticina*. Thc model is based on scenarios of leaf wetness and temperature and predicted onset of disease is one month earlier, because of raise in temperature during last five decades (Caubel et al. 2017). Heer (2014) in another study on wheat leaf rust reported that rise in temperature had a positive with the fungus' survival based on leaf wetness and humidity. However, increased oxygen levels had a negative relationship while the effect of increased carbon dioxide concentration varied in susceptible cultivars. It is very difficult to predict wheat rust if, the parameters viz., leaf wetness, carbon dioxide, temperature, and oxygen in the model, it would be difficult to predict the severity of wheat leaf rust. However, well-developed models are limited to few major food crops and certain plant disease (Newbery et al. 2016). Most of the models of crop diseases under climate change scenario are focused on major crops like rice and wheat. Effects of parameters of climate change on vegetable diseases of root crops and tubers, oil seeds, legume crops have been neglected. Sweet potato, cassava, cowpea and beans are staple food of poor people in under developed countries. Under developed countries more likely to face the negative impact of changing climate earlier than people of developed countries. There is a need of research on these minor crops and their diseases. Most important parameters responsible for rapid outbreak of crop diseases are globalization of trade, inadequate quarantine measures, uniform genetic background of major crops, emergence of new pathogen races, intensive agriculture practices and changing and conducive weather parameters. Outbreak of coffee rust in Colambia was attributed to change in climatic conditions (Bebber et al. 2016). In the past five decades, the progress of innovations in agriculture improved breeding methods and integrated crop management modules have increased vegetable production to meet the needs of growing population in spite of the emerging threats from new races/strains of plant pathogens.

During recent days crop disease models become more popular and are successfully integrated in the disease management tactics (Donatelli et al. 2017). Prediction of impacts of changing climatic parameters on plant diseases have been becoming more accurate due to advanced prediction modules. The future research should focus on development of novel crop management strategies to address the adverse effect of climate change on plant diseases and to mitigate the threats on future production of agricultural outputs.

Impact of climate change is very significant on diseases of vegetable crops. Multiple implications on crop disease epidemiology, survival of inoculum, rate of progress of disease and duration of epidemic are expected due to changing climatic scenario. Changes in the spectra of various crop diseases are also being anticipated. Abiotic stresses associated with the extremes of environmental conditions may also predict to increase, and interactions between the abiotic and biotic stresses due climate climatic extremes might cause significant damage to vegetable production. Plant breeding and genetic engineering techniques are expected to improve drought and other stress tolerance attributes of crops. There is a lot of scope for introduction of new varieties and cultivars, but more effective technologies are required to detect new pathogens races. Due to perennial nature of forest trees and fruit crops adaption of new technologies for pest and disease problems which arising from extremities in weather parameters is very slow. Adaptations in agricultural practices/ forest conservation have been occurring over the decades, but it should be accelerated because of drastic change in climatic parameters.

The consequences of climate-change have been led to increased attention; there is increased concern over the adverse effects of high temperature associated with higher concentration of carbon di-oxide ($CO_{2)}$ in the atmosphere. There is a 30% rise in the CO_2 level since the mid of nineteenth century (Barnola et al. 1995). Winter temperature is expected to increase than in summer months, increase in frost-free days in northern latitude compared to southern, increase in minimum temperatures, changes in rainfall patterns and reoccurrence of extreme weather events. The extreme climatic parameters include high and fluctuating temperatures, severe wind, severe drought, ice storms, and rain/hail storms. Precipitation may increase due to increased evapo-transpiration caused by high temperatures. Plant health should be a rationale because of our environmental and economic reliance on agricultural and horticultural crops. Agriculture and horticulture being a major sector of world economy, a keen attention has been required on effects of changing climate on crop health and food production.

Agriculture and forest crop plants are well adapted to typical climatic parameters, but highly prune to extreme climatic conditions. Change in annual mean temperature of 1 to 2 °C is rarely considered as problem. But, even very small change in value of mean involves a change in the magnitude and frequency of extreme climatic conditions. For example, severe drought has a probability of occurrence is once in every 30 years, may change to once in four-to-five-year event (Smit 2000). Changes in long term climatic parameters, or shifts in extreme weather conditions, may result in adverse impact on crop pest and diseases (Smit and Brklacich 1992). Agricultural crops may expect to benefit from elevated CO_2 levels and warm temperature in the atmosphere. Meanwhile the crops might be stressed by drought and extreme weather parameters. Stressed plants/ crops are often more vulnerable to pest and diseases. The influence may be positive or negative, depending on the parameters that require to cause infection/disease. Climate change that occurs in relatively short period may lead to serious consequences on forest health due to perennial nature of trees.

Warm/ drier climate conditions can affect forest trees directly (Colombo and Buse 1998). Indirect impacts on forest distribution patterns are possibly even more than the direct impact. In case of biotic disturbances, effect of climate change on insect population density and distribution and associated damage on forest trees are the kay factors to be considered (Volney and Fleming 2000; Harrington et al. 2001). Many diseases in response to biotic and abiotic stresses, which make plants more vulnerable to infection. In forests, diseases also interact with abiotic stresses and insect pests which result in declines of forest (Manion and Lachance 1992). Most important environmental parameters which play key role in occurrence of plant disease epidemics are moisture and temperature (Agrios 1997). In temperate climate, the most of plant pathogens are in inactive state during the winter due to low temperatures. While, some pathogen/ diseases are favored by low temperature, and some diseases are favored by higher temperature.

15.2 Effect of Changing Climate on Soil Flora and Fauna

Increased carbon dioxide (CO_2) concentration in the atmosphere has major consequences on functioning of various ecosystems and carbon cycling. Deposition level of Nitrogen, temperature and concentration of CO_2 are major parameters influencing soil microbial population and microbial communities (Garrett et al. 2006). Short-term and long-term changes in the abiotic factors not only affect plant growth and crop productivity, but also affect the the microbial population of the habitat. Any change in rhizosphere and phyllosphere microflora and fauna, affects growth of the plants and ability of plant to tolerate aggressive strains of pathogens.

15.3 Effect of Climate Change on Geographical Distribution of Vegetable Diseases

Pole ward shift of agroclimatic zones occurs as temperature increases due to which many plant pathogens will move to new geographic regions, where they may come in contact with new potential/susceptible hosts. Facultative parasites having wide host range would mostly come under this group (Savile and Urban 1982). Chakraborty et al. (1999) reported that more aggressive races of plant pathogens having wide host range like Sclerotinia, Sclerotium, Rhizoctonia, and other necrotrophic fungus can able to migrate from natural crop habitat to vegetable crop habitat. Increasing temperature resulting in rise of bacterial and viral diseases of vegetable crops.

Boag et al. (1991) quoted the examples of plant parasitic nematodes viz., *Longidorus* and *Xiphinema*. According to them, the distribution of *Longidorus* and *Xiphinema* is directly correlated to July month soil surface temperature. Another study on climate change in Finland, Carter et al. (1996) revealed that cereals cropping area will expand by 2050 due to global warming. Increased CO_2 concentration would lead to higher crop yields. Robinet et al. (2011) reported that warmer temperature leads to increase in transmission rate of invasive plant pathogens. Changing climatic parameters have significant influence on distribution and growth of different plant populations (Coakley et al. 1999).

15.4 Implications of Changing Climate on Diseases of Vegetable Crops

Climate change causes significant impact on crop diseases. Changing scenario of temperature, ozone, rainfall, carbon dioxide and ultraviolet (UV) radiations influences vegetable crop pest/ diseases (Coakley et al. 1999; Harvell et al. 2002). Effects of changing climatic factors on plant diseases are intuitive to scientists, but prediction of their ill effects on crop diseases needs much knowledge about the environmental factors and their effect on physiology/biology of host and host-parasite interactions. Therefore, much studies are needed on effects of increasing temperature, increased CO_2, UV radiation and atmospheric ozone on crop diseases and insect vector populations. Warmer temperature affects cell to cell movement, gene expression of virus and host responses in case of viral diseases (Amari et al. 2021).

Concentration of atmospheric CO_2 is projected to increase (355 ppm (*v/v*) to 710 ppm) by the end of 2050 AD (Manning and Tiedemann 1995). Higher concentration of CO_2 will cause increased biomass of plant due to rapid increase in photosynthesis. Increase in CO_2 concentration may probably have little effect on some pathogens. Many soil-borne pathogens can able to tolerate more than 10–20-fold increases in concentration of CO_2 in the atmosphere, and are slightly stimulated by the elevated CO_2 levels (Manning and Tiedemann 1995). Effect of increased level of CO_2 on plant pathogens may be due to change in physiology and anatomy of host like more accumulation of carbohydrates, reduced concentration of nutrients, more synthesis of waxes, changes in epidermal cells and increase in fiber content, and increase in number of mesophyll cells (Chakraborty et al. 1998). Many references are available on slow decomposition of organic matter at higher concentration of CO_2 (Coakley et al. 1999). Increased biomass, slow decomposition, and increased survival rate of pathogens due to mild winters may result in increased load of primary inoculum to infect crop plants. Two common types of trends have been emerged on the effect increased concentration of CO_2 on host-pathogen interaction/s; (i) establishment of pathogen at initial stage may delayed due to modification/ change in aggressiveness of pathogen (or) susceptibility of host, (ii) increase in fecundity of plant pathogens. The combination of more humid microclimate and increased fecundity of pathogen within dense canopy of crop/s associated with increased concentrations of CO_2 may become pre disposing factor for severe infection.

Paul et al. (1998) reported that any variation in the intensity of Ultraviolet (UV) radiations resulted in depletion of the ozone layer, it had not received much attention in relation to crop diseases. UV light stimulates sporogenesis in fungal plant pathogens. It can also reduce viability of fungal spores during survival and dispersal of spores. Increase in UVB-radiations enhances sporulation in fungi, but normal day light also contains UV radiation which stimulate sporulation in light-dependent fungal pathogens (Manning and Tiedemann 1995). Ozone concentration during summer is increased at the rate of 0.5% every year since 1980 to 1998, even rural areas often experiencing the great increase on ozone concentration. Previously, it was thought that higher concentration of ozone did not coincides with weather parameters which are suitable to cause infection by fungal/bacterial pathogens, thus, implied minimum risk. But now it is well known that ozone induces metabolic activities in plants which may persist in crop plants for few days to over several months.

Insects acts as a vector for many crop diseases, and the potential impact of climate change on insect pests of agricultural crops has been studied by Porter et al. (1991). He reported that temperature is one of the most important climatic parameters which affects insect population. Temperature extremes (very low) could greatly increase insect mortality, while warm winters can increase the survival ability of many over-wintering insect species, and also allow changes in prevalence and distribution. High temperatures during summer and spring results in rapid development rates results in more increase of insect populations (Porter et al. 1991). Increases in demand for food grains as a result of plant disease epidemics due to extreme weather conditions resulted due to increase in development rates of pest populations (Rhoades 1985). Most likely the pest outbreaks occur with stressed plants due to reduced immunity of host plant. Non-linear climate change resulted in 5.5 percentage point decline of crop diseases in arid climate and 6/8 percentage point increase in plant disease in temperate regions (Joan Dudney et al. 2021).

15.4.1 Impact of Climate Change on Fungal diseases of Vegetable Crops

The major soil-borne fungi include *Fusarium, Pythium, Sclerotium, Sclerotinia, Botrytis, Rhizoctonia, Verticillium* and *Phytophthora*. Over wintering/ over summering spores produced by these fungi can persist and survive in soil for years. Therefore, warm winter and low soil moisture could be expected with changing climatic scenario which expected to have minimum implication on survival of these fungi. Soil invader fungi over winters in/ on infected plant tissues in perennial plants and on crop debris in annual crops. This group of fungi include *Colletotrichum, Cercospora, Venturia, Phomopsis, Alternaria* and *Septoria* species. Warmer winters could increase survival ability of these fungal pathogens. Some introduced fungal pathogens cannot able to survive during winters in soil, and their source of initial inoculum for such fungi comes from infected seed/planting material or air-borne conidia. This group of fungi include rusts, downy mildews etc. Mild winter months may allow overwintering of some such fungal pathogens. Early migration of spores' results in early infection, therefore longer epidemics may occur. Polycyclic diseases are much affected by changing climatic scenario than monocyclic diseases. In polycyclic diseases caused by *Phytophthora, Peronospora, Colletotrichum* etc. every disease cycle allows inoculum to multiplies many times in a compounding rate. Therefore, increased duration of the crop growing season expected to increase number of pathogen generation and diseases cycle. Meanwhile, monocyclic diseases like root rots, smuts and vascular wilts will become more severe due to severe drought conditions. Moisture, in the form of precipitation, humidity, dew, fog etc. is one of the most critical parameters which influences development of epidemics caused by plant pathogens. High moisture levels in the atmosphere and soil, and rain splashes stimulate the disease development, fungal sporulation, facilitate spore detachment and dissemination, and spore germination (Agrios 1997). Prolonged exposer of host to high moisture conditions may results in occurrence/development of epidemics. Particular Fungus can grow at particular range of temperatures. But activity may be highest at optimum temperature range. Whenever optimum temperatures and adequate moisture is prevalent, polycyclic pathogens can grow rapidly and completes their life cycle very quickly (within few days). Present scenario of changing climatic conditions of warm and dry summer months, crop diseases caused by fungal pathogens might expected to be decreased. Thus, disease progress expected to be in slower rate during warm and dry summers.

For polycyclic diseases there will be reduction in number of disease and a slow development of monocyclic diseases may result in decreased primary/initial inoculum for the next cropping season. In contrast, few fungal disease of crop plants such viz., *Sphaerotheca, Fusarium, sclerotioum, Podosphaera, Uncinula* etc. can thrive well in warm and dry weather conditions (Trigiano et al. 2004) and such diseases will increase and may cause epidemics in future under the changing climatic scenarios.

In case of forest diseases, many native microflora and fauna may cause damage in large area. Root rots caused by *Armillaria* spp., canker and decay fungi are belongs to this category. Drier and warmer climatic conditions can reduce the incidence/severity of diseases like white pine blister rust which needs cool and wet conditions for causing infection and disease development (Marosy et al. 1989). Root rot caused by *Heterobasidion annosum* will increase under warm and dry climatic conditions, and it will be restricted within a range of cooler climatic conditions. Facultative plant pathogens like *Armillaria* spp., *Verticillium* spp. and pine wood nematode (*Bursaphelenchus xylophilus)* would be benefitted under drought stress conditions (Schoeneweiss 1975). Frequent occurrence of ice/hail and wind storms will increase the vulnerability of forests to diseases and lead to mortality and decay.

15.4.2 Impact of Climate Change on Bacterial Diseases of Vegetable Crops

On perennial plants, bacteria, *Erwinia amylovora* (on apple) overwinter / over summer in infected plant tissues, this primary inoculum is spread from plant to plant during subsequent season. On annual crops, bacteria like *Pseudomonas syringae* pv. *Phaseolicola* survives in/on the crop debris in soil. Vector-borne bacterial diseases like *Erwinia* sp. survives in insect vectors, and these vectors act as primary inoculum for the next growing season. Some introduced bacterial plant pathogens such as *Xanthomonas campestris* pv. *Vesicatoria* and *Pseudomonas syringae* pv. *Tomato,* can survive in/on infected seed/planting material, crop debris in soil and weeds. Short and mild winters have a very limited impact on these pathogens. However, growth and multiplication of vector-borne bacterial diseases will increase. Plenty of water/moisture enhances multiplication and dissemination of bacterial pathogens. Under wet and humid climatic conditions, infected crop tissues exude masses of bacterial ooze which spread to other healthy hosts through irrigation/rain water and insects. Hence, the warm and dry summers expected due to climate change could reduce bacterial diseases in vegetable crops. Bacteria enters into their host plant through natural openings and wounds (Agrios 1997; Trigiano et al. 2004), thus the expected increase in intensity and frequency of summer storms and heat waves with high wind velocity, rainfall, and hail storm will increase wounding of plant tissues and increased moisture will facilitate the spread of bacteria. Bacterial wilt of vegetables has been causing epidemics in many parts of the world due to increasing winter and summer temperatures.

15.4.3 Impact Climate Change on Viral Diseases of Vegetable Crops

Amount of inoculum of viruses is greatly influenced by survival of virus in its hosts (alternate, main host, volunteer plants vectors, planting material) during winter. Mode of transmission by vector also influence the virus diseases significantly. If, vector transmits the virus persistent manner the chances of epidemics are more. Rate of accumulation, accumulation feeding period, transmission feeding periods also influence the diseases development (Irwin et al. 2000). Most of the viral diseases are favored by the higher temperature which is the result of climate change (Wu et al. 1993). In recent years, many new virus diseases are evolving frequently and causing considerable crop losses in terms of quality and quantity. Aphids' population is expected to be increased due to mild winter temperatures which favors growth and multiplication rates of viruses leading to more disease incidence and severity (Alonso-Prados et al. 2003). Mild winter months also influence survival ability of weed hosts. Increases in intensity and frequency of summer storms and high rainfall, wind velocity, rain, hailstorm increases the chances of wounding of plants and results in increased transmission rate of viruses. Hence, the climate change has the positive correlation with viral diseases of vegetable crops. Increase in insect vector population and their early appearance due to increase in winter temperature resulted in more and more epidemics of viral diseases in sugarbeet and potato.

15.4.4 Impact of Climate Change on Nematode Diseases of Vegetable Crops

Under optimum environmental conditions nematode takes 2–4 weeks to complete its life cycle. Temperature greatly influences nematode population in the soil. Cooler soil temperatures slow down the nematode growth and development. Viability of eggs of root knot nematode decreases under warm winter months. Increasing temperature due to climate change might increase the nematode development and thus results in increase in number of generations per season/year. Dry conditions also cause water stress in host plants infected with nematode. Changing climatic conditions thus enhances the nematode disease incidence and severity in crops. Increasing global temperature has been resulted in shortening of life cycle of plant pathogenic nematodes and thus contributing to rise in incidence and severity of root knot nematode in many vegetable crops.

15.4.5 Impact on Climate Change on Phytoplasma Diseases of Vegetable Crops

Primary source of inoculum of phytoplasma is overwintering of phytoplasma in the host plants (Agrios 1997). By feeding on infected plants leaf hoppers acquires phytoplasma and when they feed on healthy plant, they transmit phytoplasma. Mild winters helps the phytoplasma to survive in infected perennial weed hosts and also in insect vectors. Hot summer months and high temperature favors the development and multiplication rate of phytoplasma cells. Thus, climate change expected to increase the threat of epidemics of phytoplasma diseases like little leaf of brinjal in vegetable crops. Leaf hoppers population has been highly favored by mild winters and hot summer months which lead to increase in incidence of little leaf of brinjal.

15.4.6 Impact of Changing Climate on Abiotic Diseases and Disease Complexes in Vegetables

Abiotic plant diseases are due to non-infectious factors or abiotic factors like temperature extremes, extreme rainfall, soil moisture extremes, nutrient deficiency, drought conditions and air pollutants (Agrios 1997; Trigiano et al. 2004). Some week pathogens like *Alternaria* spp. (saprogens) cause infection on stressed crop (Manion and Lachance 1992). Due to climate change, chances of new combinations of host-stress-pathogen increased in forest trees. In temperate regions, stressed plants often more vulnerable to freezing injury during winter season. Vegetable diseases associated with biotic and abiotic stress complex are unique and critical area of consideration to assess the impact of changing climate on crop diseases (Millers et al. 1989). In Northern Hemisphere the onset of crown die-back in 1925, 1937, and 1981 is a best example to the relationship between changing climate and crop diseases (Auclair et al. 1992). Root knot nematode + Fusarium wilt complex and Root knot nematode +Pseudomonas wilt complex diseases of tomato and cucurbits are being emerging as a great importance due to change in rainfall pattern, shift in monsoon rains and increasing atmospheric temperatures.

15.4.7 Impact of Changing Climate on Historical Emerging Diseases of Vegetables

An emerging disease is a newly identified /recognized disease which appeared in a location and could develop very fast in terms of incidence and severity (Daszak and Hyatt 2003). If left it unchecked in its early stage, it could result in severe epidemic proportions. Introduction of highly hazardous plant pathogen to a new area exhibits higher risk of food production and security (Strange and Scott 2005). Plant disease epidemics leads to hunger and suffering of human beings. Lete blight of potato (*Phytophthora infestans* (Mont.) de Bary) is a one such devastating plant disease caused death and migration of millions of people of Irland (Agrios 2005). Similarly, brown spot of rice is another epidemic caused by *Cochliobolus miyabeanus* which was badly affected West Bengal in 1945 (Padmanabhan 1973). Monoculturing and favorable environmental parameters, facilitates the occurrence of plant disease epidemics which might result in human deaths and migration due to starvation and crop losses (Strange and Scott 2005). Co-volution of host and pathogens presumably results in new race/s emergence. New plant species have been constantly introducing to different countries/ locations, and grown as monoculture. Varieties are being selected based on their productivity, without considering susceptibility to new/major plant pathogens. The favorable environmental conditions act as pre disposing factor for development of new plant disease emergence and development of epidemics (Oliver and Solomon 2008). Disease epidemic is a cumulative effect of host and pathogen interactions. Soil microflora and fauna also affect the host-pathogen interaction. The implications of changing climate vary with different host-pathogen systems. Due to adverse climatic changes/conditions crop plants need higher application rates of pesticide, thus maximizing input cost of farmers and also increasing the chances of resistance development to fungicides because of heavy dose of application (Juroszek and von Tiedemann 2011). Vegetable crops are more flexible in its adoption and cultural practices as compared to horticultural crops. Persistence of agro-chemicals is highly dependent on weather parameters in the rhizosphere and phyllosphere. The efficacy of chemicals also depends on intensity, frequency and duration of precipitation. Temperature has direct correlation with degradation of pesticides, physiology of plant; indirectly affects the penetration, persistence, translocation and modes of action of pesticides (Coakley et al. 1999).

15.5 Mitigation of Effects of Climate Change on Diseases of Vegetables

In recent years, development of extreme weather parameters like droughts, hurricanes, ice storms, severe floods have been reported from globally and locally which are due to changing climatic conditions. Recently, Standing Committee on Agriculture and Forestry prepared the report i.e., '*Climate Change: We Are at Risk*' in which they revealed about the influence of changing climate on agriculture/ forestry and rural population. Due to rapid climate change, there is an acceleration in cultivation of new crops and cultivars. The adaptation of cultivars of vegetable and fruit crops in particular geographical locations and soil is co-established over many decades. But, during present scenario the cultivars or varieties which performs better across various locations and environmental conditions are being used for commercial cultivation. Such adaptation of appropriate crop genotypes is very successful for several years and is this practice of selection of adapted cultivars is being continued under expected changing climate scenario. But the concern is, whether this practice is effective under extreme and erratic weather events which are predicted with rapid changing scenario of climate. Emphasis should be given for breeding crop plants for stress tolerance (drought/ flood/ extreme temperature events etc.). Tolerance of crop plants to stress conditions might result in production of healthy crop/s with higher yields. Advanced techniques that can enhance development of host plant tolerance/ resistance to various biotic/ abiotic diseases are the key strategies to facilitate such adaptation.

Drastic changes in climatic parameters in relatively short period of time is expected to be accompany with the emergence and introduction of new pest and diseases of vegetable crops. It is also important to develop adequate and efficient diagnostic tools /technical staff/ skills to develop to mitigate the uncertainties which may occur due to climate change. McKenney et al. (2003) explained about the changes in rainfall pattern and temperature to predict the risk of dissemination and distribution of new pest and diseases, and such models might contribute to assess the risk of climate change and distribution and spread of vegetable diseases. Extreme drought conditions may result in increased use of irrigation. There is need to develop technologies for efficient utilization of water resources in arid regions. Advanced irrigation systems (drip irrigation) also contribute to reduce humidity and leaf wetness in the cropping system which in turn help to overcome many foliar diseases. Due to perennial nature of fruit trees its highly necessary to consider the impacts of climate change. Therefore, critical consideration is necessary while selecting plant species, breeding method, selection of geographical area and management practices. Viral diseases like mosaic and leaf curl become more common and extensive due to the predisposing factors like drought stress and warmer winter temperatures. Drought stress affecting areas could not be selected for raising commercial vegetable crops like tomato, chilli, potato, brinjal etc. Commercially cultivated vegetable crops are mostly belonging to annuals and short duration. Therefore, appropriate selection of varieties/cultivars and agronomic package of practices will play a vital role to mitigate the adverse effects of changing climate on vegetable diseases. Undoubtedly over the period of time, there will be big changes might occur in the cropping system, has occurred during past few decades. However, it is expected the accelerated changes in cultivation practices due to changing climatic patterns.

The continuous efforts from plant breeders, agronomists, entomologists, plant pathologists and plant biotechnologists are very essential for success production of vegetable crops. With the changing scenario of climate, it become necessary to develop suitable crop improvement and management strategies through pest risk analysis, crop improvement and management tools and techniques. Climate change will be predicted to occur very rapidly than ever seen before, therefore the research strategies developed must meet the future challenges. Actions to overcome the pest/disease epidemic situations and efforts to reduce the threats by developing and implementing regulatory strategies to prevent the epidemics caused by plant pathogens with high potential to cause crop losses. Many International and National organizations to provide information regarding probable climate changes and their impacts and also the mitigation measures among governments, scientists, farmers and public are very crucial to reduce the risk of emerging plant pathogens.

Sanitary and Phytosanitary measure are regulations worldwide to combat emerging and introduced pests which affect vegetable production. Quarantine measures are the example for prevention of movement of pest and diseases to country/state where the pest/disease not known to be existed. Quarantine measures also delays introduction of pest/pathogens into new area through pest rick analysis and regulatory measures. International quarantine measures, prevent the spread of exotic pest and diseases. While, domestic quarantine reduces the spread of new pest and diseases from one state to another, and also control disease that already introduced in few areas of the country. Phytosanitary measures also include surveillance at airports, sea ports and borders by trained persons to locate seeds, vegetables, fruits and any other plant materials. Such programs have been found very successful to stop the spread of pest and diseases from other countries. In spite of such regulatory measures, plant pathogens manage to spread to new crop areas via humans, insect vectors and other climatic factors. It is the responsibility of researchers to focus on risks associated with climate change on vegetable crops and collaborations to formulate policies, designing of phyto-sanitary guidelines and development of suitable and rapid diagnostic protocols which could strengthen government policies. Diagnostic protocols play vital role in accurate diagnosis of crop diseases. Traditional methods coupled with advanced molecular and serological methods will facilitate the rapid detection of new plant pathogens or their variants (Phillips et al. 2013).

15.6 Crop Protection Strategies Under Changing Climatic Scenario

Changing climatic parameters can even having the implications on efficacy of fungicides applied to manage the crop diseases. Efficacy of agro-chemicals may also affect due to change in onset, intensity and duration of rainfall. High and heavy rainfall pattern may also wash-off the fungicides applied on the crop plants. Temperature influences on pesticide degradation and also influence on physiology and morphology of plants thus affecting penetration, translocation and efficacy of pesticide/s (Elad and Pertot 2014).

Increase in thickness of cuticular wax on leaf surface causes slow uptake of fungicides by the host plant. Dense crop canopy due to increased CO_2 can affect the penetration of systemic fungicides, its translocation within the plant system and mechanism of inhibition of pathogen (Bowes 1993). Plant surface roughness and wax layer on the cuticle have a negative correlation with retention of fungicides (Hunsche et al. 2006). Increase in canopy size might have a negative relation with spray coverage and which may also lead to dilution of active ingredients in the fungicide formulation. In case of biological control of plant disease is concerned minimum work has been done on effect of climate change. Few reports suggests that increased concentration of CO_2 decreases nitrogen levels and nitrogen availability in soil. A few studies suggested that high temperature and low humidity increases the efficiency of *Trichoderma harzianum* in soil (Elad et al. 1993). Due to variability in interaction among soil microorganisms, it is very difficult to simulate implication of changing climatic parameters on plant disease in suppressive soils (Davelos et al. 2004). Climate change may have positive or negative implications on vegetable diseases. Regularly evaluated/tested disease management strategies are needed to be incorporation against diseases under changing climate conditions. Many simulation modules are also developed against some major diseases of vegetable crops. Todays need is to collaborate of all the relevant disciplines together for mitigation of this global issue. More and more rational and holistic approaches are need to be developed to overcome the impact of climate change on diseases of vegetable crops.

15.7 Conclusion

Limited research has been done on impact of changing climate on vegetable crop diseases under and also on management of crop diseases under changing climatic scenario. However, limited reports are available for few countries, states, locations, crop/s and particular diseases which are concern with food safety and security. Emphasis to be shifted from assessment of impact to development of adaptation and mitigation strategies. Efficacy of new disease management tactics, including disease-resistant or tolerant cultivars to be evaluated under changing climate scenario and also include climate scenarios in future research works which aimed to develop new pest management tactics. Pest risk analyses should be performed based on host-pathogen interaction/s to understand the effect of imminent changes in climatic parameters.

Exotic crop pest and diseases causes huge losses in vegetable and fruit crops. Due to international trade and commerce many exotic diseases spreading quickly to the areas where they were not a problem in the past. It's a need to stay prepared to combat the introduced and exotic pest and diseases in case of their entry into a new geographic area or to new cropping system. Pest risk assessment and developing strategies to combat exotic pest/ diseases is gaining much scope at federal as well as state levels. Preparation of research design/ models on these emerging diseases before the entry into new area, funding and involvement of administrators and policy makers might help in reducing the impact of exotic pest/ diseases. This may be the result of climate change, intensification of particular cultivars, cropping patterns which exerts pressure on pest population. It indicates the importance to undertake frequent and continuous studies on a particular individual disease/s, even for the diseases which are of minor importance in at present scenario. Particularly, the data on status of major strains/races of pathogens needs to be updated to identify the effective resistant genes for the prevailing strains/races of pathogens. Such studies are of great significance in designing strategies whenever the pathogen appears in its epidemic form. In addition, the advanced molecular and serological tools are needs to be utilized for the investigation of unanswered mysterious questions of host-parasite interaction.

References

Agrios GN (1997) Plant pathology, 4th edn. Academic, SanDiego, Calif

Agrios GN (2005) Fitopatología, 2nd edn. Limusa Noriega, Editorial, p 838

Alonso-Prados JL, Luis-Arteaga M, Alvarez JM, Moriones E, Batlle A, Lavina A, Garcia-Arenal F, Fraile A (2003) Epidemics of aphid-transmitted viruses in melon crops in Spain. Eur J Plant Pathol 109:129–138

Amari K, Huang C, Heirlein M (2021) Potential impact of global warming on virus propagation in infected plants and agricultural productivity. Front Plant Sci 12:649768

Auclair AA, Worrest RC, Lachance D, Martin HC (1992) Climatic perturbation as a general mechanism of forest dieback. In: Manion PD, Lachance D (eds) Forest decline concepts. American Phytopathological Society (APS) Press St. Paul, Minn, pp 38–58

Barnola JM, Anklin M, Porheron J, Raynaud D, Schwander J, Stauffer BTI (1995) CO_2 evolution during the last millennium as recorded by Antarctic and Greenland ice. Tellus Ser.B. Chem Phys Meteorol 47:264–272

Bebber DP, Castillo AD, Gurr SJ (2016) Modelling coffee leaf rust risk in Colombia with climate reanalysis data. Phil Trans R Soc B 371:20150458. https://doi.org/10.1098/rstb.2015.0458

Boag B, Crawford JW, Neilson R (1991) The effect of potential climatic changes on the geographical distribution of the plant-parasitic nematodes *Xiiphinema* and Longidorus in Europe. Nematologica 37:312–323

Bowes G (1993) Facing the inevitable: plants and increasing atmospheric CO_2. Annu Rev Plant Physiol Plant Mol Biol 44:309–332

Burdon JJ, Zhon J (2020) Climate change and disease in plant communities. PLoS Biol 18(11):e3000940

Carter TR, Saariko RA, Niemi KJ (1996) Assessing the risks and uncertainties of regional crop potential under a changing climate in Finland. Agricult Food Sci Finland 5:329–350

Caubel J, Launay M, Ripoche D, Gouache D, Buis S, Huard F, Huber L, Brun F, Bancal M (2017) Climate change effects on leaf rust of wheat: implementing a coupled crop disease model in a French regional application. Eur J Agron 90:53–66. https://doi.org/10.1016/j.eja.2017.07.004

Chakraborty S, Newton AC (2011) Climate change, plant diseases and food security: an overview. Plant Pathol 60:2–14. https://doi.org/10.1111/j.1365-3059.2010.02411.x

Chakraborty S, Murray GM, Magarey PA, Yonow T, O'Brien RG, Croft BJ, Barbetti MJ, Sivasithamparam K, Old KM, Dudzinski MJ, Sutherst RW, Penrose LJ, Archer C, Emmett RW (1998) Potential impact of climate change on plant diseases of economic significance to Australia. Australas Plant Pathol 27:15–35. Paul ND, Ayres PG, Rasanayagam S, Royle DJ (1998)

Chakraborty S, Perrott R, Ellis N, Thomas MR (1999) New aggressive Colletotrichum gloeosporioides strains on Stylosanthess Cabra detected by virulence and DNA analysis. Plant Dis 83:333–340

Coakley SM, Scherm H, Chakraborty S (1999) Climate change and plant disease management. Annu Rev Phytopathol 37:399–426

Colombo SJ, Buse LJ (1998) The impacts of climate change on Ontario's forests. Ontario Forest Research Institute, Ontario Ministry of Natural Resources, Sault Ste Marie, Ont., Forest Research Information Paper No. 143

Daszak P, Cunningham AA, Hyatt AD (2003) Infectious disease and amphibian population declines. Divers Distrib 9(2):141–150

Davelos, A. L., Kinkel, L. L. And Samac, D. A. (2004). Spatial variation in frequency and intensity of antibiotic interactions among Streptomycetes from prairie soil. Appl Environment Microbiol,70: 1051–1058

Donatelli M, Magarey RD, Bregaglio S, Willocquet L, Whish JPM, Savary S (2017) Modelling the impacts of pests and diseases on agricultural systems. Agric Syst 155:213–224. https://doi.org/10.1016/j.agsy.2017.01.019

Dudney J, Willing CE, Das AJ, Latimer AM, Nesmith JCB, Battles JJ (2021) Non linear shifts in infectious rust disease due to climate change. Nat Communicat. 10.1038

Eastburn DM, McElrone AJ, Bilgin DD (2011) Inuence of atmospheric and climatic change on plant–pathogen interactions. Plant Pathol 60:54–69. https://doi.org/10.1111/j.1365-3059.2010.02402

Elad Y, Pertot I (2014) Climate change impacts on plant pathogens and plant diseases. J Crop Improv 28:99–139

Elad Y, Zimand G, Zaqs Y, Zuriel S, Chet I (1993) Biological and integrated control of cucumber grey mold (Botrytis cinerea) under commercial greenhouse conditions. Plant Pathol 42:324–332

Garrett KA, Dendy SP, Frank EE, Rouse MN, Travers SE (2006) Climate change effects on plant disease: genomes to ecosystems. Annu Rev Phytopathol 44:489–509

Garrett KA, Nita M, De Wolf ED, Esker PD, Gomez-Montano L, Sparks AH (2011) Plant pathogens as indicators of climate change, Chapter 21. In: Letcher TM (ed) Climate change (2nd edn): observed impacts on planet earth. Elsevier BV, pp 325–338. https://doi.org/10.1016/B978-0-444-63524-2.00021

Harrington R, Fleming RA, Woiwod IP (2001) Climate change impacts on insect management and conservation in temperate regions: can they be predicted? Agric For Entomol 3:233–240
Harvell CD, Mitchell CE, Ward JR, Altizer S, Dobson AP, Ostfeld RS, Samuel MD (2002) Climate warming and disease risks for terrestrial and marine biota. Science (Washington, D.C.) 296:2158–2162
Heer S (2014) Rust fungi and global change. New Phytol 201:770–780. https://doi.org/10.1111/nph.12570
Hunsche M, Bringe K, Noga G (2006) J Pest Manag Sci
Irwin ME, Ruesink WG, Isard SA, Kampmeier GE (2000) Mitigating epidemics caused by non-persistently transmitted aphid-borne viruses: the role of the pliant environment. Virus Res 71:185–211
Juroszek P, von Tiedemann A (2011) Potential strategies and future requirements for plant disease management under a changing climate. Plant Pathol 60:100–112
Manion PD, Lachance D (eds) (1992) Forest decline concepts. American Phytopathological Society (APS) Press, St. Paul, Minn
Manning WJ, Tiedemann AV (1995) Climate change: potential effects of increased atmospheric carbon dioxide (CO2), ozone (O3) and ultraviolet-B (UV-B) radiation on plant diseases. Environ Pollut 88:219–245
Marosy M, Patton RF, Upper CD (1989) A conducive day concept to explain the effects of low temperatures on the development of *Scleroderris* shoot blight. Phytopathology 79:1293–1301
McKenney DW, Hopkin AA, Campbell KL, Mackey BG, Foottit R (2003) Opportunities for improved risk assessments of exotic species in Canada using bioclimatic modeling. Environ Monit Assess 88:445–461
Millers I, Shriner DS, Rizzo D (1989) History of hard wood decline in the eastern United States. US Dep Agric For Serv Gen Tech Rep:NE-126
Newbery F, Qi A, Fitt BDL (2016) Modelling impacts of climate change on arable crop diseases: progress, challenges and applications. Curr Opin Plant Biol 32:101–109. https://doi.org/10.1016/j.pbi.2016.07.002
Oliver RP, Solomon PS (2008) Recent fungal diseases of crop plants: is lateral gene transfer a common theme? Mol Plant-Microbe Interact 21(3):287–293
Padmanabhan SY (1973) The great Bengal famine. Annu Rev Phytopathol 11(1):11–24
Pautasso M, Doring TF, Garbelotto M, Pellis L, Jeger MJ (2011) Impacts of climate change on plant diseases—opinions and trends. Eur J Plant Pathol 133:295–313. https://doi.org/10.1007/s10658-012-9936-1
Phillips AJL, Alves A, Abdollahzadeh J, Slippers B, Wingfield MJ, Groenewald JZ, Crous PW (2013) The Botryosphaeriaceae: genera and species known from culture. Stud Mycol 76:51–167
Porter JH, Parry ML, Carter TR (1991) The potential effects of climatic change on agricultural insect pests. Agric For Meteorol 57:221–240
Rhoades DF (1985) Offensive–defensive interactions between herbivores and plants: their relevance in herbivore population dynamics and ecological theory. Am Nat 125:205–238
Robinet C, Van Opstal N, Baker R, Roques A (2011) Applying a spread model to identify the entry points from which the pine wood nematode, the vector of pine wilt disease, would spread most rapidly across. Eur Biol Invasions 13:2981–2995
Savile DBO, Urban Z (1982) Evolution and ecology of *Puccinia graminis*. Preslia 54:97–104
Schoeneweiss DF (1975) Predisposition, stress, and plant disease. Annu Rev Phytopathol 13:193–211
Smit WH (2000) Ice and forest health. North J Appl For 17:16–19
Smit B, Brklacich M (1992) Implications for global warming for agriculture in Ontario. Can Geogr 36:75–80
Solankey SS, Kumari M, Kumar M, Silvana N (2021) The role of research for vegetable production under a changing climate future trends and goals. In: Solankey SS (ed) Advances in Research on Vegetable Production Under a Changing Climate, Vol, 1. Springer, Switzerland, pp 1–12

Strange RN, Scott PR (2005) Plant disease: A threat to global food security. Annu Rev Phytopathol 43:83–116

Stratospheric ozone depletion (1998) UVB radiation and *Septoria tritici* infection of wheat. [online]. In: 7th International Congress of Plant Pathology. 9–16, vol. 1. Compiled by British Society for Plant Pathology, Birmingham. Abstr. 4.2.6S. Available from http://www.bspp.org.uk/icpp98, Edinburgh

Trigiano RN, Windham MT, Windham AS (eds) (2004) Plant pathology: concepts and laboratory exercises. CRC Press, New York

Volney WJA, Fleming RA (2000) Climate change and impacts of boreal forest insects. Agric Ecosyst Environ 82:283–294

Wu G, Kaper JM, Tousignant ME, Masuta C, Kuwata S, Takanami Y, Pena L, Diaz-Ruiz JR (1993) Tomato necrosis and the 369 nucleotide Y satellite of cucumber mosaic virus: factors affecting satellite biological expression. J Gen Virol 74:161–168

Chapter 16
Impact of Climate Change on Postharvest Quality of Vegetables

K. Prasad, S. K. Singh, Panchaal Bhattocharjee, Joy Rudrapaul, Udit Kumar, Sudheer Kumar Yadav, Shubham Maurya, and Araghya Mani

Abstract The postharvest value of vegetables depends on its quality at the time of harvest. This postharvest quality is attributed by synthesis of plant pigments such as chlorophylls, carotenoids, xanthophylls, bio active photochemical and complex carbohydrates. This process of pigments and carbohydrate synthesis occur through photosynthetic activity, which is intricately linked to the environmental/pre-harvest conditions (climate) during production and harvesting stage. There are several preharvest factors such as genetic and environmental which affect quality of vegetables. The climate at production stage is one of the major factors which affects the quantity and quality of produced vegetable. The changing climatic conditions such as elevated temperature, irregular rainfall pattern, elevated carbon dioxide concentration, irregular weather event, salinity, draught, biotic stress *etc.* directly and indirectly affects the vegetable quality by altering duration of maturation, respiration, transpiration, development of bio-chemical, flavour components and nutritional value *etc.* The alteration in postharvest quality due to changing climate may be both

K. Prasad (✉)
Department of Horticulture, Tirhut College of Agriculture (TCA), Dholi,
Dr. Rajendra Prasad Central Agricultural University, Pusa, Bihar, India

S. K. Singh
Department of Plant Pathology, Post Graduate College of Agriculture (PGCA),
Dr. Rajendra Prasad Central Agricultural University, Pusa, Bihar, India

P. Bhattocharjee
Department of Horticulture, Anand Agricultural University, Anand, Gujarat, India

J. Rudrapaul · U. Kumar · S. K. Yadav · S. Maurya
Department of Horticulture, Post Graduate College of Agriculture (PGCA),
Dr. Rajendra Prasad Central Agricultural University, Pusa, Bihar, India

A. Mani
Department of Horticulture, School of Agriculture, Lovely Professional University,
Phagwara, Punjab, India

S. S. Solankey, M. Kumari (eds.), *Advances in Research on Vegetable Production Under a Changing Climate Vol. 2*, Advances in Olericulture,
https://doi.org/10.1007/978-3-031-20840-9_16

positive or negative. This chapter put light on the influence of climate change and its triggering factors on growing conditions, quality, harvest maturity and nutritional value of vegetables.

Keywords Climate change · Pre-harvest factors · Postharvest quality · Vegetables

16.1 Introduction

Vegetables are rich in minerals, vitamins and dietary fibres which play a noteworthy role in individual's diet, its consumption in recommended diet reduce the risk of many diseases, stroke, heart diseases and diabetes (Ali et al. 2017). In recent times, under changing climatic situation, crop production is facing several challenges these includes reducing yield, decline in postharvest quality and increasing insect and disease infestation (Moretti et al. 2010; Nemeskéri and Helyes 2019; Leisner 2020).

The climate has changed numerous times, varying from the ice ages to the warming period. The rise in air temperatures has been reported over the last several decades and related impact on the cultivation of various food crops, including vegetables became well evident (Adams et al. 1990; Nemeskéri and Helyes 2019).

In developing countries vegetables are the main source of livelihood for most of communities because vegetables are loaded with several vitamins, carbohydrate, salts and proteins (Solankey et al. 2021). The production of vegetables and its quality perspective has been one of the significant divisions to be studied in terms of the potential impacts of climate change on the post-harvest quality due to its popularity and demand around the world (Moretti et al. 2010; Nemeskéri and Helyes 2019; Ali et al., 2020).Vegetable quality is influenced by many pre-harvest factors including the alterations in the growing climate components such as temperature, precipitation, concentration of CO_2, CH_4, long-term water shortages, unsuitable soil conditions, drought, desertification,solar radiation, ozone, and UV radiation *etc.* (Beverly et al. 1993; Ali et al. 2017; Leisner 2020; Ali et al. 2020).

Climatic components such as carbon dioxide (CO_2) level in the atmosphere is increasing at faster rate than the previous years and no sooner may reach higher levels (Hansen et al. 2013). Similarly, temperatures throughout the globe have been increasing gradually from about last 40 years (Anderson 2013; Bales 2013), as it is estimated that at the end of twenty-first century, the raise in temperature will be by 2.2 °C–3.7 °C and the concentration of carbon dioxide will be 420–935 ppm. Due to increase in temperature and raise in CO_2 levels have direct influence on the length and intensity of drought on global level could be observed. In addition to that there might be alteration of rainfall in dry and wet regions, and between wet and dry seasons in different parts of globe (Medellin-Azuara et al. 2011; Deschenes and Greenstone 2012; Leisner 2020).

It is critical to crucial to understand how climatic changes affect the post-harvest handling, storage, and quality of the commodities. It is well knowledge that various environmental conditions during the preharvest stages significantly affect and change the postharvest physiology and associated quality features of many edible plant products during storage. Because of the diverse climatic circumstances during the preharvest stage, the harvested products contain different level of secondary metabolites, nutraceutical substances, mycotoxins, nutrient status, ethylene biosynthesis, respiration rate, and postharvest shelf- or storage-life potential. Additionally, by impacting food quality, climate change indirectly impacts the markets and companies that deal with produced products (Ali et al. 2020). Bisbis et al. 2018 are obserbed that Climate change will most likely cause higher temperatures and higher atmospheric CO_2 concentrations. Both will have an influence on yield and product quality of vegetables. These influences and interactions can be either positive (þ) or negative (–) (Fig. 16.1).

Apart from these, there are several others climatic factors which affects quality of vegetables and have been discussed in detail in this chapter so as to assess the potential impact of climate on vegetable production system.

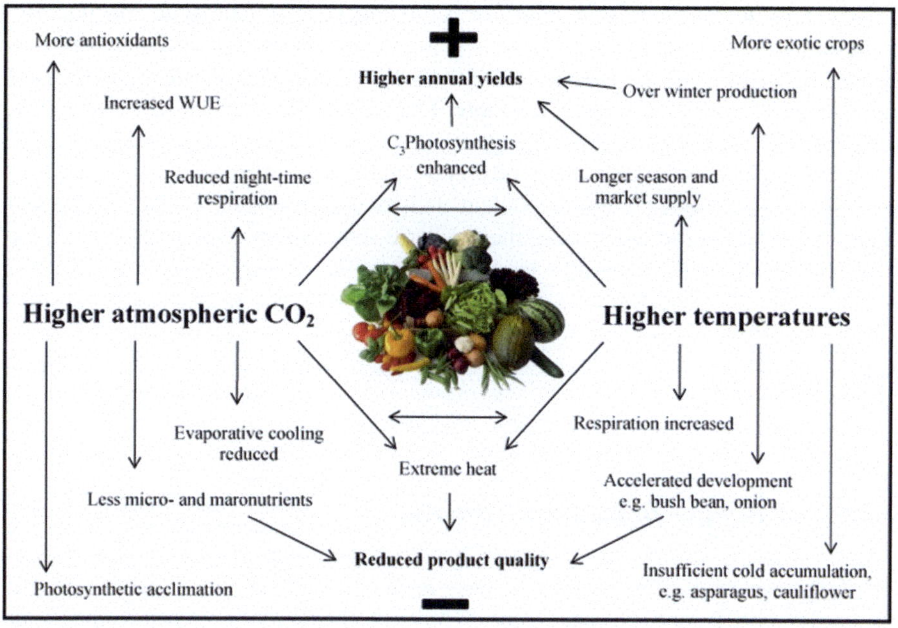

Fig. 16.1 Schematic representation of climate change are influence on post harvest quality of vegetables (Bisbis et al. 2018)

16.2 Components of Climate Change Triggering the Postharvest Quality of Vegetables

16.2.1 Elevated Temperature

Since the starting of nineteenth century to 1980 a rapid rise in the global temperature was recorded in each 13.5 years; since 1981, it was found to be increased to every 3 years (NASA 2021). The mean temperature risen at a rate of 0.07 °C till 1880 and thereafter since 1981 it increased to double at a rate of 0.18 °C and in 2019 the mean temperature increased by 0. 95 °C and the current rise in mean temperature in 2020 was 1.02 °C. Out of all warmest years, the five warmest years was recorded since 2015 whereas top ten warmest years occurred since 2005 (NASA 2021).

Temperature plays significant role in shaping the pre and post-harvest quality of many vegetables (Bindi et al. 2001). During growth and development of vegetables, elevated temperatures can have an effect on photosynthesis, respiration, membrane stability, hormones level and accumulation of metabolites. Most of the physiological processes in vegetable crops functions normally in temperatures varying from 24° to 27°, if temperature goes above or below this range it causes detrimental disturbance in the physiology of crop influencing the postharvest quality of many vegetables (Bewley 1997; Nemeskéri and Helyes 2019). It is also reported that the heat tolerant genotypes of tomato have more TSS content as compare to susceptible genotypes (Solankey et al. 2017).

There might be several positive effects of high temperature such as higher antioxidant activity, high phenolics content and total soluble solids (Ackerman and Stanton 2013). On the other hand, there will be some undesirable alteration on the qualities which will eventually lead to decrease in the postharvest quality for instances the high temperatures and long days cause flowering in lettuce and spinach but the produce is of inferior quality. Adaptation of certain varieties is the mitigation technique to overcome this flowering trigger, known as 'bolting.' (Abou-Hussein 2012).

16.2.2 Changes in Rainfall Patterns

The long-term rainfall pattern trends have been changed and since nineteenth century and mostly from twentieth century. The rainfall pattern varies with year to year such as 1901–1930 was a dry period and thereafter 1931–1960 was recorded as a wet period, whereas 1961–1990 was recorded as a dry period and 1991–2020 a wet period (NASA 2021).

The rainfall patterns have been observed in terms of precipitation amount. Higher precipitation has been experienced in some regions of the world, such as North America, South America, Northern Asia and Central Asia. On the other hand, in some areas, such as southern Africa, Mediterranean region, and some parts of southern Asia, less rainfall was observed (Guhathakurta and Rajeevan 2008).

Spatially and temporally, precipitation has become extremely uneven. The change in the rainfall pattern have a harmful impact on the production and quality of vegetables during development for example if high rainfall occurs during the growing season, those vegetables which need less water may not come up resulting in degradation of fresh and processing quality (Pena and Hughes 2007; Lopez et al. 2011).

16.2.3 Elevation of Carbon Dioxide Concentration

Over the last 171 years, it was observed that the carbon dioxide (CO_2) concentrations increased by about by 47% more than the pre-industrial era of 1850. Thereafter since twentieth century the rise in the atmosphere carbon dioxide (CO_2) concentrations was much higher, during 2005 the carbon dioxide (CO_2) concentrations in atmosphere were recorded 380 ppm, in 2007, 382 ppm, in 2010, 388 ppm, in 2013, 395 ppm, in 2016, 402 ppm, in 2019, 410 ppm, and the current concentrations of carbon dioxide (CO_2) in January 2021 was recorded up to 415 ppm (NASA 2021).

Elevated carbon dioxide (CO_2) concentrations can increase the photosynthesis, biomass production, and increase the use efficiency of light, water, and nutrient and thereby increase the overall yield (Panja and Adhikary 2018).

Carbon dioxide CO_2 concentration is increasing gradually which plays an important role in global warming that effects the vegetable production and its postharvest quality. According to some of the studies conducted in the recent years concluded that CO_2 might have both positive and negative role in maintain the postharvest quality of vegetable crops (Panja and Adhikary 2018; Nemeskéri and Helyes 2019).

16.2.4 Extreme Weather Events

When the concentration CO_2increases, in one hand the globe gets heat up on the other hand many extreme weather events takes place as its side effect, such as, heavy droughts, change in rainfall pattern, flood etc. may become more common, change in rainfall pattern not only delay time of harvesting but also the occurrence of rains just before harvesting can cause serious cracking and splitting of many vegetables such as tomato, curd of cauliflower which results in low production along with deteriorating postharvest quality (Nemeskéri and Helyes 2019; Panja and Adhikary 2018).

Vegetables become very susceptible especially during the formation bulb, leafy head, tuber and curd. For many of the vegetables such as leafy vegetables drought decreases water content and reduces the postharvest quality and storage life (Nemeskéri and Helyes 2019).

Vegetables are vulnerable to extreme contrasts in temperature and soil moisture fluctuations, combination of these, contribute to a diminishing yield. By limiting photosynthesis, modifying metabolism, and inducing changes in enzymatic behaviour, environmental change has an effect on physiological and biochemical processes. (Tkemaladze and Makhashvili 2016; Bisbis et al. 2018).

16.3 The Impact of Climate Change Factors on Postharvest Quality of Vegetables

16.3.1 Effect of Changing or Elevated Temperature

The increasing temperature throughout the globe is a result of climate change, and is greater than before. Various vegetable crops and their broad diversity makes it complicated to sum up the possible effects of climate change on the postharvest quality of vegetables (Silva et al. 2017).

The vegetable grown under high temperature and high relative humidity will lead to some unwanted physiological changes which may results in detrimental quality such as blotchy ripening, sun scald in tomato, higher solanin accumulation in potato *etc.* in addition to that vegetables which are harvested under high temperature will have higher physiological heat (8–10 °C) than that of harvested at normal temperatures which results in the higher respiration and metabolic rate, leading to higher percentage of weight loss and browning in some vegetables (Wheeler et al. 2000).

Fruit colour very important in measuring the marketable quality of tomato. 25–30 °C. is the most favourable temperature for development of lycopene in tomato if the temperature reaches 27 °C lycopene starts degrading at above and at 40 °C lycopene completely destroyed (Kalloo et al. 2001). High temperature influences the development of red colour in ripe chilli fruits. The temperature alteration interferes the ripening of melon fruits that indirectly affects the sweetness (Arora et al. 1987).

Tomato grown at 40 ± 2 °C under polytunnel recorded a decrease quality parameter in terms of marketability such as total lycopene content due to degradation of lycopene at high temperature, total sugar and antioxidant compounds whereas the quality parameters such as total phenolics content, total flavonoids, total soluble solid and acidity were not influenced by high temperature (Lokesha et al. 2019).

Global warming would have a bigger effect on the open field culture than control cultivation of tomatoes. While it is unclear about the extent of future impacts, methods may be used to forecast such impacts. GCMs (Global Climate models) are valuable instruments for analyzing the potential repercussions on plant species (Silva et al. 2017).

Similarly, in lettuce which is generally grown in areas having cool days and nights if they are exposed to high temperature the heads will develop fewer tender leaves along with bitter taste (Peirce 1987). Tip burnt lettuce are prone to soft rot after harvesting and during storage, cabbage and lettuce leaves affected by sun scald are more vulnerable to post harvest decay (Moretti et al. 2010).

In tomato fruit, it has been observed that fruit size gets reduced due to heat stress leads to decrease in the surface to volume ratio increases more moisture loss throughout the storage period. Carrot exposed to preharvest heat stress may lead to faster water loss during storage period and may perhaps make many vegetables more vulnerable to chilling injury (Toivonen and Hodges 2011).

It is well evident that cauliflower must be exposed to adequate vernalisation to begin the formation of a curd. For a strong curd formation, temperature of 14 °C is most favourable. In tropical climate, cauliflower may fail to form a curd and flowering due to high temperature, where as in places where very low temperature prevails, few cultivars may have need of shield from low temperature effects (Putland and Deuter 2011).

Antioxidants of vegetable get altered by higher temperatures. Vegetables grown in warmer days and night temperature may have higher antioxidant activity than vegetables grown under low temperature (Moretti et al. 2010). Tomato grown under higher temperatures just prior to harvest inhibits the ethylene production and softening. Similarly, cucumber grown under high temperature have no influence on vitro amino cyclopropane carboxylate oxidase activity which affects the post-harvest quality of tomato fruits (Picton and Grierson 1988; Table 16.1).

16.3.2 Effect of Carbon Dioxide (CO_2) Concentration

Carbon dioxide (CO_2) is considered as one of the fundamental components to crop. For production of carbohydrate, higher concentration of CO_2 is supposed to be related directly (Bisbis et al. 2018; Dong et al. 2020). It is evident now that with increase in the carbon dioxide concentration, temperature and water availability is likely to be changed which create a complex situation where quality retention may be a challenge (Dong et al. 2020). For instance, reduction of plant growth, physiological parameters and alteration of some of the fruit qualities such as total soluble solid, acidity etc. (Daymond et al. 1997; Shaw et al. 2002; Dong et al. 2020). Due to decreased stomatal conductance, high CO2 often enhances vegetable yields and WUE, although long-term exposure may also cause growth depressions (Bisbis et al. 2018).

Only few selected crops such as lettuce, potato, and tomato have been studied in reference with the effect of carbon dioxide (CO_2) on postharvest quality of vegetables.

In context to vegetable such as tomato, increase in CO_2 have no effect on the quality of in terms of appearance, while on the other hand increase in CO_2 above 1000 $\mu L.L^{-1}$ shown an increase in the ascorbic acid and was found to be increasing with increase in the Carbon dioxide level (Kimball and Mitchell 1981). This increased carbon dioxide (CO_2) lifts up the consistency and taste of growing

Table 16.1 Exposure of vegetable crops to high temperatures resulting in physiological disorders affecting the post-harvest quality

Crop	Symptoms	Reference
Beans	Brown spots on pods and later on spots form water-soaked area	Bhat et al. (2017)
Cabbage	Outer leaves become bleached, papery and becomes more prone to decay	Bhat et al. (2017)
Tomato	Sunburn	Bhat et al. (2017)
Potato	Dark-gray to black discoloration at the center of the tuber	Bhat et al. (2017)
Lettuce	Heads will develop a bitter taste along with less tender leaves	Peirce (1987)
Lettuce and Cabbage	Sun scald leaves are more susceptible to rot.	Moretti et al. (2010)
Tomato	At 40 °C lycopene completely destroyed	Kalloo et al. (2001)
Cole crops	High temperature causes bolting	Kondinya et al. (2014)
Lettuce	Increase in bitter compounds	Wien (1997)
Cauliflower	Delaying head induction	Bisbis et al. 2018
Lettuce/Broccoli, Chinese Cabbage	Tipburn/Loose heads	"
Tomato	Green shoulders, Fruit coloration	"
Tomato, pepper	Blossom-end rot, Fruit cracking	"
Onion	Bulb splitting	"
Broccoli	Flower bud size, Hollow stem	"
Cauliflower, Broccoli	Bracting	"
Bean, Asparagus	Fibres	"
Sweet corn	Tip fill (cob)	"
Carrot	Terpenes	"
Tomato, Broccoli	Anthocyanin, Flavonols, Phenols, Glycosinolates	"
Pea, tomato, cabbage, melon, sweet corn	Sugar content	"
Sweet corn	Starch	"

tomatoes by increasing the concentration of fructose, glucose, and total soluble sugars in the fruit (Li et al. 2007; Zhang et al. 2014).

While in case of lettuce, the increased CO_2 concentration in atmosphere modifies the taste of lettuce by rising in the accumulation of soluble sugar by 27.1% as confirmed in a number of studies. Under salinity stress or high light intensity lettuce plants exposed to elevated carbon dioxide have privileged on its antioxidant metabolism (Jin et al. 2009; Becker and Klaring 2016). Surge in carbon dioxide leads to increase the ascorbic acid by 7.1% and total antioxidant by 82.0% (Baslam et al. 2012; Pérez- Lopez et al. 2018).

In tubers, elevated carbon dioxide can surge the tuber malformation which will further affect the postharvest quality of stored potato. Increase in Carbon dioxide level (CO_2) in the atmosphere result in decreasing the tuber quality due to increase in the sugar level which may lead to higher proportion of browning (Bisbis et al. 2018; Leisner 2020).

In context to internal quality and sensory attributes, the increased CO_2 level in the atmosphere can lead to decrease in postharvest qualities and sensory evaluation value of many vegetable crops, with noteworthy negative impacting case of potato, in higher carbon dioxide (CO_2) concentration, decrease in protein content, potassium and calcium levels in potato tuber can be observed (Moretti et al. 2010). On the other hand, elevated carbon dioxide (CO_2) increases soluble sugars and starch and maintained the organic acid level (Kumari et al. 2013).

Some research has shown, increased that increased CO_2 has the ability to lower the nutritional value of crops by lowering the mineral content in the plants (McGrath and Lobell 2013; Loladze 2014; Smith and Myers 2018; Medek et al. 2017; Table 16.2).

16.3.3 Effect of Ozone (O_3) Concentration

In addition to carbon dioxide, rise in the O_3 level as well contribute to global warming and climate change (IPCC 2014). Due to the fact that O_3 is simply absorbed through stomata it leads to development of reactive oxygen species and free radicals that are harmful to the plant cells which results in slow plant growth (Kumari et al. 2013).

The detrimental effects of ozone on postharvest quality especially for leafy vegetables is very high and are not direct but in the form of bronzing, leaf chlorosis, premature senescence and stressed leaf tissues, less biomass production and, eventually, decreases the postharvest quality in terms of appearance, overall acceptance and colour developments (Felzer et al. 2007; Kumari et al. 2013).

Some experiments were conducted to see the influence of ozone on vegetable quality. Positive results were observed in case of seedless cucumbers and broccoli, as shelf-life seedless cucumbers and broccoli can be extended by using Ozone at the rate of 0.04 μL / L (Skog and Chu 2001). Quality characteristic and sensory score were assessed on tomato after they were ozone exposed with a concentration varying from 0.005–1.0 μmol / mol this study revealed that soluble sugar, firmness of fruit, percentage of weight loss, ethylene production, vitamin C, citric acid, and total phenol were not considerably affected by ozone (Tzortzakisa et al. 2007).

A research conducted in potato with an AOT40 exposure to 12.5 μL/L × h in OTCs revealed that paste made from potato tubers was extra viscous in nature. Potato plants treated with an AOT40 with value of 27.1 μL/L × h, revealed that the starch granules of tuber were less resistant to swelling and unwanted glycoalkaloid

Table 16.2 Influence of elevated CO_2concentration /climate change on post-harvest quality of vegetables

Crop	Quality Parameters (QP)	Effect of CO_2 on QP	Reference
Leafy vegetables	Total soluble sugar	Increased	Jin et al. (2009)
Stem vegetables	Total soluble sugar	Decreased	Jin et al. (2009)
Leafy vegetables	Total antioxidant capacity	Increased	Dong et al. (2018)
Leafy vegetables	Ascorbic acid	Increased	Dong et al. (2018)
Root vegetables	Ascorbic acid	Decreased	Dong et al. (2018)
Fruit vegetables	Antioxidant capacity	Decrease	Dong et al. (2018)
Lettuce and cabbage	Sugars	Increased	Becker and Klaring (2016)
Lettuce, Celery,	Nitrate content	Decreased	Jin et al. (2009)
Lettuce, Spinach	Macronutrients (N, P, K, S, Mg)	Decreased	Giri et al. (2016)
Lettuce, Spinach	Micronutrients (Cu, Zn)	Decreased	Giri et al. (2016)
Potato	Browning and acryl amide formation during processing,	Increased	Högy and Fangmeier (2009)
	Glucose, fructose, sugar concentrations, protein, potassium, calcium and visual quality	Reduced	Ali et al. (2020)
	Shelf life with antioxidant enzymes and sensory quality		
Oyster Mushroom	Browning index, Flavor and antioxidant enzymes	Enhanced	Li et al. (2013), Zhang et al. (2015)
Button Mushroom		Decreased Increased	Lin et al. (2017)

content was found to be in higher amount which causes bitter taste in potato. O_3 exposure of potato plants resulted in decrease in reducing sugar and starch concentrations whereas increasing in ascorbic acid (Vorne et al. 2002). Exposure of water melon to O_3 decreased soluble solids 4% to 8% (Gimeno et al. 1999; Table 16.3).

16.3.4 Effect of Salinity

Under climate change situation, increase in soil salinity can be observed in arid region of the globe because of higher rates of evapotranspiration where as in the coastal regions due to saltwater intrusion, in. extreme soil salinity can alter the

Table 16.3 Impact of high ozone(O_3) concentration on post-harvest quality of vegetables

Crop	Quality Parameters (QP)	O_3 Effect on QP	Reference
Potato; carrot	Sucrose	No significant effect	Hildebrand et al. (2008)
Carrot	Isocumarin	Increased	Hildebrand et al. (2008)
Cucumber	Firmness	Increased	Skog and Chu (2001)
Broccoli	Color	Increased	Skog and Chu (2001)
Potato	N, P	No significant effect	Heagle et al. (2003)
Potato	K, Mg	Increased	Piikki et al. (2003)
Potato	Nitrate	Increased	Vorne et al. (2002)
Potato	Ascorbic acid	Decreased	Vorne et al. (2002)
Potato	Malic acid	Decreased	Piikki et al. (2003)
Tomato	Ascorbic accid	No significant effect	Tzortzakisa et al. (2007)
)Potato	Citric acid	Increased	Vorne et al. (2002)
Tomato	Citric acid	No significant effect	Tzortzakisa et al. (2007)
Potato	Reducing sugars	Decreased	Vorne et al. (2002)
Potato	Starch	No significant effect	Vorne et al. (2002)
Bell pepper	Glucose, fructose, and total phenolic contents	Increased	Glowacz et al. (2015)
	Weight loss during storage		
Cucumber/ zucchini	Disease incidence, weight loss, and maintained fruit firmness	Reduced	Glowacz et al. (2015)
Chili	Flavonoid contents	Reduced	Glowacz and Rees (2016)
Peppers		Increase	
Hot pepper	Soluble solid content and shelf life	Increased	Sachadyn-Król et al. (2016)
	Nutritional quality		
Carrot		Increased	Souza et al. (2018)
Swiss Chard		Reduced	Elvira et al. (2021)

production, productivity and quality of many vegetable crops (Machado and Serralheiro 2017).

Increase in TSS, acidity and flavour were observed when tomato was exposed to salinity stress (Beverly et al. 1993; Weston and Barth 1997).Increase in salinity may have an effect on the reverse osmosis which results in of loss of water from plant cells leads to loss in the quality many vegetables especially in leafy vegetables crops such as amaranthus, palak and spinach, (Pena and Hughes 2007).In chilli salinity stress decreases dry matter production, leaf area, but increases leaf area ratio where as in cucurbits high salt concentration leads to a decrease in fresh and dry weight. Decrease in water and total chlorophyll content was also found to be associated with the high salt concentration (Baysal et al. 2004).

It is well evident that soluble solids, sugars, acidity and pH are some vital quality parameters for both fresh and processing tomatoes, other parameter such as shelf life and taste are also important in determining the overall quality fruits. These quality parameters are highly influenced by the salinity stress. Higher NaCl concentration is likely to increase the consumer preference since high NaCl results in higher sweetness and flavor, but the fruit produced seems to be a little hard (Magan et al. 2008). It also alters the acidic taste of tomato because in high salt concentration ascorbic acid tends to be increased along with total soluble solids and titratable acidity on the other hand fruit firmness tends to be decreased (Azarmi et al. 2010). Internal browning of potato tubers is the effect of oxidation by PPO. The potato plants exposed to high saline condition will have higher percentage of internal tissue browning, probably due to presence of higher concentrations of chlorogenic acid and PPO in the tubers. Along with PPO and chlorogenic acid there might be involvement of some enzymes in browning of internal tissue of potato (Kirk et al. 2006).

Gurgel et al. (2010) confirmed increase in soluble solids content with increase of salinity in melon fruits thereafter with the increase in the storage period soluble solids was found to be gradually decreased.

According to Medeiros de et al. (2010) excess salinity in soil leads plants to absorb less water and less nutrients, as a result a larger period for vegetative growth lasting for longer time in plant, which may lead to excessive enzymatic activity, which is an indication of lower levels of some physiochemical compound such as sugars. The fruit weight of chilli is negatively correlated with salinity, as the salinity increased fruit weight was found to be decreased along with thickness of pulp on the other hand dry matter and total soluble solids increased. The hue of fruit was changed considerably with more orange colour. Sodium chloride stress showed slight raise in total phenolic content (10%) and significant rise in carotenoids (40%). Plants grown with high salinities produced fruit having high ascorbic acid compared to plants grown in normal soil condition (Giuffrida et al. 2014; Table 16.4).

16.3.5 Drought and Waterlogging

Due to climate change, unpredictable climate situation throws a huge challenge to human being as it results in the incidence of situations like flooding and drought. The vegetable crops are susceptible to both flood and drought. Drought stress results in reduction in photosynthesis, protein synthesis and nucleic acid metabolism and thus growth, development, post-harvest quality and storage life of many droughts sensitive vegetables (Bray et al. 2000). On the other hand, heavy rain fall could result in water logging situation in some of the areas due to poor drainage system.

Table 16.4 Impact of high salinity on postharvest quality of vegetables

Crop	Quality Parameters (QP)	Effect of salinity on QP	Reference
Tomato	Soluble solid	Increased	Zhang et al. (2016)
Tomato	Titratable acidity	Increased	Zhang et al. (2016)
Tomato	Sweetness and Flavour	Increased	Zhang et al. (2016)
Tomato	Firmness of fruit	Decreased	Azarmi et al. (2010)
Potato	Internal tissue browning	Increased	Kirk et al. (2006)
Potato	Proline concentration	Increased	Kirk et al. (2006)
Melon	Soluble solids content	Increased	Gurgel et al. (2010)
Chilli	Fruit weight	Decreased	Giuffrida et al. (2014)
Chilli	Total phenol content	Increased	Giuffrida et al. (2014)
Chilli	Carotenoids	Increased	Giuffrida et al. (2014)
Gherkin	Length and diameter of fruits,	Decreased	de Morais et al. (2018)
	pH, soluble solids	Increased	
Potato	Content and SS/TA ratio Enzymatic and non-enzymatic antioxidant activities	Enhance	Chourasia et al. (2021)

Water logging may result in lower rates of gas exchange, which will restrict the respiration of the roots resulting in alteration of the physiology, yield and the post-harvest quality of the vegetables (Patel et al. 2014).

Drought stress during vegetative growth decreases the accumulation of carbohydrate. Starch content of vegetable grains is highly influenced by drought. The effect of drought in starchy tuber crops such as cassava, potato is negatively correlated with starch concentration (Vandegeer et al. 2013; Ballmer et al. 2012). Under the severe drought situation, the reducing sugar was found to be increased with the intensity of drought in cucumber whereas total phenol content was negatively correlated with drought stress (Farag et al. 2019).

A lot of studies have confirmed that drought stressing tomato (Bartholomew et al. 1991; Sivakumar and Srividhya 2016) results in fast decrease in the quantity of 'Rubisco small subunit transcripts', which may point towards the decline in the synthesis of soluble protein, on the other hand a slight increase in vitamin C, TSS, lycopene was reported, on the other hand during the growing period, High rainfall in spinach resulted in the reduction of storage life by 40% (Weston and Barth 1997; Table 16.5).

Table 16.5 Impact of drought on post-harvest quality of vegetables

Crop	Quality Parameters (QP)	Effect of drought on QP	Reference
Tomato	Fruit weight	Decreased	Nahar et al. (2011)
Tomato	Total soluble solids,	Increased	Nahar et al. (2011)
Tomato	Ascorbic acid	Increased	Sivakumar and Srividhya (2016)
Tomato	Lycopene	Increased	Sivakumar and Srividhya (2016)
Broccoli	Polyphenolic concentrations	Increased	Fortier et al. (2010)
Lettuce	Total phenolic content and antioxidant capacity	Increased	Oh et al. (2010)
Tomato	Carotenoids	Increased	Favati et al. (2009)
Chilli	β-Carotene and capsanthin	Increased	Marín et al. (2009)
Tomato	Regulated deficit irrigation improved yield, antioxidants, bioactive compounds, and lycopene	Improve	Bogale et al. (2016)
Lettuce	Higher content of carotenoids, chlorophylls, caffeic acid, monocaffeoyl tartaric acid, malercyl quercetin glucoside, quercetin-3-O-glucuronide, and greater total antioxidant activity at harvest.	Improve	Paim et al. (2020)

16.3.6 Ultraviolet Radiations

UV irradiation has a beneficial impact on physico-chemical quality. UV treatmentreduced firmness loss and colour development of in tomatoes during storage and handling is highly desired. Although UV irradiation reduces the loss of vitamin C, the increased vitamin E loss remains a concern (Mditshwa et al. 2017). The application of a low UV-C dose can trigger some desirable reactions in fruits and vegetables, which lead to beneficial effects, including modulation of ripening and senescence, extending shelf life, or increasing health-promoting, compounds, elicitation of nutraceutical compounds as a defense mechanism and the induction of cross-stress tolerance. In addition, the UV-C radiation has successfully demonstrated its potential on microbial decontamination on food products without altering physicochemical and sensorial properties when used at proper doses. UV treatments have been shown to induce the accumulation of phytochemicals, including ascorbic acid, carotenoids, glucosinolates, and, more frequently, phenolic compounds. (Naradisorn 2021; Darré et al. 2022). Induction of cross-stress resistance and

Table 16.6 Effect of ultraviolet (UV) radiations on postharvest quality of edible plant products

Crop	Irridiation	Quality traits	Effect of UV on Quality traits	References
Garlic	UV-C	Microbial count,	Reduced,	Park and Kim (2015)
		Antioxidant/flavonoids and quercetin	Increased	
Amaranth	UV-C	Weight loss,	Decreased	Gogo et al. (2017)
		Lignin and cellulose/hemicellulose contents	Increased	
Cucumber	UV-C	Brightness and maintained quality	Increased	Imaizumi et al. (2018)
Broccoli	UV-B	Quality of broccoli florets and glucosinolates	Increased	Lu et al. (2018)
Tomato	UV-C	Phenolic/flavonoid and antioxidant activities	Increased	Liu et al. (2018), Esua et al. (2019)

synergistic responses: It is currently known that biotic and abiotic stress responses use common signals, pathways, and triggers (Charles 2019). This overlap includes common changes in cellular redox status, reactive oxygen species, hormones, protein kinase cascades, and calcium gradients as common elements and helps to explain cross-tolerance phenomena (Atkinson and Urwin 2012). Some quality traits are depicted in Table 16.6.

16.3.7 Biotic Stresses on Vegetable Crops

The growth rates of weeds, parasites and insect pests can be changed by the alteration of temperature, rainfall pattern, moisture levels and carbon dioxide elevation. The diseases caused by bacteria, fungi, viruses and nematodes could be strongly affected by climate change. (Nopsa et al. 2014). These biotic stresses react to climate change in a multiple manner starting from interactions among hosts, pathogen, and potential vectors to damaging the crop to its highest potential. High concentration of CO_2 may increase canopy density leading in less penetration of light, air circulation, which alters the explosion, survival of many pathogens (Pannga et al. 2013). Thus, climate change can create a new environment for emergence of certain weeds, insect pest and disease which may outbreak in the new environment which can results in decreasing the productivity, decline in the postharvest quality and storage life even in total failure of crop (Pannga et al. 2013; Abewoy 2018).

Climate change through global warming enhances the influence of pest and disease on vegetable crops in much larger region. Climate change significantly lower the yields or complete crop loss, lower the quality, and further enhance pest and disease outbreaks, rendering vegetable production unprofitable (Abewoy 2018).

Due to climate change, the regions which earlier were not favourable witnessing the increase in the frequency of outbreaks of field and storage pests and disease (Epstein 2001) resulting in more losses of produce. Elevated temperatures might cause shorter life cycles of insect pests and diseases, which may further stimulate reproduction process and expand the population of field and storage pests and diseases (Epstein 2001; Abewoy 2018). Due to occurrence of high temperature, some crops may get too dry reducing the quality and shelf life (Stathers et al. 2013).

High moisture percentage in the seed or other economical parts of vegetables are more prone to storage insect pest and disease. The increase of fungal disease on some of the vegetable will lead to produce aflatoxin or other mycotoxin which will deteriorate the heath of consumer and increase the post-harvest loss of the stored vegetables and its product (Hodges et al. 2011; Abewoy 2018).

16.3.8 Postharvest Losses of Vegetables

The quality of vegetables is determined by crop production methods and postharvest handling (Prasad et al. 2017; Prasad et al. 2018a). After harvesting, vegetable crops and their produced become more prone to biotic and abiotic deterioration (Prasad et al. 2018). The rate at which this deterioration takes place is significantly determined by environmental surroundings especially climate change (Allen et al. 2018; Prasad et al. 2018).

The increases in the temperatures and rainfall will not be same throughout the storage period. Instead, there will be a great variation in the duration, intensity and the frequency of rainfall and on the other hand the temperatures are likely to surge in every part of globe (FAO 2008). It is also expected that there will be more weather extremes like drought, flood. Unpredictable precipitation can dampen down the crops before they are ready to harvest which may lead in disease infestation on the field and later on may transfer into the storage room. Higher temperature is favourable for growth of many fungi and mould which produces mycotoxin that are hazardous to human being. At very high temperature some vegetable crops get too dried on the other hand unfavourable damp weather during the drying can lead to poor quality of stored vegetables (Stathers et al. 2013). Van Gogh et al. (2013) observed that non-optimal temperature regulation in the cold chain is the most significant source of postharvest losses in a report on developed and emerging economies.

16.4 The Possible Mitigation Strategies

More emphasis is needed on breeding facilities and capital investment on developing new vegetable varieties which can withstand changing climate and bears high postharvest quality. It should be noted that substantial investment has been made in supplying genome assemblies for a wide range of specialty crops, as enumerated in recent reviews (Chen et al. 2019; Hao et al. 2020).

The majority of sequenced genomes for vegetable crops have come from the Solanaceae, Cucurbitaceae, Cruciferae, and Leguminosae families, with broad scale resequencing efforts ongoing for at least five major vegetable crops (Cucumis *sativus, Brassica oleracea, Citrullus lanatus, Solanum lycopersicum,* and *Capsicum* spp.) (Hao et al. 2020). Crop responses to climate change could be bolstered or hastened with gene editing techniques, and crops could be biofortified to provide sufficient food quality content to an increasing population (Leisner 2020).

Genome editing methods including clustered frequently interspaced short palindromic repeats (CRISPR) schemes, for example, have been used to engineer higher nutritional value and stress resistance in wild tomato relatives (Khan et al. 2019).

Recently, efforts have been directed toward biotechnological techniques for a biofortified ion accumulating cassava, which accumulates considerably more zinc and potassium. The research involving individual climatic parameters along with their combined effect on diverse range of vegetable crop's growth and quality is need of modern world to mitigate future challenges in scientific manner (Narayanan et al. 2019).

16.5 Conclusion

Climate change has turned into a serious concern for vegetable sector, as most of the time consecutive years of unpredictable weather, increasing temperature, draught, irregular rainfall pattern led to disturbed crop cycle and affects postharvest quality. This result in reduced and irregular supply of raw material to the consumer as well as processing factories. The adverse effects of climate change on vegetable quality includes abnormal vegetable growth, pests and disease attack, premature senescence, altered biochemicals, altered physiology, physiological disorders *etc*. This adverse effect of climate change on vegetable quality is lowering down the returns of vegetable growers.

The vegetable breeding (conventional as well as modern breeding approaches), research on climatic parameters, genomics, biotechnological tools, marker technology, development of varieties having resistance to adverse/diverse climatic conditions and developing varieties having high postharvest quality and rich in biochemicals and postharvest shelf-life attributes are some of the research areas where researchers are working on so that to minimize the adverse effect of climate change on vegetable quality.

References

Abewoy D (2018) Review on impacts of climate change on vegetable production and its management practices. Adv Crop Sci Technol 6(1):10–30

Abou-Hussein SD (2012) Climate change and its impact on the productivity and quality of vegetable crops. J Appl Sci Res 8:4359–4383

Ackerman F, Stanton E (2013) Climate impacts on agriculture: a challenge to complacency 13-04. Global Development and Environment Institute, Tufts University, Medford

Adams RM, Rosenzweig C, Peart RM, Ritchie JT, McCarl BA, Glyer JD, Curry RB, Jones JW, Boote KJ, Allen JLH (1990) Global climate change and US agriculture. Nature 345:219–224

Ali S, Liu Y, Ishaq M, Shah T, Ilyas A, Din IU (2017) Climate change and its impact on the yield of major food crops: *evidence from Pakistan*. Foods 6:39

Ali S, Ejaz S, Anjum MA, Nawaz A, Ahmad S (2020) Impact of climate change on postharvest physiology of edible plant products. In: Plant ecophysiology and adaptation under climate change: mechanisms and perspectives I. Springer, Singapore, pp 87–115

Allen MR, Shine KP, Fuglestvedt JS, Millar RJ, Cain M, Frame DJ, Macey AH (2018) A solution to the misrepresentations of CO_2-equivalent emissions of short-lived climate pollutants under ambitious mitigation. Clim Atmos Sci 1(1):1–8

Anderson M (2013) Climate change, floods and adaptation. Presented at the California Department of Food and Agriculture Climate Change Adaptation Consortium, Monterey

Arora SK, Partap PS, Pandita ML, Jalal I (1987) Production problems and their possible remedies in vegetable crops. Indian Hort 32(2):2–8

Atkinson NJ, Urwin PE (2012) The interaction of plant biotic and abiotic stresses: from genes to the field. J Exp Bot 63(10):3523–3543

Azarmi R, Taleshmikail RD, Gikloo A (2010) Effects of salinity on morphological and physiological changes and yield of tomato in hydroponics system. J Food Agric Environ 8(2):573–576

Bales RC (2013) Managing forests for snowpack storage and water yield. Presented at the California Department of Food and Agriculture Climate Change Adaptation Consortium, Monterey

Ballmer T, Hebeisen T, Wüthrich R, Gut F (2012) Potential for drip irrigation in potato production under changing climatic conditions. Agrarforschung Schweiz 5:244–251

Bartholomew DM, Bartley GE, Scolnik PA (1991) Abscisic-acid control of rbcS and cab transcript ion intomato leaves. Plant Physiol 96:291–296

Baslam M, Garmendia I, Goicoechea N (2012) Elevated CO2 mayimpair the beneficial effect of arbuscular mycorrhizal fungi on the mineral and phytochemical quality of lettuce. Ann Appl Biol 161:180–191

Baysal G, Tipirdamaz R, Ekmekci Y (2004) Effects of salinity on some physiological parameters in three cultivars of cucumber (*Cucumis sativus*). Progress in cucurbit genetics and breeding research. Proceedings of Cucurbitaceae. The 8th EUCARPIA Meeting on Cucurbit genetics and Breeding, Olomouc, Chech Republic

Becker C, Klaring HP (2016) CO_2 enrichment can produce high red leaf lettuce yield while increasing most flavonoid glycoside and some caffeic acid derivative concentrations. Food Chem 199:736–745

Beverly RB, Latimer JG, Smittle DA (1993) Preharvest physiological and cultural effects on postharvest quality. Academic, In Postharvest Handling, pp 73–98

Bewley JD (1997) Seed germination and dormancy. Plant Cell 9(7):1055–1066

Bhat MA, Ahsan H, Husain S (2017) Climate change and its impact on food quality. Int J Pure App Biosci 5(3):709–725

Bindi M, Fibbi L, Miglietta F (2001) Free air CO2 enrichment (FACE) of grapevine (*Vitis vinifera* L.): II. Growth and quality of grape and wine in response to elevated CO_2 concentrations. Eur J Agron 14:145–155

Bisbis MB, Gruda N, Blanke M (2018) Potential impacts of climate change on vegetable production and product quality – a review. J Clean Prod 170:1602–1620

Bogale A, Nagle M, Latif S, Aguila M, Müller J (2016) Regulated deficit irrigation and partial root-zone drying irrigation impact bioactive compounds and antioxidant activity in two select tomato cultivars. Sci Hortic 213:115–124

Bray EA, Bailey-Serres J, Weretilnyk E (2000) Responses to abiotic stresses. In: Buchanan BB, Gruissem W, Jones RL (eds) Biochemistry and molecular biology of plants, pp 1158–1203

Charles F (2019) Current challenges of physical treatments to control quality and postharvest diseases of fresh fruits and vegetables. In IOP Conference Series: Earth and Environmental Science (Vol. 309, No. 1, p. 012003). IOP Publishing

Chen F, Song Y, Li X, Chen J, Mo L, Zhang X, Lin Z, Zhang L (2019) Genome sequences of horticultural plants: past, present, and future. Hortic Res 6(1):1–23

Chourasia KN, Lal MK, Tiwari RK, Dev D, Kardile HB, Patil VU, Kumar A, Vanishree G Kumar, D Bhardwaj V and Meena J.K (2021) Salinity stress in potato: understanding physiological, biochemical and molecular responses. Life, 11(6), p.545

Darré M, Vicente AR, Cisneros-Zevallos L, Artés-Hernández F (2022) Postharvest ultraviolet radiation in fruit and vegetables: applications and factors modulating its efficacy on bioactive compounds and microbial growth. Foods 11(5):653

Daymond AJ, Wheeler TR, Hadley P, Ellis RH, Morison JIL (1997) The growth, development and yield of onion (*Allium cepa* L.) in response to temperature and CO_2. J Hort Sci 72:135–145

de Morais DL, Aroucha EMM, de Oliveira FDA, de Medeiros JF, de Paiva CA, Nascimento LV (2018) Impact of salinity on quality and post-harvest conservation of gherkin (*Cucumis anguria* L.). J Agric Sci 10(4)

Deschenes O, Greenstone M (2012) The economic impacts of climate change: evidence from agricultural output and random fluctuations in weather: reply. Am Econ Rev 102:3761–3773

Dong J, Gruda N, Lam SK, Li X, Duan Z (2018) Effects of elevated CO_2 on nutritional quality of vegetables: a review. Front Plant Sci 9:924

Dong J, Gruda N, Li X, Tang Y, Zhang P, Duan Z (2020) Sustainable vegetable production under changing climate: the impact of elevated CO2 on yield of vegetables and the interactions with environments-a review. J Clean Prod 253:119920

Elvira S, Sanz J, Gonzalez-Fernandez I, Bermejo-Bermejo V (2021, December) Ozone effects on the quality of Swiss Chard. Peri-urban crops a case study. In Biology and Life Sciences Forum (Vol. 11, No. 1, p 16). MDPI

Epstein PR (2001) Climate change and emerging infectious diseases. Microb Infect 3:747–754

Esua OJ, Chin NL, Yusof YA, Sukor R (2019) Effects of simultaneous UV-C radiation and ultrasonic energy postharvest treatment on bioactive compounds and antioxidant activity of tomatoes during storage. Food Chem 270:113–122

FAO (2008) Food and agricultural organisation, climate change and food security: a framework document. http://www.fao.org/docrep/010/k2595e/k2595e00.html. Rome, Italy

Farag MI, Behera TK, Munshi AD, Bharadwaj C, Jat GS, Khanna M, Chinnusamy V (2019) Physiological analysis of drought tolerance of cucumber (*Cucumis sativus*) genotypes. Indian J Agric Sci 89(9):1445–1450

Favati F, Lovelli S, Galgano F, Miccolis V, Di Tommaso T, Candido V (2009) Processing tomato quality as affected by irrigation scheduling. Sci Hortic Amsterdam 122:562–571

Felzer BS, Cronin T, Reilly JM, Melillo JM, Wang X (2007) Impacts of ozone on trees and crops. ComptesRendus Geosci 339(11–12):784–798

Fortier E, Desjardins Y, Tremblay N, Belec C, Cote M (2010) Influence of irrigation and nitrogen fertilization on broccoli polyphenolics concentrations. Acta Hortic 856:55–62

Gimeno BS, Bermejo V, Reinert RA, Zheng Y, Barnes JD (1999) Adverse effects of ambient ozone on watermelon yield and physiology at a rural site in eastern Spain. New Phytol 144:245–260

Giri A, Armstrong B, Rajashekar CB (2016) Elevated carbon dioxide level suppresses nutritional quality of lettuce and spinach. Am J Plant Sci 7(1):246–258

Giuffrida F, Graziani G, Fogliano V, Scuderi D, Romano D. Leonardi C (2014) Effects of nutrient and NaCl salinity on growth, yield, quality and composition of pepper grown in soilless closed system. J Plant Nutr, 37(9):1455–1474

Glowacz M, Rees D (2016) Exposure to ozone reduces postharvest quality loss in red and green chilli peppers. Food Chem 210:305–310

Glowacz M, Colgan R, Rees D (2015) Influence of continuous exposure to gaseous ozone on the quality of red bell peppers, cucumbers and zucchini. Postharvest Biol Technol 99:1–8

Gogo EO, Opiyo AM, Hassenberg K, Ulrichs C, Huyskens-Keil S (2017) Postharvest UV-C treatment for extending shelf life and improving nutritional quality of African indigenous leafy vegetables. Postharvest Biol Technol 129:107–117
Guhathakurta P, Rajeevan M (2008) Trends in the rainfall pattern over India. Int J Climatol 28(11):1453–1469
Gurgel MT, Oliveira FHT, Heyi HR, Fernandes PD, Uyeda CA (2010) Qualidadepos-colheita de variedades de melõesproduzidos sob estressesalino e doses de potássio. Rev Brasileira de CienciasAgrárias 5:398–405
Hansen J, Mki S, Russell G, Kharecha P (2013) Climate sensitivity, sea level, and atmospheric carbon dioxide. Phil Trans R Soc A 371:2012–2094
Hao N, Han D, Huang K, Du Y, Yang J, Zhang J, Wen C, Wu T (2020) Genome-based breeding approaches in major vegetable crops. Theor Appl Genet 133(5):1739–1752
Heagle AS, Miller JA, Pursley WA (2003) Atmospheric pollutants and tracegases growth and yield responses of potato to mixtures of carbon dioxide andozone. J Environ Qual 32:1603–1610
Hildebrand PD, Forney CF, Jun S, Lihua F, McRae KB (2008) Effect of a continuous low ozone exposure (50 nL L^{-1}) on decay and quality of stored carrots. Posthar Biol Tech 49(3):397–402
Hodges RJ, Buzby JC, Bennett B (2011) Postharvest losses and waste in developed and less developed countries: opportunities to improve resource use. J Agric Sci 149:37–45
Högy P, Fangmeier A (2009) Atmospheric CO2 enrichment affects potatoes: 2. Tuber quality traits. Eur J Agron 30(2):85–94
Imaizumi T, Yamauchi M, Sekiya M, Shimonishi Y, Tanaka F (2018) Responses of phytonutrients and tissue condition in persimmon and cucumber to postharvest UV-C irradiation. Postharvest Biol Technol 145:33–40
IPCC (2014) Climate Change 2014: synthesis report. In: Core Writing Team, Pachauri RK, Meyer LA (eds) Contribution of Working Groups I, II and III to the Fifth Assessment Report of the Intergovernmental Panelon Climate Change. IPCC, Geneva, pp 40–54
Jin C, Du S, Wang Y, Condon J, Lin X, Zhang Y (2009) Carbon dioxide enrichment by composting in greenhouses and its effect on vegetable production. J Plant Nutr Soil Sci 172:418–424
Kalloo G, Benarjee MK, Tiwari RN (2001) Tomato. In: Thumbraj S, Singh N (eds) Vegetables tuber crops and spices. Directorate of Information and Publication in Agriculture, Indian Council of Agricultural Research, New Delhi, pp 10–28
Khan MZ, Zaidi SS, Amin I, Mansoor S (2019) A CRISPR way for fast-forward crop domestication. Trends Plant Sci 24(4):293–296
Kimball BA, Mitchell ST (1981) Effects of CO_2 enrichment, ventilation, and nutrient concentration on the flavor and vitamin content of tomato fruit. J Plant Nutr Soil Sci 172:418–424
Kirk WW, Da Rocha AB, Hollosy SI, Hammerschmidt R, Wharton PS (2006) Effect of soil salinity on internal browning of potato tuber tissue in two soil types. Am J Potato Res 83(3):223–232
Kondinya A, Palash S, Pandit MK (2014) Impact of Climate Change on vegetable cultivation - a review. Int J Environ Agric Biotech 7(1):145-1 55
Kumari S, Agrawal M, Tiwari S (2013) Impact of elevated CO_2 and elevatedO_3 on Beta vulgaris L.: pigments, metabolites, antioxidants, growth and yield. Environ Pollut 174:279–288
Leisner CP (2020) Climate change impacts on food security-focus on perennial cropping systems and nutritional value. Plant Sci 293:110412
Li F, Wang J, Chen Y, Zou Z, Wang X, Yue M (2007) Combined effects of enhanced ultraviolet-B radiation and doubled CO2 concentration on growth, fruit quality and yield of tomato in winter plastic greenhouse. Front Biol China 2:414–418
Li P, Zhang X, Hu H, Sun Y, Wang Y, Zhao Y (2013) High carbon dioxide and low oxygen storage effects on reactive oxygen species metabolism in *Pleurotus eryngii*. Postharvest Biol Technol 85:141–146
Lin Q, Lu Y, Zhang J, Liu W, Guan W, Wang Z (2017) Effects of high CO2 in-package treatment on flavor, quality and antioxidant activity of button mushroom (*Agaricus bisporus*) during postharvest storage. Postharvest Biol Technol 123:112–118

Liu C, Zheng H, Sheng K, Liu W, Zheng L (2018) Effects of postharvest UV-C irradiation on phenolic acids, flavonoids, and key phenylpropanoid pathway genes in tomato fruit. Sci Hortic 241:107–114

Lokesha AN, Shivashankara KS, Laxman RH, Geetha GA, Shankar AG (2019) Effect of high temperature on fruit quality parameters of contrasting tomato genotypes. Int J Curr Microbiol Appl Sci 8:1019–1029

Loladze I (2014) Hidden shift of the ionome of plants exposed to elevated CO_2 depletes minerals at the base of human nutrition. elife 3:e02245

Lopez MAH, Ulery AL, Samani Z, Picchioni G, Flynn RP (2011) Response of chilli pepper (*Capsicum annuum* L.) to salt stress and organic and inorganic nitrogen, growth and yield. Trop Subtrop Agroecosyst 14:137–147

Lu Y, Dong W, Alcazar J, Yang T, Luo Y, Wang Q, Chen P (2018) Effect of preharvest CaCl2 spray and postharvest UV-B radiation on storage quality of broccoli microgreens, a richer source of glucosinolates. J Food Compos Anal 67:55–62

Machado RMA, Serralheiro RP (2017) Soil salinity: effect on vegetable crop growth. Management practices to prevent and mitigate soil salinization. Horticulturae 3:30

Magan JJ, Gallardo M, Thompson RB, Lorenzo P (2008) Effects of salinity on fruit yield and quality of tomato grown in soil-less culture in greenhouses in Mediterranean climatic conditions. Agric Water Manag 95:1041–1055

Marín A, Rubio JS, Martinez V, Gil MI (2009) Antioxidant compounds in green and red peppers as affected by irrigation frequency, salinity and nutrient solution composition. J Sci Food Agric 89:1352–1359

McGrath JM, Lobell DB (2013) Reduction of transpiration and altered nutrient allocation contribute to nutrient decline of crops grown in elevated CO_2 concentrations. Plant Cell Environ 36(3):697–705

Mditshwa A, Magwaza LS, Tesfay SZ, Mbili NC (2017) Effect of ultraviolet irradiation on postharvest quality and composition of tomatoes: a review. J Food Sci Technol 54(10):3025–3035

Medeiros de PRF, Duarte SN, Dias CTS, Silva MFD (2010) Tolerância do pepino a salinidadeemambienteprotegido: Efeitossobrepropriedadesfísico-químicas dos frutos. (Cucumber salinity tolerance under protected environment: effect on the fruits physicochemical properties). Irrigation 15:301–311

Medek DE, Schwartz J, Myers SS (2017) Estimated effects of future atmospheric CO_2 concentrations on protein intake and the risk of protein deficiency by country and region. Environ Health Persp 125(8):087002

Medellin-Azuara J, Howitt RE, Duncan J, MacEwan Lund JR (2011) Economic impacts of climate related changes to California agriculture. Clim Chang 109:387–405

Moretti CL, Mattos LM, Calbo AG, Sargent SA (2010) Climate changes and potential impacts on postharvest quality of fruit and vegetable crops: a review. Food Res Int 43(7):1824–1832

Nahar K, Ullah SM, Islam N (2011) Osmoticadjustment and quality response of five tomato cultivars *(Lycopersicon esculentum Mill.)* following water deficit stress under subtropical climate. Asian J Plant Sci 10:153–157

Naradisorn M (2021) Effect of ultraviolet irradiation on postharvest quality and composition of foods. In: Food losses, sustainable postharvest and food technologies. Academic, pp 255–279

Narayanan N, Beyene G, Chauhan RD, Gaitán-Solís E, Gehan J, Butts P, Siritunga D, Okwuonu I, Woll A, Jiménez-Aguilar DM, Boy E (2019) Biofortification of field-grown cassava by engineering expression of an iron transporter and ferritin. Nat Biotechnol 37(2):144–151

NASA Global Climate Change: vital sign of the planet (2021). https://climate.nasa.gov/vital-signs/carbon-dioxide. Washington, DC, United States

Nemeskéri E, Helyes L (2019) Physiological responses of selected vegetable crop species to water stress. Agronomy 9(8):447

Nopsa JFH, Sharma ST, Ann GK (2014) Climate change and plant disease. In: van Alfen N (ed) Encyl Agri Food Sys, pp 232–243

Oh MM, Carey EE, Rajashekar CB (2010) Regulated water deficits improve phytochemical concentration in lettuce. J Am Soc Hortic Sci 135:223–229

Paim BT, Crizel RL, Tatiane SJ, Rodrigues VR, Rombaldi CV, Galli V (2020) Mild drought stress has potential to improve lettuce yield and quality. Sci Hortic 272:109578

Panja P, Adhikary KVS (2018) Abiotic stresses and its Management for Quality Production of vegetables. In: Research trends in horticulture sciences, vol 1. Akinik Publications, New Delhi, pp 37–56

Pannga IB, Hanan J, Chakraborty S (2013) Climate change impacts on plant canopy architecture: implications for pest and pathogen management. Eur J Plant Path 135:596–610

Park MH, Kim JG (2015) Low-dose UV-C irradiation reduces the microbial population and preserves antioxidant levels in peeled garlic (*Allium sativum* L.) during storage. Postharvest Biol Technol 100:109–112

Patel PK, Singh AK, Tripathi N, Yadav D, Hemantaranjan A (2014) Flooding: abiotic constraint limiting vegetable productivity. Adv Plant Agric Res 1:96–103

Peirce LC (1987) Vegetables: characteristics, production and marketing. Wiley, New York

Pena R, Hughes J (2007) Improving vegetable productivity in a variable and changing climate. SAT e J 4(1):1–22

Pérez- Lopez U, Sgherri C, Miranda-Apodaca J, Micaelli F, Lacuesta M, Mena-Petite A (2018) Concentration of phenolic compounds is increased in lettuce grown under high light intensity and elevated CO_2. Plant Physiol Biochem 123:233–241

Picton S, Grierson D (1988) Inhibition of expression of tomato-ripening genes at high temperature. Plant Cell Environ 11:265–272

Piikki K, Vorne V, Ojanperä K, Pleijel H (2003) Potato tuber surgars, starch andorganic acids in relation to ozone exposure. Potato Res 46(2):67–79

Prasad K, Jacob S, Siddiqui MW (2017) Fruit maturity, harvesting and quality standards. In book: springer- Preharvest modulation of postharvest fruit and vegetable quality. Elsevier Publications. Apple Academic Press, pp 41–69. https://doi.org/10.1016/B978-0-12-809807-3.00002-0

Prasad K, Guarav A, Preethi P, Neha P (2018) Edible coating technology for extending market life of horticultural produce. Acta Sci Agric 2:55–64

Prasad K, Siddiqui, MW, Sharma RR, Gaurav AK, Neha P and Kumar N (2018a) Edible coatings and their effect on postharvest fruit quality. n.d.Springer- Postharvest Biology and Technology Book Series. Apple Acedmic Press,. pp. 201–264

Putland D, Deuter P (2011) The effects of high temperatures on vegetable production and the rapid assessment of climate risk in agriculture. APEC Workshop on Collaboration on the Promotion of Indigenous Vegetables for Coping with Climate Change and Food Security. Asia-Pacific Economic Cooperation

Sachadyn-Król M, Materska M, Chilczuk B, Karaś M, Jakubczyk A, Perucka I, Jackowska I (2016) Ozone-induced changes in the content of bioactive compounds and enzyme activity during storage of pepper fruits. Food Chem 211:59–67

Shaw MR, Zavaleta ES, Chiariello NR, Cleland EE, Mooney HA, Field CB (2002) Grassland responses to global environmental changes suppressed by elevated CO2. Science 298:1987–1990

Silva RS, Kumar L, Shabani F, Picanço MC (2017) Assessing the impact of global warming on worldwide open field tomato cultivation through CSIRO-Mk3·0 global climate model. J Agric Sci 155(3):407–420

Sivakumar R, Srividhya S (2016) Impact of drought on flowering, yield and quality parameters in diverse genotypes of tomato (*Solanum lycopersicum*L.). Adv Hortic Sci 30(1):3–11

Skog LJ, Chu CL (2001) Effect of ozone on qualities of fruits and vegetables in cold storage. Can J Plant Sci 81:773–778

Smith MR, Myers SS (2018) Impact of anthropogenic CO_2 emissions on global human nutrition. Nat Clim Chang 8(9):834–839

Solankey SS, Shirin A, Pallavi N, Meenakshi K (2017) Effect of high temperature stress on morpho-biochemical traits of tomato genotypes under polyhouse condition. *Indian.* J Ecol 44(Special Issue-5):497–502

Solankey SS, Kumari M, Kumar M, Silvana N (2021) The role of research for vegetable production under a changing climate future trends and goals. In: Solankey SS et al (eds) Advances in research on vegetable production under a Changing Climate, vol 1. Springer, Cham., ISBN 978-3-030-63497-1, pp 1–12

Souza LPD, Faroni LRDA, Heleno FF, Cecon PR, Gonçalves TD, da Silva GJ, Prates LHF (2018) Effects of ozone treatment on postharvest carrot quality. LWT 90:53–60

Stathers T, Lamboll R, Mvumi BM (2013) Postharvest agriculture in changing climates: its importance to African smallholder farmers. Food Secur 5(3):361–392

Tkemaladze GS, Makhashvili KA (2016) Climate changes and photosynthesis. Ann Agrar Sci 14(2):119–126

Toivonen PM, Hodges DM (2011) Abiotic stress in harvested fruits and vegetables. Abiotic stress in plants-mechanisms and adaptations. InTech, China, pp 39–58

Tzortzakisa N, Borlanda A, Singletona I, Barnes J (2007) Impact of atmospheric ozone-enrichment on quality-related attributes of tomato fruit. Postharvest Biol Technol 45(3):317–325

Van Gogh JB, Aramyan LH, Van der Sluis AA, Soethoudt JM, Scheer FP (2013) Feasibility of a network of excellence postharvest food losses: combining knowledge and competences to reduce food losses in developing and emerging economies. Wageningen UR-Food & Biobased Research, Wageningen

Vandegeer R, Miller RE, Bain M, Gleadow RM, Cavagnaro TR (2013) Drought adversely affects tuber development and nutritional quality of the staple crop cassava (*Manihot esculenta* Crantz). Funct Plant Biol 40:195–200

Vorne V, Ojanperä K, De Temmerman L, Bindi M, Hogy P, Jones M (2002) Effects of elevated carbon dioxide and ozone on potato tuber quality in the European multiple-site experiment 'CHIP-project'. Eur J Agron 17(4):369–381

Weston LA, Barth MM (1997) Preharvest factors affecting postharvest quality of vegetables. HortScience 32(5):812–816

Wheeler TR, Craufurd PQ, Ellis RH, Porter JR, Prasad PV (2000) Temperature variability and the yield of annual crops. Agric Ecosyst Environ 82:159–167

Wien HC (1997) The physiology of vegetable crops. Cab International, Oxon/NewYork

Zhang Z, Liu L, Zhang M, Zhang Y, Wang Q (2014) Effect ofcarbon dioxide enrichment on health-promoting compounds and organoleptic properties of tomato fruits grown in greenhouse. Food Chem 153:157–163

Zhang L, Gao J, Hu H, Li P (2015) The activity and molecular characterization of a serine proteinase in *Pleurotus eryngii* during high carbon dioxide and low oxygen storage. Postharvest Biol Technol 105:1–7

Zhang P, Senge M, Dai Y (2016) Effects of salinity stress on growth, yield, fruit quality and water use efficiency of tomato under hydroponics system. Rev Agric Sci 4:46–55

Index

S. S. Solankey, M. Kumari (eds.), *Advances in Research on Vegetable Production Under a Changing Climate Vol. 2*, Advances in Olericulture, https://doi.org/10.1007/978-3-031-20840-9

U

V

W

Z

The manufacturer's authorised representative in the EU is Springer Nature Customer Service Centre GmbH, Europaplatz 3, 69115 Heidelberg, Germany. If you have any concerns regarding our products, please contact ProductSafety@springernature.com

Printed and bound by CPI Group (UK) Ltd, Croydon, CR0 4YY
30/09/2026
02225315-0009